small-sample confidence interval for μ $\quad \bar{X} \pm t_{\alpha/2}$

small-sample confidence interval for $(\mu_1 - \mu_2)$ $\quad (\bar{X}_1 - \bar{X}_2) \pm t_{\alpha/2}\, s \sqrt{\dfrac{1}{n_1} + \dfrac{1}{n_2}}$

Student's t (single mean) $\quad t = \dfrac{\bar{X} - \mu_0}{s/\sqrt{n}}$

Student's t (comparing two means) $\quad t = \dfrac{\bar{X}_1 - \bar{X}_2}{s \sqrt{\dfrac{1}{n_1} + \dfrac{1}{n_2}}}$

Mann–Whitney U test $\quad U = n_1 n_2 + \dfrac{n_1 (n_1 + 1)}{2} - T_1$

chi-square test $\quad \chi^2 = \sum \dfrac{(O - E)^2}{E}$

proportional reduction in error $\quad \text{PRE} = \dfrac{E_1 - E_2}{E_1}$

measures of association:

   phi coefficient $\quad \phi = \sqrt{\dfrac{\chi^2}{n}}$

   contingency coefficient $\quad C = \sqrt{\dfrac{\chi^2}{n + \chi^2}}$

   lambda $\quad \lambda_r \text{ or } \lambda_c = \dfrac{E_1 - E_2}{E_1}$

   gamma $\quad \gamma = \dfrac{E_1 - E_2}{E_1}$

   Spearman's ρ $\quad \rho = 1 - \dfrac{6\Sigma d^2}{n(n^2 - 1)}$

   Kendall's W $\quad W = \dfrac{12 SS_R}{k^2 n(n^2 - 1)}$

   Pearson's r $\quad r = \dfrac{\Sigma z_x z_y}{n}$

least-squares regression line $\quad \hat{Y} = a + bX$

STATISTICS:

A TOOL FOR THE SOCIAL SCIENCES

STATISTICS:

William Mendenhall
UNIVERSITY OF FLORIDA

Lyman Ott
UNIVERSITY OF FLORIDA

Richard F. Larson
CALIFORNIA STATE UNIVERSITY, HAYWARD

A
Tool for the
SOCIAL
SCIENCES

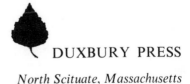
DUXBURY PRESS

North Scituate, Massachusetts

Duxbury Press
A Division of Wadsworth Publishing Company, Inc.

L.C. Cat. Card No.: 73-84305
ISBN 0-87872-053-7
Printed in the United States of America

Statistics: A Tool for the Social Sciences by William Mendenhall, Lyman Ott, and Richard F. Larson was edited and prepared for composition by Service to Publishers, Inc. Interior design was provided by Mr. David Earle and the cover was designed by Mr. Oliver Kline. The type was set by J.W. Arrowsmith Ltd., and the book was printed and bound by Kingsport Press, Inc.

1 2 3 4 5 6 7 8 9 10—78 77 76 75 74

CONTENTS

PREFACE

This book is for students in a one-quarter or one-semester statistics course who wish to gain an appreciation of the role of statistics in the social sciences. We use the term "appreciation" deliberately: the intent is to give the reader an overview of the role that statistics plays in social science investigations. This should better equip him to read and interpret articles in social science journals and to utilize statistical methods in his own social science research.

Equal time is given to the two major statistical divisions: *descriptive statistics* and *inferential statistics*; in this, the book differs from many other social science statistics books currently on the market. Included in the text is the standard treatment of descriptive statistics—so essential for the presentation and interpretation of social science data—but also a thorough (but elementary) introduction to the concepts of statistical inference.

To fulfill our teaching objective, we have reduced the dependence of the material on mathematics, particularly probability theory. The mathematical background required for the course is therefore quite elementary in nature, requiring only an understanding of the use of mathematical formulas.

The salient features of this text are its organization, its continuity, and the strong attempt we have made to show the relevance of statistics to social science investigations. Chapter topics are split into two clearly identified groups. One deals with the description of social science data. The other is concerned with statistical inference based on sampling. Definitions, formulas, and important rules for data presentation are set apart from the body of the text in boxes. Terms and discussions of major importance are shown in color. These devices, it is hoped, will make the book very easy to use.

Continuity is maintained by relating each topic to one of the two objectives of statistics—description or inference—as well as to preceding chapters and sections. Chapter introductions and summaries also assist in establishing the continuity and serve to relate the statistical concepts to practical problems encountered in the social sciences. Further relevance to real-world problems is provided by worked examples in the body of the text, many of which have been extracted from current social science literature. And interspersed throughout the chapters and at the ends of chapters are an unusually large number of practical exercises.

The subject matter can be used in a variety of ways. A one-quarter course presently taught for social science majors at the University of Florida provides a balanced offering of topics concerned with both descriptive and inferential statistics. These include Chapters 1 through 6, selected portions of Chapters 7 and 8, and Chapters 9 through 11. A course heavily tilted toward descriptive statistics could include the same chapters but with reduced emphasis on Chapters 6 through 8.

The authors wish to acknowledge the helpful comments of the reviewers, most particularly Norman R. Kurtz, Brandeis University; Susan Schwarz, Dartmouth College; and Robert P. Snow, Arizona State University. Thanks are due also to A. Hald, The Biometrika Trustees, The Chemical Rubber Company, and F. J. Massey, Jr., for permission to reprint tables. We are also indebted to our many typists, including Cathy Kennedy, Ellen Evans, and Carol Storer. And last but not least, we acknowledge the patience, encouragement, and forbearance of our families.

1

WHAT IS STATISTICS?

1.1 PRELUDE

*W*hat is statistics? Is it the addition of numbers? Is it graphs, means, medians, and modes, batting averages, percentages of passes completed, percent unemployed—in general, tedious descriptions of society and nature by means of a set of figures? Or is statistics a modern and scientific method for penetrating the unknown, that is, an exciting way of acquiring new information about the world in which we live? True, it involves numbers, and an understanding of the theory of statistics requires a basic knowledge of mathematics. But the concepts behind statistics—what it is attempting to do and how we go about it—are reasonable, easily explained, and understandable. The purpose of our text is to answer the questions: What is statistics? How does it work? How can it be applied in the social sciences? In general, we seek to provide the student with an understanding of the basic concepts of statistics and to tell him how he can use statistics as a tool. We shall do so while keeping in mind that many students in the social sciences possess relatively little mathematical training and often lack self-confidence in this area.

What is statistics? Statistics can be viewed from several perspectives. Because this text is written for students in the social sciences, we will write about statistics from a social science perspective. In so doing we need to consider two types of statistics, descriptive statistics and inferential statistics. Both are important tools in the social sciences.

What is descriptive statistics? The social scientist is often confronted with a mass of data that needs to be described. He might, for example, have specific bits of information, such as the total family income of each student in the university. If he wishes to describe the university in terms of the total family income of each student, the listing of 25,000 family incomes would be cumbersome indeed. Descriptive statistics provides us with the techniques that enable us to describe the university *concisely* in terms of the total family income of its students. We might also want to describe the drug habits of the residents of New York State, occupational status of the residents of Chicago, the religious beliefs of the residents of Tieton, or the preschool preparation of first graders currently attending Lincoln Grade School. In every instance, the researcher (it is assumed) has information on each subject in the population and merely wishes to describe a characteristic of New York State, the city of Chicago, the village of Tieton, or the first grade at Lincoln.

What is inferential statistics? Consider this story: The presidential elections of 1960 and earlier were laced with suspense. The polls closed at 7:00 or 8:00 P.M., but television viewers were treated to an exciting evening of uncertainty that, in most cases, did not end until early morning, when all the votes had been counted. Well-known political analysts exhibited a great lack of respect for the electronic monster—the computer—that had begun invading their areas of expertise, and they left no doubt in the minds of television viewers that man's judgment could never be replaced by the calculations of a machine. Yet in 1964 a Johnson victory over Goldwater was confidently forecast before midnight; and even the skeptical news analysts seemed convinced of the "monster's" predictions. An era ended in 1968 when, as viewers sat before their television sets, Nixon's victory was predicted early in the evening. The same dubious news analysts who had slighted computer calculations in 1960 and 1964 were forecasting the outcome of the 1968 election with great certainty. Conclusions in some states were forecast as early as a half-hour after the closing of the polls. Many viewers turned away from their television sets in disgust. The great race, the exciting evening they had anticipated, was over before it had begun, terminated by the great success of the electronic computer.

Our view of the evening was different. Knowing that computers only follow human instructions, we recognized the rapid and successful prediction of the election victor as an achievement of statistics. So what is inferential statistics?

The best way to answer this question is to consider a few examples. Suppose that you wish to estimate the proportion of all migrant workers in a particular state who have completed six years of public school education. Suppose, further, that you have obtained a fairly complete list of these workers from the state records for workmen's accident compensation and you have to decide how you will collect the pertinent information. One method, extremely costly and time consuming, would be to interview each of the migrant workers. But an easier and more efficient approach would be to randomly sample a thousand workers from the large body of all migrant workers and to contact each of these persons individually. You could then use the proportion of workers in the sample who possess six years of education to *estimate* the proportion of all migrant workers in the state who possess the same characteristic. We shall show later that the sample proportion will be quite close to the proportion for all migrant workers in the state. In addition, we shall be able to place a limit on the difference between the estimate and the true value of the proportion.

A second example is that of the Gallup, Harris, and other public opinion polls. How can these pollsters presume to know the opinions of more than 100 million Americans? They certainly cannot reach their conclusions by contacting every person in the United States. Instead, they randomly sample the opinions of a small number of citizens, often as few as 900, to *estimate* the reaction of every person in the country. The amazing result of this process is that the proportion in the sample who hold a particular opinion will match very closely the proportion of voters holding that opinion in the complete population from which the sample was drawn. Some students may find this assertion difficult to believe; convincing supportive evidence will be supplied in subsequent chapters.

As a third example, suppose that we wish to estimate the average number of children in families that live in the urban area of a large city. One way to do this would be to interview every family in the city, but this method has obvious disadvantages. It would be very costly to conduct so many interviews, and it would be incomplete and subject to inaccuracies because some families would not be at home when the interviewer made his visit. A second way, and one that employs the concepts of statistics, would be to select a random sample of 400 families from the city and interview each family. Families not at home would be revisited until they were interviewed. Then we would *estimate* the true average number of children in the family for the entire city by calculating the average for the sample of 400 families. This method of determining the average number of children per family is much less costly than sampling the entire city and, as we shall show later, it is very accurate. The difference between the sample estimate and the true average number of children per family is usually quite small.

Let us now try to identify from the examples the characteristics common to all inferential statistical problems. First, each example involved making an observation or measurement that could not be predicted with certainty in advance. an unpredictable (random) manner. We cannot say in advance whether a particular migrant worker selected from the state list will possess six years of education, whether a randomly selected voter will vote for Jones, or exactly how many children a particular family will have.

Second, each example involved sampling. A sample of migrant workers was selected from the state list, a sample of people was taken from the entire voting population of the United States, and a sample of 400 families was obtained from the total number of families in the city.

Third, although not obvious, each example involved the collection of data or measurements, one measurement corresponding to each element of the sample. Each migrant worker is interviewed and is assigned a score: $X = 1$ if he has had as many as six years of education and $X = 0$ if he has not. The total number of migrant workers with at least six years of education in the sample is the sum of these measurements. Scores, representing ratings, are measurements. Similar measurements are obtained when we sample voter intentions; and the sampling of the number of children in the 400 households yields 400 measurements, one for each family.

Finally, each example exhibits a common objective. That is, the purpose of sampling is to obtain information and to make an inference about a much larger set of measurements called the population. For the estimation problem concerning migrant workers, the population of interest is the large set of 1s and 0s corresponding to those who have and those who have not had at least six years of education. Similarly, the population for the voter problem would be the set of 1s and 0s corresponding to the 100 million or more voters in the United States. Each voter would be assigned a 1 if he intended to vote for candidate Jones and a 0 if he did not. The objective of sampling is to estimate the proportion of eligible voters who favor Jones, that is, the proportion of 1s in the population. The population associated with the household survey is the number of children per household recorded for all families in the city. The objective is to infer the mean number of children per family. In other words, the researcher wishes to make an inference about a large set of families based on information contained in the sample of 400 families. Note again that populations are collections of measurements or scores and are not collections of people (in accordance with the usual connotation of the term). Note also that populations may exist in fact or may be imaginary. The populations for the three examples exist even though we do not actually possess the complete collection of 1s and 0s corresponding to the two educational categories for migrant workers, or the entire set of voters favoring or opposing Jones. In contrast, if we sample the pulse rates of a set of heart patients, the population of pulse rates measured on all heart patients in the future is a product of the physician's imagination and thus is an imaginary population.

When we group the four characteristics—random observations, sampling, numerical data, and a common inferential objective—we might think to define *inferential statistics* as a theory of information. Information is obtained by experimentation or, equivalently, by *sampling*. The data are employed to make an inference about a larger set of measurements, existing or imaginary, called a *population*. Inferences about populations are usually expressed as estimates of or as decisions about one or more characteristics of the population.

Relevant definitions are as follows:

Definition 1.1

A *population* is the set representing all measurements of interest to the researcher.

Definition 1.2

A *sample* is a subset of measurements selected from the population of interest.

> **Definition 1.3**
>
> *Inferential statistics* is a theory of information concerned with the acquisition of data, with sampling, and with the use of the data in making inferences about a population.

> **Definition 1.4**
>
> The *objective of inferential statistics* is to make an inference about a population based on the information contained in a sample.

To summarize, social scientists are concerned with both descriptive and inferential statistics. Descriptive statistical techniques are employed primarily to describe large masses of data where the social scientist has information on all measurements in a population. In contrast, inferential statistical techniques are used when the scientist possesses only a sample of data and wishes to make an inference about the population from which it was selected.

1.2   WHY STUDY STATISTICS

There are two good reasons for taking an introductory course in statistics. First, every citizen is exposed to manufacturers' claims for products, the results of consumer and political polls, and the published achievements of scientific research. Most of these results involve inferences based on sampling. Some of the inferences are valid, others are invalid. Some are based on samples of adequate size, others are not. Yet the published results bear the ring of truth. Some people say that statistics can be made to support almost anything (particularly statisticians). Others say it is easy to lie with statistics. Both statements are true. It is easy to use statistics to distort the truth, purposely or unwittingly, when presenting the results of sampling to the uninformed. Your primary reason for studying statistics, then, is to learn how to evaluate published numerical findings: when to believe them, when to be skeptical, and when to reject them.

A second reason to study statistics is that, as a social scientist, you will be required to interpret the results of sampling (or surveys or experimentation), to employ statistical methods of analysis to make inferences in your work, or to describe a particular population of interest.

For example, you may be asked to estimate, on the basis of a sampling of hospital records, the proportion of hospital patients whose bills are "forgiven." Or you might wish to compare the bill forgiveness rates for two hospitals. Do the different rates in the two samples really imply a difference in the forgiveness rates for the two hospitals, or is the observed difference in results due simply to random variation in the records from day to day? Or you might wish to estimate, based on a random sample of 200 criminals selected from those who have "graduated" from a rehabilitation program, the frequency of reincarceration of criminals, broken down by social class.

Statistics is essential to the social, biological, and physical sciences because they all make use of observations of natural phenomena, through sample surveys or experimentation, to develop theories and to test hypotheses. Statistics has therefore become one of the major tools of the scientific method and is necessary for the development of knowledge in all fields of science. Although we are primarily interested in the role statistics plays in the social sciences, it is important to see that statistics is used in almost all areas of science, business, and industry. In business, sample data are utilized to forecast sales and profit. They are used in engineering and manufacturing to monitor product quality. And sampling of accounts is a new and useful tool to assist accountants in conducting audits.

1.3 THE INTERPLAY BETWEEN STATISTICS AND THE SOCIAL SCIENCES

In terms of research, what do social scientists do? Social scientists, particularly anthropologists, political scientists, and sociologists, devote their consulting and research time to two major methodological issues. The first concerns the acquisition of sample data relative to substantive issues. Sample surveys and experiments cost money and yield information. By varying the survey or experimental procedure—where we select the data and how many observations we take from each source—we can vary the cost, quality, and quantity of information in the experiment. Rather simple modifications in the selection of data often can reduce the cost of the sample to 1/100 or less of the cost of conventional sample procedures. Knowledge of statistics and statistical techniques can assist us in making these sampling decisions.

Social scientists are also called upon to select the appropriate method of inference for a given sample survey or experimental design. Social science inference makers, like all others, vary in their ability: some are good, some are bad, and some seem to be best for most occasions. In any case, the inference maker will either predict or make a decision about some characteristic of the population.

The foregoing discussion points to the most important contribution of statistics to the social sciences. Anyone can devise a method to make inferences based on sample data. The major contribution of the theory of statistics is in evaluation of the "goodness" of the inference. If, for example, we estimate the percentage of criminals reincarcerated after a rehabilitation program as 30 percent, we would like to know how far this estimate lies from the true percentage. Is the error 5, 10, or 30 percent? Similarly, when predicting a future occurrence based on sample data, we seek an upper limit to the error of the prediction. In reaching a decision concerning a characteristic of the population, we should be aware of the probability of reaching an incorrect conclusion.

To summarize: Social scientists use statistics, first, to design surveys and experiments that minimize the cost of obtaining a specified quantity of information. Second, they seek the best inference makers for a given sampling situation. Finally, we imply that one can measure the goodness of an inference and that all designs and inference-making procedures are selected to achieve an inference of specified quality at minimum cost.

1.4 A NOTE TO THE STUDENT

We think with words and concepts. Thus statistics, like a foreign language, requires the memorization of new terms and ideas. Commit to memory all definitions, theorems, and concepts. At the same time, you must focus on the broader aspects of statistics. What is it? How does it work? What are some of the more important applications? Do not let details obscure the broader characteristics of the subject, which identify the teaching objective of this text.

QUESTIONS AND PROBLEMS

1   What is statistics?

2   Distinguish between descriptive statistics and inferential statistics.

3   What is a population?

4   What is a sample?

5   Write a brief essay stating in your own words *why* people in the social sciences need to understand statistics. Could you make a similar argument for most people in an industrial society?

REFERENCE

Tanur, J. M., F. Mosteller, W. H. Kruskal, R. F. Link, R. S. Pieters, and G. R. Rising. *Statistics: A Guide to the Unknown.* San Francisco: Holden-Day, Inc., 1972.

2

MEASUREMENT

2.1 INTRODUCTION

We stated in Chapter 1 that statistics is concerned with two types of problems: description of large masses of data and the making of inferences about a population based on information contained in a sample. Implicit in this statement is the fact that we can actually measure elements of interest.

For example, it is easy to classify people according to racial categories; or to measure birth rates, population sizes, or family incomes. It is more difficult to assign a measure of social status to a person, to give an exact measure of a person's intelligence, or to measure the "religiosity" of a person. Consequently, before discussing data description and statistical inference, we need to consider the types of measurements encountered in the social sciences and to learn how to conduct them.

2.2 CONSTANTS AND VARIABLES

Statistics is a theory of information aimed at inference. Basic to this theory is the process of using information in a sample to make inferences about a population. Suppose that we plan to conduct a study to determine the cost of obtaining a marriage license at a county court house. To do this we send a student to inquire about the license fee. Note here that it would take only one observation (the result of one student inquiry) to ascertain the cost of a license, because the marriage fee is constant over a given period of time. Different observations from different studies would yield the same fee.

Definition 2.1

When observations on the same phenomenon remain constant in successive trials, the phenomenon is called a constant.

Consider the following extension of the previous study. We plan to obtain a sample of marriage fees charged by counties in a state and use this information to estimate the average fee for the entire state. Ten students are assigned to contact 10 different counties and each one is to obtain the stated marriage fee. Although all counties perform essentially the same function, each of the 10 students observed a different fee.

Definition 2.2

When observations on the same phenomenon vary from trial to trial, the phenomenon is called a variable.

We have seen that the marriage license fee charged in a county court house is a constant (for a given point in time), whereas the county marriage license fee in a state is a variable, because it may vary from county to county.

Social scientists are concerned primarily with the observation of variables. No particular theory of information is necessary for making inferences when the observed values remain constant from trial to trial. When the observations vary, however, statistics can be a valuable tool in formulating techniques for making inferences.

Definition 2.3

A quantitative variable is one whose observations vary in magnitude from trial to trial.

Examples of quantitative variables include

1 Family income.
2 Population size.
3 Birth rate.
4 Divorce rate.
5 Social cohesion.
6 Religiosity.
7 Age.

Quantitative variables can be further classified as one of two types: discrete or continuous.

Definition 2.4

When observations on a quantitative variable can assume only a countable number of values, the variable is called a discrete variable.

Typical examples of discrete variables include

1   The number of children for each household.
2   The number of voters in a sample favoring a political candidate.
3   The number of deaths for each year due to lung cancer.
4   The number of accidents for each year at a given intersection.

Note that for each of these examples we are able to count the number of values the variable can assume.

Definition 2.5

When observations on a quantitative variable can assume the infinitely large number of values in a line interval, the variable is called a continuous variable.

For example, the average amount of electricity consumed per household per month in Rochester, New York, can assume any of an infinitely large number of values. During a particular month, the average might be 90.5 kilowatt-hours, 105.126, 118.77789, and so on. Another example of a continuous quantitative variable is the age of freshmen U.S. Senators when they take the oath of office. Certainly a senator could assume every possible gradation of age between the regulatory lower limit of 30 years and the outer limits of life. Other typical continuous variables are height, weight, and length of life of persons receiving Medicare. From a practical standpoint, we treat measurements as if they were discrete. Weight, as just indicated, is continuous, but in determining someone's weight we count by ounces if we are dealing with an infant, and we count by pounds if we are dealing with an adult.

Not all variables encountered in the social sciences are quantitative. For example, "occupation" differs from one person to another, but the variable, "occupation," cannot be quantified. We can attach a number (called a score) to a person, assigning a 1 if a carpenter, a 2 if a plumber, etc. (or follow another procedure), but this would be only for purposes of identification. No quantitative significance is attached to such a number. Susceptibility to quantification distinguishes between quantitative and qualitative variables.

Definition 2.6

A *qualitative variable* is one whose observations are not subject to a *quantitative interpretation.*

Sex, religious affiliation, marital status, political party affiliation, voter registration, and employment status all provide examples of qualitative variables. Note that none of these qualitative variables can be arranged in order of magnitude. One political party affiliation cannot be considered as "greater" or "more" than an affiliation with another party. This inability to rank the attributes for qualitative variables does not hold for quantitative variables.

In our discussion of qualitative and quantitative variables, we have assumed that we are able to "measure" the variable of interest. However, not all variables can be measured with the same sophistication or scale. We consider next the different scales of measurement.

EXERCISES

1   Racism is one of the major concerns of social scientists. In doing research would this variable be more likely to be treated in a quantitative or in a qualitative manner? Explain your answer.

2   In the following list of concepts or terms of interest to the social scientist, indicate the terms which, in your judgment, refer to quantitative variables. Explain your choices.

infant mortality rate	extent of industrialization
type of economy	social institutions
caste systems	marriage forms

3   Of the terms listed in Exercise 2, list those which refer to qualitative variables.

4   Are there some variables in Exercise 2 which, in actual practice, can be treated as either qualitative or quantitative? Which ones? Explain your answer.

5   Infants die one at a time, yet we consider infant mortality rates as continuous variables. Explain this apparent inconsistency.

6 Categorize each of the following variables as qualitative or quantitative:

> penal codes   assessed valuation of homes
> voter concern   income tax bracket
> political parties   occupation
> campaign contributions

2.3 SCALES OF MEASUREMENT FOR QUALITATIVE AND QUANTITATIVE VARIABLES

The result of mensuration (the process of measuring) is called a measurement. Measurements may be in common units such as pounds, feet, dollars, or number of persons. Or measurements may be scores or values assigned to objects or quantities that represent a rating or evaluation: for example, you might rate the degree of interaction between two social groups on a scale of 1 to 10. Or you might score individuals on the basis of sex, 1 if a female, 0 if a male.

Social scientists and statisticians generally agree that there are four basic scales of measurement. These vary in the degree to which they quantify a variable. They are called nominal, ordinal, interval, and ratio scales. As you will see, a particular type of variable can be measured on one or perhaps more than one scale, but no single scale is dominant in the social sciences. Because statistical techniques and procedures have been developed for use with particular scales of measurement, it is essential that we understand the scales and be able to determine which is best for use with the variable (or variables) that we wish to measure.

Definition 2.7

A nominal scale defines specific categories by name. These categories are called levels of the scale.

All qualitative variables are measured on a nominal scale. For example, the qualitative variable "political party affiliation" could be divided into three levels—"Republican," "Democrat," and "Other"— so that all observations on potential voters could be "measured" or categorized as being in one of these levels. Sometimes nominal scales are labeled with numbers. For purposes of identification, we could assign the number 1 to a Republican, 2 to a Democrat, and 3 to an individual with any other party affiliation. We must remember, however, that we have used numbers to indicate levels on a nominal scale for the qualitative variable, party affiliation, but we cannot assign an order of magnitude to the various levels.

Definition 2.8

An ordinal scale *incorporates all properties of a nominal scale and the additional property that observations can be ranked from low to high.*

It should be pointed out that, although we can rank observations on an ordinal scale from low to high, we cannot assign a distance between the ranks. (On an ordinal scale a rank denotes the level of the variable.) For example, a crude prestige ranking of occupations such as the one in Table 2.1 would serve as an ordinal scale. Note, however, that we have no way of determining the distance between levels of prestige for the occupations in Table 2.1.

Table 2.1 *Prestige Ranking of Occupations*

Occupation	Rank
White collar	3
Blue collar	2
Laborer	1

Other typical examples of variables often measured on an ordinal scale are degrees of beauty, alienation, and the categorization of intelligence from mentally deficient to genius. Any variable that can be measured on an ordinal scale (i.e., levels can be ranked) can also be measured on a nominal scale (i.e., levels are not ranked). The three levels of occupational prestige ranking in Table 2.1 (white collar, blue collar, and laborer) therefore would also serve as a nominal scale.

Definition 2.9

An interval scale *incorporates all properties of an ordinal (and hence nominal) scale and the additional property that we can specify distances between levels on the scale.*

IQ tests provide an example of an interval scale. Suppose that eight sixth graders are administered an IQ test with the results given in Table 2.2. Not only can we rank the IQs for the eight individuals from lowest to highest but we can also state exact distances, measured in score units on an IQ test, between the students. Thus student 7 intellectually resembles student 6 more than he does student 8. Other examples of quantitative variables that can be measured by interval-level scales are the social science variables that employ techniques of measurement in which the intervals appear to the social scientist to be of equal width, such as anomie, group morale, social attitudes, and social distance.

Table 2.2 *Student IQ Results*

Student	IQ
1	150
2	128
3	126
4	125
5	122
6	120
7	110
8	75

One undesirable feature of the interval scale is that the origin on the scale is undetermined; that is, we do not know where 0 is located. For example, a zero IQ test score does not correspond to zero intelligence. In fact, we do not know the exact intelligence level implied by a zero test score.

Not knowing where the origin is located on a scale means that you cannot form valid ratios of observations. For example, using the data in Table 2.2, we might be tempted to form the ratio

$$\frac{\text{IQ student 1}}{\text{IQ student 8}} = \frac{150}{75} = 2$$

This result, however, *does not* permit us to say that student 1 is twice as intelligent as student 8. Some scales do permit valid examination of ratios. These are the ratio scales.

Definition 2.10

Ratio scales incorporate all the properties of interval (and hence nominal and ordinal) scales and the additional property that levels on the scale may be expressed as ratios.

Typical examples of ratio scales are age, birth and death rates, and divorce rates. For example, a country with a birth rate of 2.4 children per married couple has twice as large a birth rate as a country with 1.2 children per married couple. Similarly, a person 60 years old is twice as old as a person 30 years old.

To summarize, there are four scales of measurement for variables: nominal, ordinal, interval, and ratio. Since no one scale of measurement dominates the social sciences, we must be familiar with all four.

Summary of Measurement Scales

Scale	Description
Nominal	No quantitative implications. Levels are identified by name.
Ordinal	Levels on the scale are assumed to be quantitative, but exact numerical values are not known. Data can be ordered (ranked) according to their relative magnitudes.
Interval	Levels can be identified with points on a line, but the origin of the scale is undetermined.
Ratio	Levels can be identified with points on a line, and the origin is known.

Note that qualitative variables are always measured on a nominal scale. Ordinal, interval, and ratio scales are appropriate for quantitative variables, but they vary in the degree to which they express the magnitude of the variable levels. Ordinal scales are the least informative because they simply show the ranks of the levels. Interval scales preserve distance between two measurements, but the location of the origin (zero) is unknown. The most informative quantitative scale, the ratio scale, possesses a known origin and is constructed so that ratios of measurements are truly meaningful. It is seen, therefore, that the four scales differ in their ability to quantify variables. The nominal scale represents the least quantification and the ratio scale possesses the most quantitative sophistication. Certain statistical techniques have been developed for each measurement scale, but we do have some leeway. Ratio data can be employed as interval data, so statistical techniques appropriate for measurement on an interval scale can also be used for variables measured on a ratio scale. Similarly, statistical techniques developed for ordinal data can be used for interval and ratio data. Techniques for nominal data can be employed for any other of the three kinds of data. However, some information about the data is lost by this process. More discussion on the techniques for each scale will be presented in succeeding chapters as we develop statistical procedures. We turn now to some examples of data that one might collect in a study.

T

2.4   EXAMPLES OF SOCIAL DATA

he results of Section 2.3 will be illustrated with several examples.

Example 2.1 *Welfare recipients in a certain city were surveyed and categorized on the basis of marital status using the classifications given in Table 2.3. Identify the scale used in measuring the data in the table.*

Table 2.3 *Marital Status*

Single
Married
Divorced
Widowed
Other

Solution *The levels of the qualitative variable, marital status, differ in name only and cannot be ranked. Hence the data would be measured on a nominal scale.*

Example 2.2 *Administrative officials at many universities require parents of students applying for financial aid to file financial reports. If a particular university required parents to check one of the income categories listed in Table 2.4, determine the scale of the measurements collected.*

Table 2.4 *Gross Annual Income Categories*

Far below average
Below average
Average
Above average
Far above average

Solution *Since we can assign a rank to the levels of the income categories but cannot determine the exact distance between levels, the appropriate scale would be the ordinal scale.*

Example 2.3 *The Neal and Seeman Powerlessness Scale was administered to five groups of college students. Each student responded to the seven-item point scale. The results—average for each group—are listed in Table 2.5. Determine the scale used in measuring these data.*

Table 2.5 *Average Powerlessness Score for Five Groups*

4.5	1.3
3.2	2.8
4.4	

Solution *The distance between two scores is preserved, but the location of the origin is unknown. For example, a zero score on the Neal–Seeman social scale does not correspond to a true zero measure of powerlessness. In fact, we do not know what a zero Neal–Seeman score implies. Hence the resulting scale is an interval scale.*

Note again that we could easily convert the present interval scale into an ordinal scale. The categories "little power," "moderate power," and "much power" would form an ordinal scale.

Example 2.4 *The* Uniform Crime Reports for the United States: 1970, *lists auto theft rates per 100,000 people for all standard metropolitan statistical areas in the United States. A sample of seven metropolitan areas was taken and the auto theft rates were as shown in Table 2.6. Determine the scale of the measurement appropriate for the auto theft rates.*

Table 2.6 *Auto Theft Rates (per 100,000 people)*

City	Auto Theft Rate
Anderson, Ind.	161.1
Boise, Idaho	276.0
Cleveland, Ohio	595.2
Detroit, Mich.	757.6
Fresno, Calif.	1426.9
Lansing, Mich.	1710.0
Madison, Wis.	220.5

SOURCE: *Uniform Crime Reports for the United States: 1970* (Washington, D.C.: Department of Justice), pp. 78–86.

Solution   *It is apparent that the appropriate scale for the data of Table 2.6 is the ratio scale, because the auto theft-rate data preserve distances between measurements and possess a known origin. Consequently, an auto theft rate of zero would indicate that no thefts had been recorded in that standard metropolitan statistical area (a rather unlikely result). Since we have a known origin, meaningful ratios of measurements can also be formed. The auto theft rate for Lansing, Michigan, is 6.2 times the rate for Boise, Idaho:*

$$\frac{1710.0}{276.0} = 6.2$$

It is interesting to note that we can easily convert the ratio scale of Example 2.4 to a scale of lesser quantitative sophistication. For example, we could categorize the auto theft rates for standard metropolitan statistical areas using the classifications given in Table 2.7. The scale appropriate for measuring data collected according to the categories of the table is the ordinal scale, because we can rank the levels of variable auto rate classification from low to high.

Table 2.7   *Auto Theft Rate Classifications*
(rates per 100,000 *people)*

Low (rates less than 200)
Medium (rates between 200 and 1000)
High (rates greater than 1000)

Now that we can adequately distinguish among the four scales of measurement, we shall consider several methods that are useful in making relative comparisons among variables.

7   One measure of fertility is the crude birth rate. Is the crude birth rate measured on a nominal, ordinal, interval, or ratio scale? Explain your answer.

8   Give an example (other than one mentioned in the text) of a nominal scale currently being used in the social sciences.

9   A researcher first ranked the faculties at three universities according to national prestige, then according to actual scholarly productivity. The two rankings

of the three universities (denoted as *A*, *B*, and *C*) are as follows:

National Prestige	Scholarly Productivity
1 *A*	1 *A*
2 *B*	2 *C*
3 *C*	3 *B*

Given this level of data, can a researcher *meaningfully* add the ranks (i.e., 2 for university *A*, 5 for *B*, and 5 for *C*) and conclude that universities *B* and *C* are equal universities? Comment.

10   What are the advantages in measuring social phenomena by a ratio scale as opposed to an interval scale?

11   Using the IQ scores contained in Table 2.2 (page 19), suggest categories for treating the data at the interval level and then at the ordinal level. Indicate which students would be placed in the various categories.

12   Search the literature in your library for any one of the following variables. Try to locate examples where infant mortality, for example, is treated at the ordinal level and then at the ratio level.

infant mortality   aggression
alienation   industrialism
group morale   popularity

2.5   COMPARISONS OF VARIABLES
EMPLOYING RATIOS, PROPORTIONS,
PERCENTAGES, AND RATES

Among the many types of data that you will encounter in the social science literature, ratios, proportions, percentages, and rates are used as relative means for comparison of two numbers in a set or between two sets of measurements. For example, we may be interested in comparing the consumer price index for this month relative to the value of the index last month. Labor unions are interested in comparing their salaries to the national average for all related industries. Similarly, the high school senior contemplating a particular college might be interested in comparing the number of girls to the number of boys attending the college.

Comparisons can be made in two ways. We can subtract the two measurements in question, or we can divide one measurement by the other. Subtraction provides us with an absolute measure of the difference between two quantities and can be applied to data measured on either an interval or a ratio scale. The second, the quotient of two numbers, provides a relative measure of difference and can be applied only to data measured on a ratio scale. Absolute measures of comparisons will be dealt with in detail later. At this point we shall consider several measures of the relative difference between two measurements.

Attendance figures for a state university at the end of the spring quarters of 1971 and 1972 showed the breakdown given in Table 2.8. A potential student concerned about the campus social life might wish to

Table 2.8 *Male–Female Enrollment at the State University for Spring* 1971 *and* 1972

	1971	1972
Male	13,000	14,000
Female	8,000	8,000
Total	21,000	22,000

compare the number of males to the number of females. For the spring quarter of 1972, the *ratio* of males to females on the state university campus was, from Table 2.8,

$$\text{male-to-female ratio} = \frac{\text{number of males}}{\text{number of females}} = \frac{14{,}000}{8{,}000} = 1.75$$

That is, there were 1.75 males to every female on campus or, equivalently, there were 175 males to every 100 females. Similarly, during the spring quarter of 1971, the male-to-female ratio was

$$\frac{13{,}000}{8{,}000} = 1.625$$

That is, there were 1.625 males for every female.

This leads us to the following formal definition:

Definition 2.11

If there are n_a elements in group A and n_b elements in group B, the quantity n_a/n_b is called the ratio *of the number of elements in group A relative to the number of elements in group B. Similarly, n_b/n_a is the ratio of the number of elements in group B relative to the number in group A.*

(Note: Elements cannot belong in both groups. They are either in group A or group B.)

Ratios are used extensively to compare the magnitude of the number of elements in one group relative to the number in a second group.

Example 2.5 *Voter registration lists in a particular county precinct listed the following numbers of Democrats and Republicans:*

$$\begin{array}{ll} Democrats & 240 \\ Republicans & 200 \end{array}$$

Compute the ratio of the number of Democrats to Republicans in the precinct.

Solution *The Democrat to Republican ratio is given by*

$$\frac{number\ of\ registered\ Democrats}{number\ of\ registered\ Republicans} = \frac{240}{200} = 1.2$$

Thus there are 1.2 registered Democrats for every Republican registered in the precinct.

We are not always interested in comparing two numbers using a ratio. Sometimes a *proportion* provides a useful comparison. For example, using the data in Table 2.8, the proportion of males at the state university in the spring of 1972 was

$$\text{proportion of males} = \frac{\text{number of males}}{\text{total number of students}} = \frac{14{,}000}{22{,}000} = .64$$

That is, .64 of the campus student body was composed of males. Similarly, the proportion of females for the spring of 1972 was

$$\text{proportion of females} = \frac{\text{number of females}}{\text{total number of students}} = \frac{8,000}{22,000} = .36$$

Definition 2.12

If there are n_a elements in group A and n_b in group B, the proportion *of the total number of elements $n_a + n_b$ which are in group A is*

$$\frac{\text{number in group } A}{\text{total number in groups } A \text{ and } B} = \frac{n_a}{n_a + n_b}$$

Similarly, the proportion in group B is $\dfrac{n_b}{n_a + n_b}$.

(Note: Elements cannot belong in both groups. They are either in group A or in group B.)

Example 2.6 *Using the data from Example 2.5, compute the proportion of Democrats and the proportion of Republicans registered in the precinct.*

Solution *Recall that there were* 240 *registered Democrats and* 200 *registered Republicans. Hence the proportion of Democrats is*

$$\text{proportion of Democrats} = \frac{240}{240 + 200} = .55$$

Similarly, the proportion of Republicans is 200/440 = .45.

Some people dislike working with fractions and decimals; rather than working with proportions, they use percentages.

Definition 2.13

If there are n_a elements in group A and n_b elements in group B, the percentage *of the total elements in group A is the proportion in group A multiplied by* 100. *That is,*

$$\text{percentage in group } A = (\text{proportion in group } A)(100)$$

$$= \frac{n_a}{n_a + n_b}(100)$$

(Note: Elements cannot belong in both groups. They are either in group A or in group B.

Example 2.7   *Compute the percentage of registered Democrats and registered Republicans using the data in Example 2.5.*

Solution   *In Example 2.6 we computed the proportions of Democrats and Republicans as .55 and .45, respectively. So the percentage of Democrats is*

$$.55 \times 100 = 55$$

and the percentage of Republicans is

$$.45 \times 100 = 45$$

The final method of comparison that utilizes the ratio of two measurements concerns rates. Common published examples include birth rates, death rates, murder rates, and divorce rates. What are rates and what do they measure?

Definition 2.14

The rate of occurrences of a particular outcome is found by dividing the actual number of occurrences by the number of possible times the outcome could have occurred.

Table 2.9   *Murder and Nonnegligent Manslaughter in* 10 *Cities in* 1970

City	Population	No. Murders and Nonnegligent Manslaughters
Akron, Ohio	679,239	37
Albuquerque, N.M.	315,774	23
Altoona, Pa.	135,356	6
Wichita, Kan.	389,352	21
Waco, Tex.	147,553	11
Syracuse, N.Y.	635,946	22
Stockton, Calif.	290,208	26
Roanoke, Va.	181,436	19
Racine, Wis.	170,838	8
Monroe, La.	115,387	17

SOURCE: *Uniform Crime Reports for the United States: 1970* (Washington, D.C.: Department of Justice), pp. 78–87.

Example 2.8 *Table 2.9 shows the manslaughter rate in each of 10 selected cities. Determine the murder and nonnegligent manslaughter rates in 1970 for each of the 10 cities listed in Table 2.9.*

Solution *Utilizing Definition 2.14 and Table 2.9, we have the murder rates in Table 2.10.*

Table 2.10 *Murder Rates for 10 Selected Cities in 1970*

City	Murder Rate
Akron, Ohio	$\dfrac{37}{679,239} = .000054$
Albuquerque, N.M.	$\dfrac{23}{315,774} = .000073$
Altoona, Pa.	$\dfrac{6}{135,356} = .000044$
Wichita, Kan.	$\dfrac{21}{389,352} = .000054$
Waco, Tex.	$\dfrac{11}{147,553} = .000074$
Syracuse, N.Y.	$\dfrac{22}{635,946} = .000035$
Stockton, Calif.	$\dfrac{26}{290,208} = .000090$
Roanoke, Va.	$\dfrac{19}{181,436} = .000105$
Racine, Wis.	$\dfrac{8}{170,838} = .000047$
Monroe, La.	$\dfrac{17}{115,387} = .000147$

Quite often rates are multiplied by a larger number, such as 100,000, to give the number of occurrences per 100,000 possible occurrences. This makes the rate easier to read and less subject to misinterpretation. Thus the rates for Example 2.8 could be multiplied by 100,000 to give the number of murders and nonnegligent manslaughters per 100,000 people. These are listed in Table 2.11.

Table 2.11 *Murder Rates (per* 100,000 *people) for* 10 *Selected Cities in* 1970

City	Murder Rate
Akron, Ohio	5.4
Albuquerque, N.M.	7.3
Altoona, Pa.	4.4
Wichita, Kan.	5.4
Waco, Tex.	7.4
Syracuse, N.Y.	3.5
Stockton, Calif.	9.0
Roanoke, Va.	10.5
Racine, Wis.	4.7
Monroe, La.	14.7

In this section we have considered comparisons of measurements by making use of the quotients of numbers. Ratios, proportions, percentages, and rates provide ways for comparing two measurements. Although we never emphasized this point, most of the examples given to illustrate these procedures utilized measurements that were made at a fixed point in time. For example, we compared the number of boys to the numbers of girls enrolled at a state university in the spring of 1972. Similarly, we compared the proportion (and percentage) of registered Democrats and Republicans for a specified precinct at a given point in time.

Not all comparisons are made at a fixed point in time. Examination of a time series deals with the observation of a variable at different points in time. A time series will show increases or decreases, or, more generally, trends and patterns, of a variable with respect to time. Section 2.6 will provide us with a brief introduction to time series.

EXERCISES

13   Use the data in the table to answer the questions.

(a) What is the proportion of upper-middle-class males to all males?

(b) What is the ratio of all males to all females?

(c) What is the percentage of lower-middle-class males and females in the total group?

(d) If 2 males of the 48 died during the study, what would be the crude death rate for males?   ．O 4η

(e) If 6 females moved to other states during the study, what would be the geographical mobility rate for females?   ．|O|η

(f) Would the data be easier to understand if the numbers were converted to percentages? Comment.

Social Class Characteristics of 107 *Students in Fourth, Eighth, and Twelfth Grades, by Sex*

Class	Males	Females	Totals
Upper	20	23	43
Upper-middle	18	16	34
Middle	5	10	15
Lower-middle	4	7	11
Lower	1	3	4
Total	48	59	107

SOURCE: T. M. Meisenhelder, "Sex Roles and Person Perception: The Evaluation of Speakers by Children," University of Florida unpublished M.A. thesis, 1972, p. 62; by permission.

2.6   MEASUREMENT OF A VARIABLE OVER TIME: TIME SERIES

*E*xamination of trends, patterns, and fluctuations of a variable observed over a period of time is a vital part of the evaluation of social change, evolution, and trends.

Definition 2.15

Observations of a particular variable at successive points in time form a time series.

For example, in studying the ratio of males to females we could examine the sex ratio, the total number of males to the total number of females multiplied by 100, over a period of many years. These data would form a time series (see Table 2.12).

Table 2.12   *Time Series: Sex Ratios for the United States for Selected Years, 1820–1970*

Year	Sex Ratio
1820	103.3
1840	103.7
1860	104.7
1880	103.6
1900	104.4
1920	104.0
1940	100.8
1950	98.7
1960	97.1
1970	94.8

SOURCE: U.S. Bureau of the Census, *Statistical Abstract of the United States: 1971*, Washington, D.C., p. 24.

Note that we can readily detect trends in the time series of Table 2.12. For example, it appears that the general pattern for the sex ratio during the entire twentieth century has been a declining trend. Obviously there would be periodic increases throughout the period 1900–1970, depending on how often you collected the data; however, the predominant trend is one of decline. Similarly, in a study of industrialization of the steel industry in Japan and the United States we could examine the time series given in Table 2.13.

Table 2.13   *Time Series: Man-Hours Needed to Produce 1 Net Ton of Raw Steel*

Year	United States	Japan
1960	10.6	20.6
1962	9.6	18.1
1964	8.3	13.4
1966	8.1	10.6
1968	7.7	8.6
1970	7.3	5.7

SOURCE: *The Congressional Record* (Washington, D.C., June 3, 1971), S8144. Figures originally from American Iron & Steel Institute and Japan Iron & Steel Federation.

Note that both time series depict a downward trend in the man-hours required to produce 1 net ton of raw steel from 1960 to 1970. Also it is readily apparent that Japan's man-hour requirements have been reduced faster than those for the United States. For example, the United States has gone from 10.6 to 7.3 in the 1960–1970 period, while Japan has dropped from 20.6 in 1960 to 5.7 in 1970.

As we have seen from Tables 2.12 and 2.13, a time series provides a convenient portrayal of the fluctuations, trends, and patterns of a variable over a period of time. These patterns will become even more apparent when we study graphic descriptions of time series in Chapter 3. Whenever possible, uniform time periods should be selected, such as every 2nd year or every 5th year.

We turn now to an evaluation of the precision of the data we measure. Just how precise are they?

2.7 ARE THE MEASUREMENTS ROUNDED?

There is a distinct difference between the actual value of a variable that we wish to measure and the value recorded by our measuring instrument. Owing to the limitations of measuring equipment and the inexactness of human measurers, observations will sometimes be imprecise. Weights of people are recorded to the nearest pound; distances between adjacent cities are recorded to the nearest mile or kilometer; and amounts of gasoline are recorded to the nearest tenth of a gallon. We have actually *rounded* the true weights, miles, and gallons. There are, then, two sets of data, the true measurements, which we rarely can obtain, and the actual observed measurements, called the rounded observations. The difference between the true and the rounded measurements is called the rounding error and is a function of the rounding procedure. Any evaluation of the precision of the measurements will be affected by the rounding procedure.

We summarize a general rounding procedure as follows:

1 Specify the rounding unit. We could round measurements to the nearest integer, nearest 100, nearest 1000, nearest tenth, nearest one thousandth, etc.

2 Round each measurement to the nearest rounding unit.

3 If the actual observation lies at the midpoint of a rounding unit, round toward the nearest even unit.

Example 2.9 *Time cards for four delinquent employees were examined for a period of 1 week. The data shown represent the number of late minutes charged to each employee.*

Employee	Minutes
1	125.5
2	108.2
3	106.5
4	104.9

Round these numbers to the nearest minute.

Solution *The rounding unit for this example is 1 minute, which means that we must round measurements up or down to the nearest minute. No problem arises for employee 2 or 4 because 108.2 rounds down to 108 and 104.9 rounds up to 105. However, for employees 1 and 3 the observations fall at the midpoint of a rounding unit, so we apply the additional rule of rounding toward the nearest even unit. These results are shown.*

Actual Measurement	Rounded Measurement
125.5	126
108.2	108
106.5	106
104.9	105

Example 2.10 *Round the murder rates in Table 2.14 to the nearest whole number.*

Solution *The rounding unit for the murder rates is 1, so we round up or down depending on the actual rate. The results are summarized in Table 2.15.*

Table 2.14 *Murder Rates per 100,000 Inhabitants in 1970 for 10 Cities from the South, North, and West*

South	Rate	North	Rate	West	Rate
Atlanta, Ga.	20.4	Albany, N.Y.	2.8	Bakersfield, Calif.	7.6
Augusta, Ga.	22.1	Allentown, Pa.	2.4	Boise, Idaho	5.3
Baton Rouge, La.	10.2	Atlantic City, N.J.	5.1	Colorado Springs, Colo.	4.7
Beaumont, Tex.	9.8	Canton, Ohio	3.0	Denver, Colo.	8.3
Birmingham, Ala.	13.7	Chicago, Ill.	12.9	Eugene, Ore.	4.2
Charlotte, N.C.	24.7	Cincinnati, Ohio	6.4	Fresno, Calif.	8.5
Chattanooga, Tenn.	15.4	Cleveland, Ohio	14.5	Honolulu, Hawaii	4.0
Columbia, S.C.	12.7	Detroit, Mich.	14.7	Kansas City, Mo.	12.6
Corpus Christi, Tex.	13.3	Evansville, Ind.	6.9	Lawton, Okla.	5.5
Dallas, Tex.	18.4	Grand Rapids, Mich.	2.8	Los Angeles, Calif.	9.4

SOURCE: *Uniform Crime Reports for the United States: 1970* (Washington, D.C.: Department of Justice), pp. 78–86.

Table 2.15   *Rounded Murder Rates for the Data of Table 2.14*

South	North	West
20	3	8
22	2	5
10	5	5
10	3	8
14	13	4
25	6	8
15	14	4
13	15	13
13	7	6
18	3	9

As mentioned previously, to determine the precision of our measurements we must know the rounding procedure. If we observed the rounded numbers 126, 108, 106, and 105 and knew that the actual numbers had been rounded to the nearest minute, we could try to reconstruct the actual measurements. Although this is not entirely possible, we can construct limits, called true limits, which would encompass the actual measurements. These true limits are presented in Table 2.16 for the rounded data of Example 2.9. Note that we form the upper true limit by adding one-half of a rounding unit to the rounded measurement. Similarly, the lower true limit is formed by subtracting one-half a rounding unit from the rounded measurement.

Table 2.16   *Reconstructing True Limits for Measurements Rounded to the Nearest Minute*

Rounded Measurement	True Limits of Actual Measurements
126	125.5–126.5
108	107.5–108.5
106	105.5–106.5
105	104.5–105.5

One additional comment·should be made regarding rounding and the precision or quality of our measurements. The more rounding we use, the less precision we have. But the key to the entire process is to know the limitations of the rounding procedure. If we know the rounding unit, we can determine the true limits for the measurements. These limits determine the precision of the rounded data.

Our discussion of rounding has focused on one procedure, that of rounding to the nearest even unit. Even though most data in the social sciences are rounded to the nearest even unit, such a procedure is arbitrary and far from universal. Western societies round age down to the last whole number; Japanese round age up to the next whole number. An American infant is not 1 year old until it has reached its 12th month of extrauterine existence. The infant remains at age 1 until completion of its 24th month. On the other hand, a Japanese infant is 1 year old at birth and becomes 2 years old after completing only 12 months of extrauterine existence.

EXERCISES

14 The data shown are the number of voters in city precincts favoring a political candidate. Round to the nearest tens.

742	637
675	435
1281	790
98	1285

15 Can discrete data be rounded? Explain your answer with reference to an example of your choice.

16 If an American youngster states that he is 11 years old, what are the true limits of his age?

17 Round the following numbers to the nearest tenth:

.05	1.20	139.85
.67	1.646	17.35

18 Round the following numbers to the nearest hundreds:

52	7861	156
34	4.3	122.92

19 Use the data of Table 2.14 (page 35) and round the murder rates for the South to the nearest tens.

S 2.8 SUMMARY

ocial, economic, and political variables can be of two types, quantitative and qualitative. A quantitative variable is one whose observations vary in magnitude from trial to trial. All others are qualitative. Quantitative variables can be subdivided into two groups: discrete, which can assume a countable number of values, and continuous.

Measurement of a variable, quantitative or qualitative, can be more difficult than it sounds. The first step is the selection of an appropriate scale of measurement: nominal, ordinal, interval, or ratio. If a variable lends itself to measurement on a ratio scale, it can also be measured by use of any of the less sophisticated scales: interval, ordinal, or nominal. The scales decrease in level of quantitative sophistication from the ratio scale to the nominal scale. Any variable that can be measured on one scale can also be measured on any scale of less quantitative sophistication. Qualitative data, not being at all quantitative, can only be measured on a nominal scale.

Statistical techniques appropriate for data generated by the four scales of measurement possess overlapping utility. Techniques appropriate for one scale of measurement will apply also to data collected from a scale possessing a higher level of sophistication. In particular, techniques appropriate for a nominal scale can be employed for any type of data.

We have also described ratios, proportions, percentages, rates, and time series and have discussed the precision of recorded measurements. In concluding, we distinguished between the actual and rounded values of a variable and noted the rules employed in rounding data to acquire manageable numbers. Similarly, we noted the loss of precision and utilized true limits as a measure of data precision.

QUESTIONS AND PROBLEMS

20   List 10 social science variables.

21   Distinguish between quantitative and qualitative variables. Does this distinction lie in the mind of the researcher or in the *real* nature of social phenomena? Explain your position.

22   Distinguish between continuous and discrete variables. Give examples of four continuous and four discrete variables not mentioned in the text.

23 Given the following social concepts, specify the most informative scale of measurement (nominal, ordinal, interval, or ratio) appropriate for the phenomenon in question. In each instance, explain your answer.

population of Sacramento, California
alienation
religiosity
religious affiliation
suicide rate
sex
employment status
social status
primitive tribes

24 In society *A* there are 15,000 men and 16,000 women.
(a) What is the ratio of men to women?
(b) What is the ratio of women to men?
(c) What is the proportion of men in the society?
(d) What is the proportion of women in the society?

25 A small preliterate society has a total population of 1116. Four natives die from flesh wounds. What is the rate of deaths from flesh wounds?

26 Why would a social scientist be interested in presenting his data in a time series? Specifically, what are the advantages?

27 Round the following numbers first to the nearest ones, then tens, and finally hundreds.

6.5	149.5
14.3	698.1
78.7	1028.7
100.5	1499.5

28 Give the true limits for the eight numbers listed in Exercise 27.

29 Why would a social scientist be interested in using "ratios"?

30 What is the difference between a proportion and a percentage?

31 What is the difference between a rate and a ratio?

32 Social variables occasionally change in definition so that a time series is not too meaningful. Suggest five variables for which the "definition" has changed or might change over time.

33 Convert the auto theft rates per 100,000 in Table 2.6 (page 22) to rates per 1000 people. Which rate is easier to understand? Comment.

34 Round the rates in Table 2.6 (page 22) to the nearest ones.

35 Assume that the rates in Table 2.6 (page 22) have been rounded to the nearest whole number and give the true limits for each rate in the table.

REFERENCES

Anderson, T. R., and M. Zelditch. *A Basic Course in Statistics*, 2nd ed. New York: Holt, Rinehart and Winston, Inc., 1968. Chapter 2.

Blalock, H. M. *Social Statistics*, 2nd ed. New York: McGraw-Hill Book Company, 1972. Chapters 2 and 3.

Champion, D. J. *Basic Statistics for Social Research*. Scranton, Pa.: Chandler Publishing Company, 1970. Chapters 1 and 2.

Mueller, J. H., K. F. Schuessler, and H. L. Costner. *Statistical Reasoning in Sociology*, 2nd ed. Boston: Houghton Mifflin Company, 1970. Chapters 2 and 3.

Palumbo, D. J. *Statistics in Political and Behavioral Science*. New York: Appleton-Century-Crofts, 1969. Chapter 1.

Steger, J. A. *Readings in Statistics for the Behavioral Scientist*. New York: Holt, Rinehart and Winston, Inc., 1971. Chapter 1.

Weiss, R. S. *Statistics in Social Research*. New York: John Wiley & Sons, Inc., 1968. Chapter 2.

3

GRAPHICAL TECHNIQUES FOR DESCRIBING DATA FROM A SINGLE VARIABLE

3.1 INTRODUCTION

We have two very good reasons for wishing to describe a set of measurements. We stated in Chapter 1 that social scientists use statistical techniques to make inferences about a population based on information contained in a sample. Since a population is usually a large set of measurements, and sets of measurements are by their very nature disorganized and varied, it is necessary to find some means of condensing and describing a set of measurements. After we complete the description, we shall be able to talk about a population or, equivalently, to phrase an inference about it.

A second reason for studying ways to describe a set of measurements is the need to condense information about large sets of social science data. The abortion figures for all counties in the United States in a given year would need to be described in condensed form before they were published. The data collected by the U.S. Bureau of the Census is a second example of data that require condensation and description prior to publication. Although census, sociological, and economic data may ultimately be employed to make inferences about the future, simple description of these large quantities of data is immediately necessary to make them available for public use as measures of the health and welfare of the nation.

The description of most things is difficult. Describe the person sitting next to you so precisely that a stranger could select the individual from a group of others having similar physical characteristics. It is not an easy task. Fingerprints, voiceprints, and photographs, all pictorial, are the most precise ways of human identification. The description of a set of measurements is also a difficult task but, like the description of a person, it can be accomplished most easily using graphic or pictorial techniques. We would agree with the old axiom "a picture is worth a thousand words," and we might add "or more."

Pictorial or graphical description is as old as the history of man. Cave drawings convey scattered bits of information about the life of prehistoric man. Similarly, vast quantities of knowledge about the life and culture of the Babylonians, Egyptians, Greeks, and Romans are brought to life by means of drawings and sculpture. Thus art was used to convey a picture of various life styles, history, and culture in all ages. Not surprisingly, it is also of value in describing a set of measurements.

Sets of measurements are described in two ways, graphically and numerically. Graphical description is the objective of Chapter 3. Numerical measures will follow in Chapter 4. The graphical methods in this chapter fall into two categories: those most suitable for qualitative variables—the pie chart, bar graph, and statistical maps— and those appropriate for quantitative variables—the frequency histogram, frequency polygon, and time charts. Examples of each of these techniques will be given so that you can see how these descriptive tools can be applied to real-life situations.

S   3.2   ORGANIZING QUALITATIVE DATA

ociologists, political scientists, and anthropologists are frequently required to organize and condense large quantities of data for public distribution. For example, data collected by the U.S. Bureau of the Census must be carefully condensed and packaged so that they will be readily understandable by the general public. Similarly, sociological and economic data that measure the health and welfare of the nation must be summarized before they can be made available for public consumption.

In this section we shall be concerned with the organization of data collected from qualitative variables. Two principles will guide us in grouping (categorizing) these data:

Principle of Exclusiveness

No observation can be classified in more than one category.

Principle of Inclusiveness

Every observation can be classified into a category.

In organizing a set of data, we require that enough categories be included so that every observation can be classified. At the same time we require that the extent of classification be so precise that no observation can be classified into more than one category. The categories listed in Table 3.1 illustrate the principles of exclusiveness and inclusiveness. Note that every person interviewed could be categorized into one of the six classifications but that no one could respond with an answer that would fall into more than one of the six classifications. Data organized according to these guidelines would be clear and unambiguous.

Table 3.1   *Data Organization Illustrating the Principles of Exclusiveness and Inclusiveness*

What is your religious affiliation?

Catholic
Jewish
Protestant
Other
No affiliation
No response

Now that we are somewhat familiar with the principles of exclusiveness and inclusiveness, let us examine a questionnaire organization that illustrates how *not* to condense your data. Although it is possible that some of the major causes listed in Table 3.2 were not included in the questionnaire, the real cause for concern with this method of data organization is that there are too many overlapping categories. Such classifications as "poverty," "lack of jobs," "poor housing," and "too much welfare" may all be indicators of the same problem. Although the questionnaire asks each respondent to choose one of the categories, he may become confused by the multiple entries available for the same response. Similarly, the categories "criminal elements," "failure of

Table 3.2   *Perceived Causes of the San Francisco Riots**

"Which of the following do you think is the most important cause of the disorders?"

Black nationalism
Poverty
Criminal elements
Lack of jobs
Failure of public officials
Powerlessness
Poor housing
Too much welfare
Police brutality
Don't know
Other
Not ascertained

* Fictitious.

public officials," and "police brutality" may again focus on the same general cause for the riots.

Being aware of these general principles of organization for qualitative variables, you can now begin to evaluate the organization of published results in newspapers, magazines, and professional journals. We shall assume that you can now organize data on a qualitative variable according to the principles of exclusiveness and inclusiveness and proceed to the problem of graphically describing qualitative data.

EXERCISES

1   Table 3.2 (page 45) contains the perceived causes of a racial disorder. Use these causes as the basis for developing your own set of mutually exclusive and inclusive categories.

2   In response to the question "What is your religious affiliation?", a researcher developed six categories. Evaluate these categories in terms of the principles of exclusiveness and inclusiveness.

Roman Catholic	Protestant
Episcopal	Jewish
Lutheran	none

3   Evaluate the following fivefold classification scheme for "race" using the principles of exclusiveness and inclusiveness:

Hindu	Negroid
Moslem	other
Caucasoid	

4   Evaluate the following elevenfold classification scheme for "major in college" using the principles of exclusiveness and inclusiveness:

political science	physics
anthropology	chemistry
social sciences	other
sociology	none
history	not ascertained
physical sciences	

3.3 GRAPHICAL DESCRIPTION
FOR QUALITATIVE DATA:
PIE CHART

*T*he *circle chart,* or *pie chart,* is one of the most common graphical procedures used to describe a set of measurements. Essentially, the pie chart partitions a group of measurements much as one might slice a pie. Then you can see the percentage of the total assigned to each category of interest. The data in Table 3.3 are described in the pie chart of Figure 3.1. They represent a summary of a survey to determine the percentage of adult males (over 18 years of age) in a particular city that holds down at least one job. One thousand adult males were interviewed in the city and classified in the three categories of Table 3.3.

Table 3.3

No Job	1 Job	More than 1 Job
122	536	342

Although one can scan the data in Table 3.3, the results are more easily interpreted using a pie chart. From Figure 3.1 we could make certain inferences about employment throughout the entire city. For example, we would

Figure 3.1 *Employment pie chart*

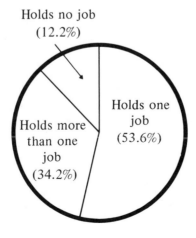

Holds no job
(12.2%)

Holds one
job
(53.6%)

Holds more
than one
job
(34.2%)

estimate that the unemployment rate (of males) for the entire city is approximately 12.2 percent.

The pie chart of Figure 3.1 is constructed by apportioning the 360 degrees of the circle in proportion to the percentage in each employment category. Thus 12.2 percent of 360 degrees, or

$$\frac{(122.2)\,(360)}{100} = 43.9$$

degrees, would be assigned to the group holding no job. Similarly, we would assign

$$\frac{(53.6)(360)}{100} = 193$$

Figure 3.2   *Quarterly inventories for 1967, 1968, and 1969*

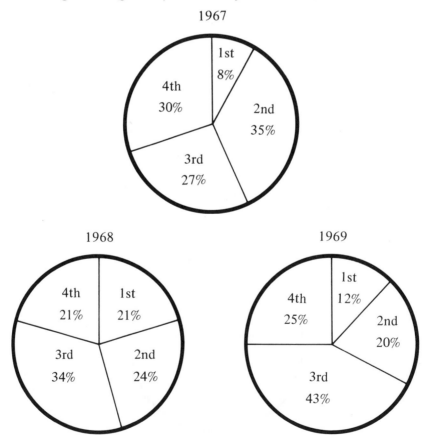

1967

1968

1969

degrees to those holding one job. The remainder of the pie would be allocated to those holding more than one job. The pie would be constructed by marking the angles with a protractor.

The three pie charts of Figure 3.2 show the quarterly inventory accumulation for all U.S. industries for the years 1967, 1968, and 1969. Each segment of a chart corresponds to the inventory for a particular quarter. Note the change in the quarterly inventory distribution over the 3-year period.

Other examples of pie charts selected from current literature are shown in Figure 3.3. The charts depict the characteristics of U.S. travelers abroad by

Figure 3.3 *Characteristics of U.S. travelers abroad.* [SOURCE: *Finance Facts*, published by National Consumer Finance Association, October 1971; by permission]

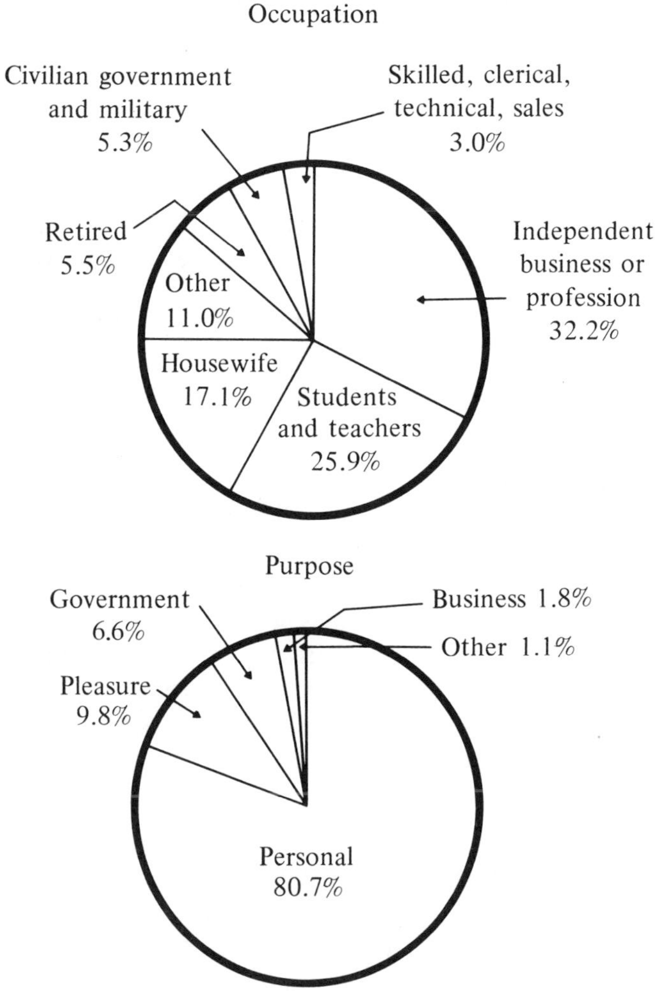

Occupation

Purpose

occupation and by purpose in 1970. Note that approximately four-fifths of the trips abroad were made for personal reasons, according to the stated purpose on passport applications.

There are several general guidelines to which one should adhere when using pie charts to describe a set of measurements:

Guidelines for the Construction of Pie Charts

1   Try to summarize the data into only a few categories. Too many categories make the pie chart difficult to interpret.

2   When possible, display the pieces of the circle chart in either ascending or descending order of magnitude.

3   Avoid using pie charts for displaying or comparing the number of measurements in each category since these charts are more difficult to interpret.

Note that pie charts are appropriate for data measured on a nominal scale and hence can be used for any data measured on an ordinal, interval, or ratio scale.

T 3.4   GRAPHICAL DESCRIPTION
FOR QUALITATIVE DATA:
BAR GRAPH

he *bar graph* provides a second procedure for displaying qualitative data in graphical form. Figure 3.4 is a bar graph that gives the estimated membership of the principal world religions in North America during 1970.

The bar graph is relatively easy to construct:

Guidelines for Constructing Bar Graphs

1   Frequencies are labeled along the vertical axis of the chart; categories of the qualitative variable are labeled along the horizontal axis.

2　Rectangles are then constructed over each category with the
height of the rectangle equal to the number of observations in the
category.

3　For clarity of presentation, a space is left between each category
on the horizontal axis.

Figure 3.4　*Estimated membership of the principal world religions in North
America in* 1970. [SOURCE: *Britannica Book of the Year, 1971*
(Chicago: Encyclopedia Britannica, Inc., 1971), p. 652; by
permission]

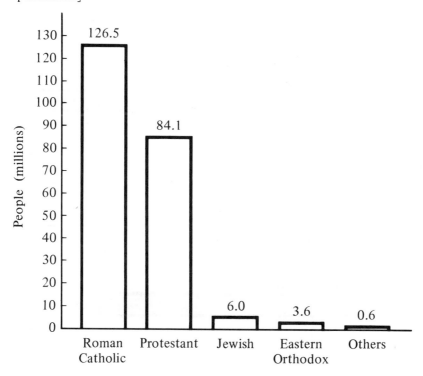

Bar graphs can also be displayed horizontally. Hence the roles of the
two axes are reversed. Figure 3.5, which displays a breakdown of expenditures
for health care by type of service and age of the recipient, illustrates this
change.

Bar charts are also used extensively to display economic (quantitative) data, especially when the variable, time, is presented in a discrete fashion. For example, universities throughout the country have had steadily increasing

Figure 3.5   *Whatever the category of health care, expenditures for the aged are highest.* [SOURCE: Adapted from *Finance Facts*, published by National Consumer Finance Association, February 1972; by permission]

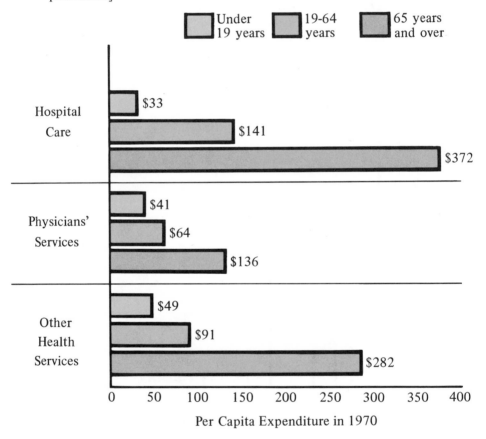

operating budgets over the years. The bar chart in Figure 3.6 illustrates increases in total operating budgets from the year 1950–1951 to 1970–1971 for the state university system in Florida.

Figure 3.6   *Cost of higher education in Florida (millions of dollars).*
[SOURCE: *St. Petersburg Times,* July 12, 1970; by permission]

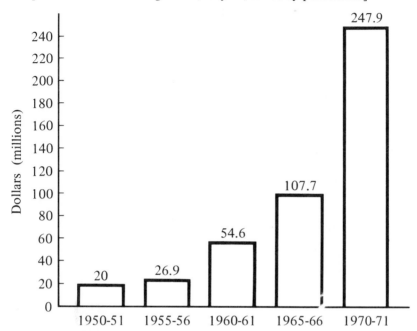

3.5   GRAPHICAL DESCRIPTION
FOR QUALITATIVE DATA:
STATISTICAL MAP

third form for graphically describing qualitative data is the statistical map. These maps are extremely useful for displaying geographical variation in such quantitative variables as income, sex ratios, marriage and divorce rates, and crime rates. Figure 3.7 presents a statistical map of average salaries of teachers in public schools by states.

As can be seen from Figure 3.7, the guiding principle of the statistical map is to graphically depict different classifications of a variable by different densities of shading. The major problem in constructing these maps is to choose the appropriate unit for shading. When we wish to display geographic variation over the entire United States, individual states provide a convenient

Figure 3.7 *Average salaries of instructional staff in public schools by states.* [SOURCE: *Finance Facts,* published by National Consumer Finance Association, November 1971; by permission]

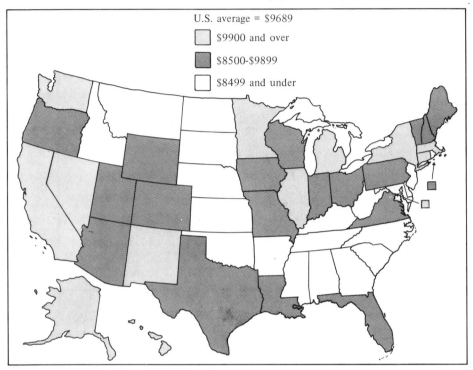

unit. Similarly, city blocks or voting precincts would provide convenient units for a statistical map of a city.

Guiding Principles in Constructing Statistical Maps

1   Select area units so they will be large enough to be easily visible to the reader, yet possess relevance to the study.

2   Use a color (or shading) code with a separate element for each category. Keep the number of categories small, say five or less, to provide a map that is easy to read.

Figure 3.8 presents a geographic breakdown by state for the increase in government jobs as a percentage of the increase in private jobs.

Figure 3.8   *Increase in government jobs as percentage of increase in private jobs by states,* 1969–1970. [SOURCE: *Finance Facts*, published by National Consumer Finance Association, December 1971; by permission]

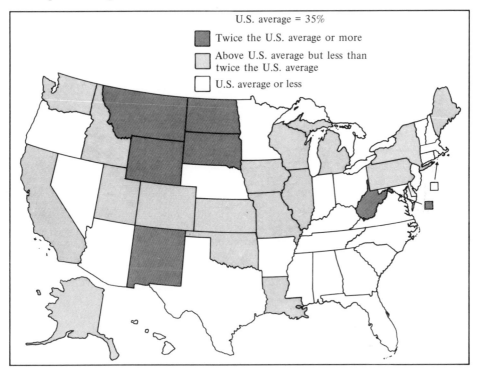

In Sections 3.2 through 3.5 we have discussed ways to organize data collected from qualitative variables and present them graphically. Although these procedures are basically designed for qualitative data, they can also be used for quantitative data. Other procedures, however, often provide more meaningful ways to organize and graph data from quantitative variables. We shall now turn to these procedures.

EXERCISES

5   Using the data in the table, construct a pie chart for quality of housing for lower-middle-class blacks.

Housing Quality by Socioeconomic Status in South County, Blacks Only

| | *Quality* | | | |
Status	*Sound*	*Unsound*	*Dilapidated*	*Totals*
Upper	13	0	0	13
Upper-middle	9	1	1	11
Middle	37	3	6	46
Lower-middle	36	38	34	108
Lower	36	39	52	127
Total	131	81	93	305

SOURCE: Gary L. Spencer, "Housing Quality and Socioeconomic Status," unpublished paper; by permission.

6 Using the data in the table, construct a pie chart for the male population by racial–ethnic groupings.

Four Racial–Ethnic Populations in the United States in 1970, by Sex (in thousands)

| *Racial–Ethnic Grouping* | *Male* | | *Female* | |
	Number	*Percent*	*Number*	*Percent*
Indian	389	36.1	404	37.3
Japanese	271	25.1	320	29.5
Chinese	229	21.3	206	19.0
Filipino	189	17.5	154	14.2
Total	1078	100.0	1084	100.0

SOURCE: U.S. Bureau of the Census, *Statistical Abstract of the United States: 1972*, Washington, D.C., p. 29.

7 Using the data in the table, construct a statistical map showing the distribution, by state, of foreign stock in the United States by leading countries of origin.

Percentage of Foreign Stock in the United States in 1970,
by State and Region

Area	Leading Countries of Origin	Area	Leading Countries of Origin
Northeast	Canada, 7.9%	North	
Maine	Canada, 13.8%	Carolina	Germany, 0.3%
New		South	
Hampshire	Canada, 13.1%	Carolina	Germany, 0.4%
Vermont	Canada, 10.4%	Georgia	Germany, 0.5%
Massachusetts	Canada, 8.2%	Florida	Cuba, 3.7%
Rhode Island	Italy, 7.7%	*East South*	
Connecticut	Italy, 7.5%	*Central*	Germany, 0.4%
		Kentucky	Germany, 0.7%
Middle Atlantic	Italy, 6.2%	Tennessee	Germany, 0.3%
New York	Italy, 7.3%	Alabama	Germany, 0.4%
New Jersey	Italy, 7.2%	Mississippi	Germany, 0.2%
Pennsylvania	Italy, 3.8%		
		West South	
East North		*Central*	Mexico, 3.7%
Central	Germany, 2.4%	Arkansas	Germany, 0.5%
Ohio	Germany, 1.8%	Louisiana	Italy, 0.8%
Indiana	Germany, 1.2%	Oklahoma	Germany, 0.8%
Illinois	Germany, 2.8%	Texas	Mexico, 6.4%
Michigan	Canada, 4.0%		
Wisconsin	Germany, 5.3%	*Mountain*	Mexico, 2.4%
		Montana	Canada, 3.0%
West North		Idaho	Canada, 1.5%
Central	Germany, 2.9%	Wyoming	Germany, 1.7%
Minnesota	Germany, 3.6%	Colorado	Germany, 2.0%
Iowa	Germany, 3.6%	New Mexico	Mexico, 3.7%
Missouri	Germany, 1.7%	Arizona	Mexico, 6.4%
North Dakota	Norway, 6.3%	Utah	United
South Dakota	Germany, 4.0%		Kingdom, 2.7%
Nebraska	Germany, 4.2%	Nevada	Italy, 1.6%
Kansas	Germany, 1.9%		
		Pacific	Mexico, 4.3%
South Atlantic	Germany, 0.9%	Washington	Canada, 4.0%
Delaware	Italy, 2.2%	Oregon	Canada, 2.5%
Maryland	Germany, 1.5%	California	Mexico, 5.6%
Virginia	Germany, 0.7%	Alaska	Canada, 2.2%
West Virginia	Italy, 1.0%	Hawaii	Japan, 13.7%

SOURCE: U.S. Bureau of the Census, *Statistical Abstract of the United States: 1972,* Washington, D.C., p. 35.

8 Using the data in Exercise 6, construct a pie chart for the female population by racial–ethnic groupings.

9 Construct a bar graph for unsound housing by socioeconomic status using the data of Exercise 5.

10 Construct a bar graph for the male population by racial–ethnic groupings for the data of Exercise 6.

11 Construct a bar graph for the female population by racial–ethnic groupings for the data of Exercise 6.

12 Using the data of Exercise 7, prepare a statistical map showing the distribution, by region (i.e., Northeast, Middle Atlantic, etc.), of foreign stock in the United States by leading countries of origin.

3.6 ORGANIZING QUANTITATIVE DATA

Categorizing quantitative data, specifically data measured on interval or ratio scales, is conducted in the same way as for qualitative variables. The only difference is that the categories for qualitative variables receive a nominal (or verbal) identification; in contrast, the categories for quantitative variables are intervals measured on an interval or ratio scale. In statistical language, these intervals are called class intervals.

The selection of classes for a quantitative variable should conform to the principles of exclusiveness and inclusiveness of Section 3.2. The intervals cannot overlap and they must be inclusive so that every measurement will fall in an interval. The following guideline expresses these two requirements in one sentence.

Guideline for Selecting Class Intervals So That They Conform to the Principles of Exclusiveness and Inclusiveness

The intervals must be chosen so that every measurement can fall in one and only one interval.

Additional criteria for selecting class intervals, chosen for convenience and ease of data interpretation, are the following:

Additional Criteria for Selecting Class Intervals

1   Intervals are chosen so that no gaps appear between them.

2   Intervals are chosen so that they possess a common width, called the class width.

Why it is desirable to satisfy the latter two criteria for selecting class intervals will become apparent when we discuss graphical methods for describing quantitative data. Now let us consider the mechanics for satisfying the criteria that we have just described.

It is easy to construct class intervals that do not overlap, are contiguous, and are of equal width. However, a point of confusion often arises concerning the endpoints of the intervals, which are called class limits. This is because intervals are sometimes constructed so that it is possible for a measurement to fall directly on the point dividing two intervals. To which interval should the measurement be assigned? Suppose that a researcher has categorized data on a quantitative variable, the ratio of the number of children to the number of bedrooms for each household in a ghetto area, by use of the following intervals: 0 to 1, 2 to 3, 4 to 5, 6 to 7, etc. The apparent class limits for these data are then 0 to 1, 2 to 3, 4 to 5, 6 to 7, etc., but the real class limits are $-.5$ to 1.5, 1.5 to 3.5, 3.5 to 5.5, 5.5 to 7.5, etc. Note that we would be undecided whether to assign the ratio 1.5 to the first or the second intervals. However, using the rounding procedures of Section 2.7, we would round to the nearest even unit. Hence 1.5 is rounded to 2. Similarly, if we observed the measurement 5.5, we would round to 6. Let us now illustrate how one organizes quantitative data.

Samples of 30 cities were selected from three regions to obtain information on crime rates. Murder rates are presented for each of the 90 cities in Table 3.4. We note from the table that the murder rates lie between 1 and 25, but it is still difficult to describe how the 90 measurements are distributed along this interval. If we had only five or six murder rates, there would not be much of a problem. With 90 rates, are most of the individual cities near 1, near 25, or are they evenly distributed along the interval of measurement? To answer these questions we shall construct a table giving the frequency distribution for the 90 observed murder rates.

We begin by dividing the range of the measurements $(25.5 - .5 = 25)$ into an arbitrary number of subintervals, called class intervals. Note that the range of a set of measurements is defined to be the difference between the upper true limit of the largest and the lower true limit of the smallest

Table 3.4 1971 *Murder Rates for* 90 *Cities Selected from the North, South, and West**

South	Rate	North	Rate	West	Rate
Atlanta, Ga.	20	Albany, N.Y.	3	Bakersfield, Calif.	8
Augusta, Ga.	22	Allentown, Pa.	2	Boise, Idaho	5
Baton Rouge, La.	10	Atlantic City, N.J.	5	Colorado Springs,	
Beaumont, Tex.	10	Canton, Ohio	3	Colo.	5
Birmingham, Ala.	14	Chicago, Ill.	13	Denver, Colo.	8
				Eugene, Ore.	4
Charlotte, N.C.	25	Cincinnati, Ohio	6	Fresno, Calif.	8
Chattanooga, Tenn.	15	Cleveland, Ohio	14	Honolulu, Hawaii	4
Columbia, S.C.	13	Detroit, Mich.	15	Kansas City, Mo.	13
Corpus Christi, Tex.	13	Evansville, Ind.	7	Lawton, Okla.	6
Dallas, Tex.	18	Grand Rapids, Mich.	3	Los Angeles, Calif.	9
El Paso, Tex.	4	Johnstown, Pa.	2	Modesto, Calif.	2
Fort Lauderdale,		Kalamazoo, Mich.	4	Oklahoma City,	
Fla.	14	Kenosha, Wis.	2	Okla.	6
Greensboro, N.C.	11	Lancaster, Pa.	2	Oxnard, Calif.	2
Houston, Tex.	17	Lansing, Mich.	3	Pueblo, Colo.	3
Jackson, Miss.	16			Sacramento, Calif.	6
Knoxville, Tenn.	8	Lima, Ohio	3	St. Louis, Mo.	15
Lexington, Ky.	13	Madison, Wis.	2	Salinas, Kans.	6
Lynchburg, Va.	18	Mansfield, Ohio	7	Salt Lake City, Utah	4
Macon, Ga.	13	Milwaukee, Wis.	4	San Bernardino,	
Miami, Fla.	16	Newark, N.J.	10	Calif.	6
				San Francisco, Calif.	8
Monroe, La.	15	Paterson, N.J.	3	San Jose, Calif.	2
Nashville, Tenn.	13	Philadelphia, Pa.	9	Seattle, Wash.	4
Newport News, Va.	8	Pittsfield, Mass.	1	Sioux City, Iowa	3
Orlando, Fla.	11	Racine, Wis.	5	Spokane, Wash.	1
Richmond, Va.	15	Rockford, Ill.	4	Stockton, Calif.	9
Roanoke, Va.	10	South Bend, Ind.	6	Tacoma, Wash.	6
Shreveport, La.	15	Springfield, Ill.	2	Topeka, Kans.	2
Washington, D.C.	11	Syracuse, N.Y.	4	Tucson, Ariz.	17
Wichita Falls, Tex.	6	Vineland, N.J.	10	Vallejo, Calif.	4
Wilmington, Del.	7	Youngstown, Ohio	7	Waterloo, Iowa	4

SOURCE: *Uniform Crime Reports for the United States: 1970* (Washington, D.C.: Department of Justice), pp. 78–94.

*Rates represent the number of murders per 100,000 inhabitants rounded to the nearest whole number.

measurement. If the number of measurements is small, we do not bother to graph them. If the number of measurements is large, we adopt a sufficient number of class intervals to provide a detailed picture of our data. As a rule of thumb, we advise using between 10 and 20 intervals for a large set of measurements.

The class intervals should be of uniform width. To determine an appropriate class width, divide the range by the number of intervals that seem appropriate for the number of measurements you wish to describe. Round the resulting number to a convenient interval width. This number is the class width or interval size, usually symbolized by the letter i. Suppose that we decide to use approximately 15 intervals for the data in Table 3.4. Then the range divided by 15 is

$$\frac{25}{15} = 1.7$$

and so an appropriate interval width is 2.

Having determined an appropriate class width, choose the first interval so that it includes the smallest observation. A convention frequently followed in the social sciences is to choose the lower apparent limit of the first interval to be either 0 or an integer multiple of the interval width. Such a procedure assures uniformity in the construction of tables and graphs. It is also important to choose subintervals so that no observation falls on a point of division between two class intervals. This eliminates any ambiguity in placing observations into class intervals. For the murder-rate example, the smallest observation is 1 and the interval width 2. An interval 0–1 with real limits of $-.5$ to 1.5 is two units wide, has zero as the apparent lower limit, and would include the values 0 and 1. The class intervals for our example, using the starting point of $-.5$, are then

−.5–1.5	7.5–9.5	13.5–15.5	19.5–21.5
1.5–3.5	9.5–11.5	15.5–17.5	21.5–23.5
3.5–5.5	11.5–13.5	17.5–19.5	23.5–25.5
5.5–7.5			

Note that no observation falls on a point of division between two class intervals.

Principles for Selecting Class Intervals

1   Decide on the number of intervals necessary to describe the measurements—usually approximately 15 intervals will do.

2   Divide the range of the measurements by the number of intervals. Round to a convenient unit. This gives the class interval width.

3   Locate the first interval so that it includes the smallest observation. The lower apparent limit of the first interval will usually be an integer multiple of the interval width or be zero.

4   If a measurement falls on a real class limit, use the rounding procedures of Chapter 2 to assign the measurement to a class interval.

Having specified the class intervals, we shall now be concerned with how the measurements or scores are distributed into the classes. Examine each of the 90 observations in Table 3.4 and keep a tally of the number falling in each of the 13 class intervals. The number falling into a given class is called the class frequency. The total of the class frequencies we call n, the sample size. For our example, $n = 90$. The tallies and class frequencies for our example are shown for each class in Table 3.5. The table portrays the frequency distribution of the $n = 90$ murder rates.

Table 3.5   *Frequency Distribution for the Data of Table 3.4*

Interval j	Class Interval	Tally	Class Frequency	Relative Frequency
13	23.5–25.5	1	1	.011
12	21.5–23.5	1	1	.011
11	19.5–21.5	1	1	.011
10	17.5–19.5	11	2	.022
9	15.5–17.5	1111	4	.044
8	13.5–15.5	╫╫ 1111	9	.100
7	11.5–13.5	╫╫ 11	7	.078
6	9.5–11.5	╫╫ 111	8	.089
5	7.5–9.5	╫╫ 1111	9	.100
4	5.5–7.5	╫╫ ╫╫ 111	13	.144
3	3.5–5.5	╫╫ ╫╫ ╫╫	15	.167
2	1.5–3.5	╫╫ ╫╫ ╫╫ 111	18	.200
1	−.5–1.5	11	2	.022
Total			90	

The relative frequency of a class interval is defined to be the frequency of the class divided by the sample size. If we let f_j denote the frequency of class *j*, then the relative frequency for class *j* is defined as follows:

Definition 3.1

The relative frequency for class j is defined to be the frequency for class j, f_j, divided by n, the sample size: (relative frequency for class j) = f_j/n

---

For example, the relative frequency for the 9th class interval, with real limits 15.5 to 17.5, can be found as follows. The sample size is $n = 90$ and the frequency for the 9th interval is $f_9 = 4$. Hence

$$\text{(relative frequency for class 9)} = \frac{4}{90} = .044$$

Having discussed the organization of data for a quantitative variable, we shall now consider graphical techniques for describing the data.

3.7 GRAPHICAL DESCRIPTION FOR QUANTITATIVE DATA: HISTOGRAM

A frequency distribution can be presented as a graph, called a frequency histogram. We mark the real class limits along the horizontal axis. Frequencies are labeled along the vertical axis. Rectangles are then constructed over each class interval with the height of the rectangle equal to the class frequency. The frequency histogram for the data in Table 3.5 is given in Figure 3.9.

Sometimes the results of a frequency table are presented graphically using a relative frequency histogram. The only difference between the frequency histogram and the relative frequency histogram is that the vertical axis is now scaled for relative frequency rather than absolute frequency. The relative frequency histogram for Table 3.5 is presented in Figure 3.10. Very little distinction is made between these two histograms since they become the same figure if drawn to the same scale. We frequently label either one as

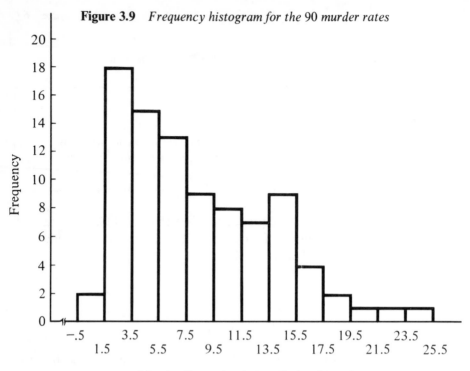

Figure 3.9 *Frequency histogram for the 90 murder rates*

Murder Rates (real class limits shown)

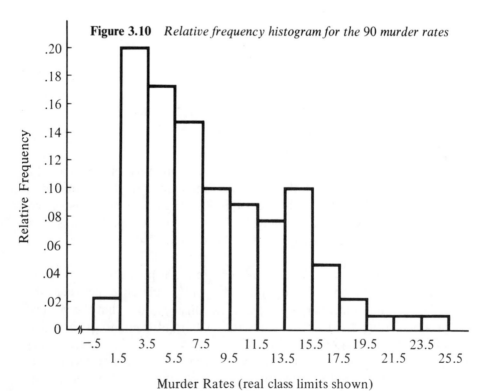

Figure 3.10 *Relative frequency histogram for the 90 murder rates*

Murder Rates (real class limits shown)

simply a histogram. To avoid graphical distortions, that is, stretching or shrinking the vertical axis to accentuate change, whenever possible histograms involving social science data are constructed so that the height of the vertical is approximately two-thirds to three-fourths of the horizontal axis.

Figure 3.11   *Histogram for n = 210 murder rates*

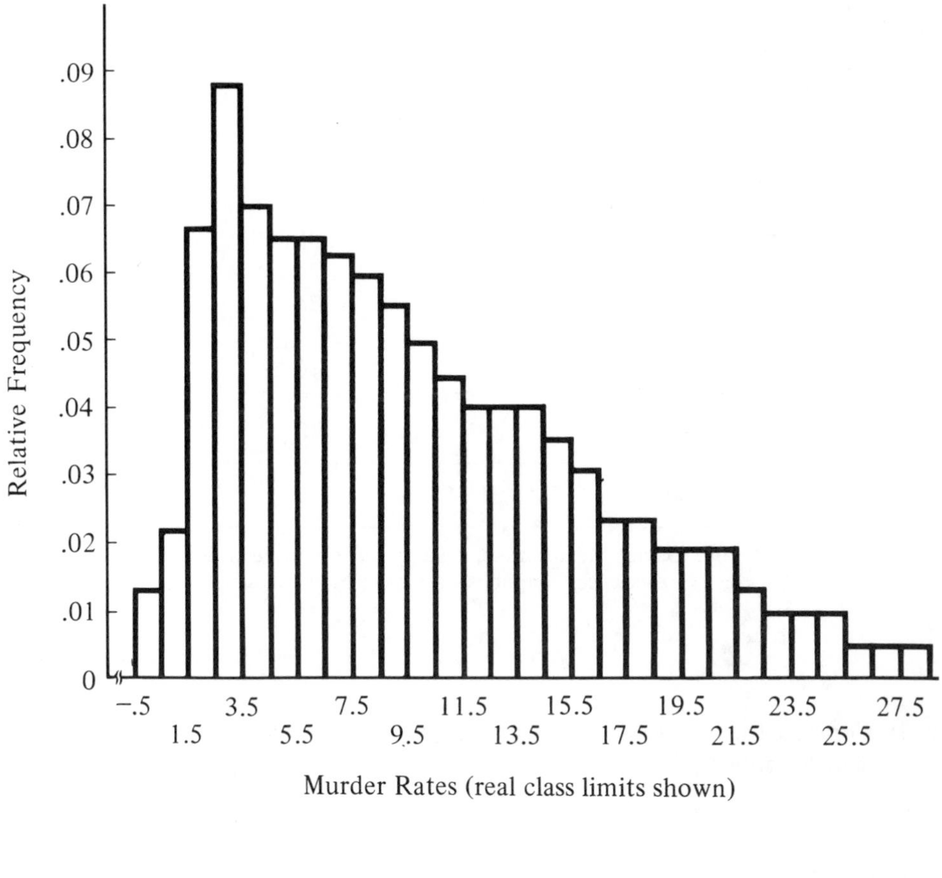

Murder Rates (real class limits shown)

Histograms for large sets of data frequently appear to be almost smooth curves. A histogram describing the set of murder rates from a sample of 210 cities is shown in Figure 3.11. We have intentionally employed more intervals than is usually the case, but we have done so to show that the relative frequency histogram begins to approach a smooth curve when both the number of measurements and the number of class intervals are increased.

13   Compute the relative frequency for each class interval for the data shown (males only).

Age at First Heterosexual (Coital) Experience, Colombia

Age	Males f	Females f
30–31	2	2
28–29	0	5
26–27	2	3
24–25	13	12
22–23	13	28
20–21	29	32
18–19	72	70
16–17	64	74
14–15	68	77
12–13	42	16
10–11	5	0
8–9	1	0
Totals	311	319

SOURCE: David M. Monsees, "Study of First Heterosexual Experience," unpublished paper; by permission.

14   Compute the relative frequency for each class interval for the data of Exercise 13 (females only).

15   Based on the frequency distribution contained in the data shown, what is the class interval width?

Degree of Job Satisfaction Among 219 Nurses

Index Score	f
68–71	4
64–67	13
60–63	24
56–59	43
52–55	38
48–51	43
44–47	33
40–43	7
36–39	7
32–35	3
28–31	1
24–27	1
20–23	2
Total	219

SOURCE: Kimball P. Marshall, "A Study of Job Satisfaction Among 219 Nurses in the Southeast," unpublished paper; by permission.

16   Based on the frequency distribution contained in the data shown, what is the class interval width? Construct a relative frequency histogram for these data.

Index of Presence of Psychiatric Symptoms in South County

Index Score	f	Percent
30–32	2	.1
27–29	3	.2
24–26	8	.5
21–23	12	.7
18–20	41	2.5
15–17	88	5.3
12–14	232	14.1
9–11	365	22.2
6–8	462	28.1
3–5	302	18.4
0–2	130	7.9
Total	1645	100.0

SOURCE: George J. Warheit, *Southern Mental Health Needs and Services Project*, NIMH 15900–05, unpublished data; by permission.

17 Set up a relative frequency column for the data shown and compute the actual relative frequencies.

Self-esteem Scores of C+ Students

Scores	f
10–11 (high)	2
12–13	10
14–15	12
16–17	16
18–19	27
20–21	37
22–23	13
24–25	12
26–27	2
28–29	4
30–31	1
32–33 (low)	1
Total	137

SOURCE: Billy L. Williams, "Self-Esteem and Patterns of Group Memberships Among High School Adolescents," University of Florida, unpublished M.A. thesis, 1973, p. 101; by permission.

18 Examine the frequency table to determine if the class intervals meet the guidelines specified in Section 3.7. Construct a frequency histogram.

Size of Ph.D. Faculties in Sociology at 111
North American Universities

Size	f
44–47	1
40–43	4
36–39	6
32–35	11
28–31	13
24–27	17
20–23	10
16–19	19
12–15	11
8–11	13
4–7	5
0–3	1
Total	111

SOURCE: Marc Petrowsky, "Departmental Prestige and Scholarly Productivity: A Replication of Previous Research," University of Florida, unpublished M.A. thesis, 1971, pp. 69–77; by permission.

3.8   COMMENTS CONCERNING HISTOGRAMS

*T*he relative frequency histogram is the graphical descriptive technique of primary interest in statistical inference. We now present some pertinent comments concerning its interpretation and relevance to inference.

First, it is important to note that the fraction of the total area under the frequency histogram over a particular interval is equal to the fraction of the total number of measurements falling in that interval. For example, the fraction of the 90 murder rates (of Table 3.5) less than or equal to 5 is 35/90. You will note in Figure 3.12 that 35/90 of the total area under the frequency histogram (shaded) lies to the left of 5.5.

The second pertinent point to make in interpreting the frequency histogram is that if a single observation is selected from the total number, the probability that it will lie in a particular interval is equal to the fraction of

Figure 3.12   *Frequency histogram showing fraction of rates below 5.5*

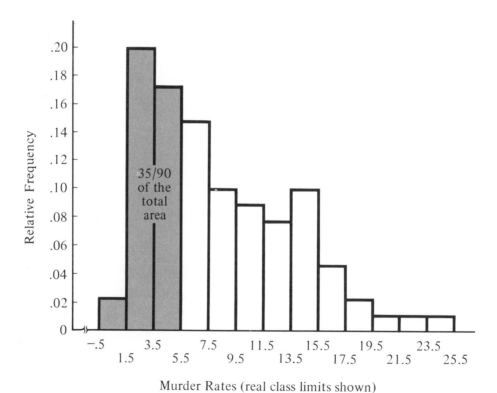

Murder Rates (real class limits shown)

the total area under the histogram over that interval. For example, if 90 cards were labeled with the respective rates in Table 3.4, then shuffled, the probability of choosing a murder rate between 3.5 and 7.5 would be 28/90, because 28 of the 90 rates fall in this interval. From Figure 3.13 we note that the area under the frequency histogram over the interval 3.5 to 7.5 (shaded) is equal to 28/90 of the total area.

Figure 3.13 *Frequency histogram showing the fraction of rates in the interval 3.5–7.5*

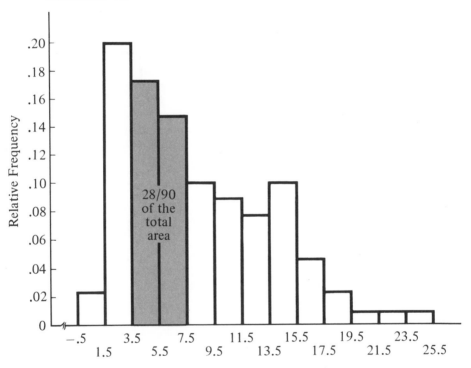

Murder Rates (real class limits shown)

The third important point to note is that one can construct a frequency histogram for any set of numerical values in a sample or a population, but our object is to describe or make inferences about a population. Although we shall seldom actually have all the measurements for the population in question and hence will not be able to physically construct the frequency histogram, we can imagine that one could be constructed and that it would possess an outline similar to that obtained for a sample. Since populations usually contain a large number of measurements, the number of classes can be made

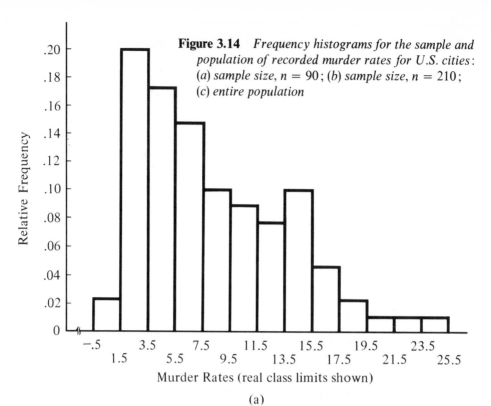

Figure 3.14 *Frequency histograms for the sample and population of recorded murder rates for U.S. cities: (a) sample size, n = 90; (b) sample size, n = 210; (c) entire population*

(a)

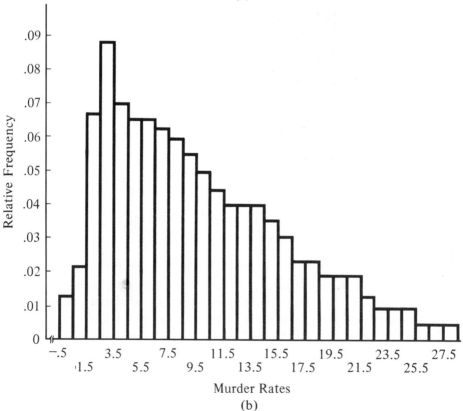

(b)

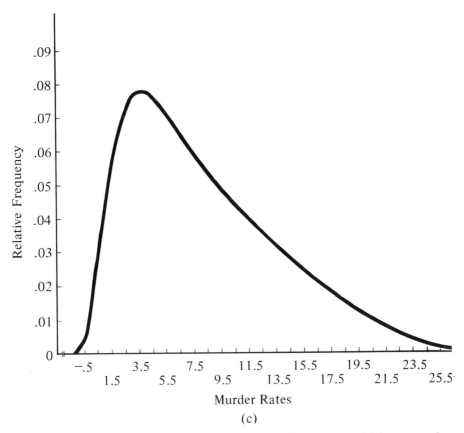

Murder Rates

(c)

rather large so that the population frequency histogram will become almost a smooth curve. The 90 rates were recorded on random samples of 30 cities from three regions. As the number of measurements in a sample increases, we can select smaller class intervals. The resulting histogram will then become more regular and tend to become a smooth curve. Figure 3.14 shows three frequency histograms for murder rates. The first is a histogram for a sample of $n = 90$ measurements, the second for a sample of $n = 210$ measurements, and the third for the entire population. Note that the scale of the relative frequency (ordinate) will change from one figure to another. Second, observe that it may be necessary to change the endpoints of the class intervals [as was done in parts (a) and (b)] so that no measurement falls on the boundary between two classes.

The frequency histogram is an excellent way to characterize a population of measurements. The area under the frequency histogram for the population tells us the proportion of the total number of measurements in the population falling in given intervals. We can also see from the frequency histogram for the population what one might expect the largest and the smallest measurements to be.

19   Using the data of Exercise 13 (page 66) plot the distribution of scores (age at first heterosexual experience) for males in the form of a frequency histogram.

20   Repeat Exercise 19 using female scores.

21   Use a frequency histogram to describe graphically the degree of job satisfaction among nurses for the data of Exercise 15 (page 67).

22   If the data of Exercise 16 (page 67), index of presence of psychiatric symptoms, were plotted in a frequency histogram and then a relative frequency histogram, would the two figures be identical? Comment.

23   Using the data of Exercise 17 (page 68), construct a relative frequency histogram.

24   If you were to plot the data of Exercise 18 (page 69) in the form of a histogram, would you use $-.5$ to 3.5 as your first interval on the horizontal axis or 0 to 3? Comment?

3.9   GRAPHICAL DESCRIPTION FOR QUANTITATIVE DATA: FREQUENCY POLYGON

*T*he frequency polygon represents an alternative way to present graphically the results of a frequency distribution. Once again the vertical axis is labeled with frequencies and the real class limits are marked on the horizontal axis. The frequency associated with each class is indicated by placing a dot over the midpoint of a class interval with the height of the dot equal to the class frequency. The dots are then joined by straight lines. The frequency polygon for the murder-rate data, Table 3.5, is presented in Figure 3.15. The reader will note that the polygon commences at the lower real limit of the first class interval and ends at the upper real limit of the last class interval. Again, to avoid distortions produced by different scales on the axes, we attempt to hold the height of the vertical axis to approximately $\frac{2}{3}$ to $\frac{3}{4}$ the length of the horizontal axis.

Frequency polygons can be used to summarize frequency data in almost any field. For example, the U.S. Weather Bureau keeps elaborate records on the number and intensity of tropical storms and hurricanes each

Figure 3.15 *Frequency polygon for 90 sampled cities*

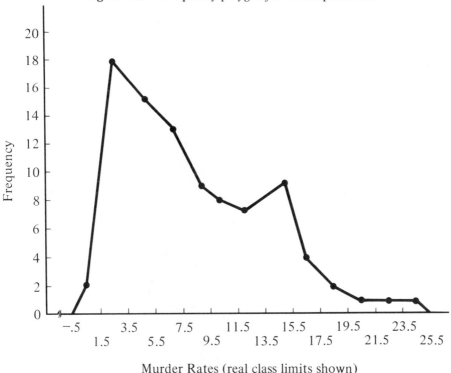

Murder Rates (real class limits shown)

year. Some of these results are presented as a frequency polygon, Figure 3.16. The total number of hurricanes and tropical storms by month during the hurricane season is displayed for the years 1887–1958.

EXERCISES

25 Plot the data contained in Exercise 15 (page 67) in the form of a frequency polygon.

26 Use both sets of data contained in Exercise 13 (page 66) to plot a frequency polygon first for the males and then for the females. Plot the two polygons on the same graph and interpret the results.

27 Do frequency polygons have the same probabilistic interpretation as relative frequency histograms?

Figure 3.16 *Monthly variation of hurricanes.* [SOURCE: *St. Petersburg Times*, July 12, 1970; by permission]

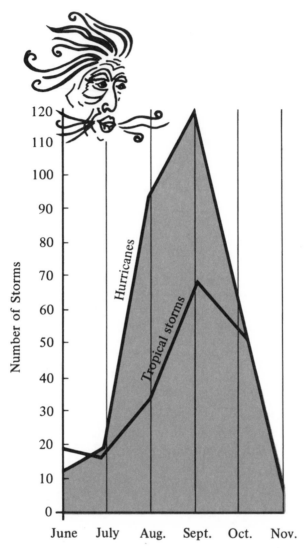

28   Plot the data contained in Exercise 16 (page 67) in the form of a frequency polygon.

29   Plot the data contained in Exercise 17 (page 68) in the form of a frequency polygon.

30   Plot the data contained in Exercise 18 (page 69) in the form of a frequency polygon.

3.10   CUMULATIVE RELATIVE FREQUENCY POLYGONS

ocial scientists are often interested in the proportion of measurements less than or equal to some specific value. For example, they might be interested in the fraction of the standard metropolitan statistical areas, Table 3.5, with murder rates less than or equal to 7.5. This quantity, called the cumulative relative frequency at 7.5, is the sum of the relative frequencies for all classes up to and including class 4, that is, the class with upper real limit equal to 7.5.

Definition 3.2

The cumulative relative frequency at class j is equal to the sum of the relative frequencies for all classes below and including the jth class.

Table 3.6   *Cumulative Relative Frequencies for the Data of Table 3.5*

Interval	Class Interval	Relative Frequency	Cumulative Relative Frequency
13	23.5–25.5	.011	.999*
12	21.5–23.5	.011	.988
11	19.5–21.5	.011	.977
10	17.5–19.5	.022	.966
9	15.5–17.5	.044	.944
8	13.5–15.5	.100	.900
7	11.5–13.5	.078	.800
6	9.5–11.5	.089	.722
5	7.5–9.5	.100	.633
4	5.5–7.5	.144	.533
3	3.5–5.5	.167	.389
2	1.5–3.5	.200	.222
1	−.5–1.5	.022	.022

*Note that the cumulative relative frequencies add to .999 rather than to 1.000. This is a common occurrence caused by rounding the relative frequencies.

For the murder-rate data, the cumulative relative frequency at the fourth class is

$$\text{(cumulative relative frequency at class 4)} = .022 + .200 + .167 + .144$$
$$= .533$$

In a similar way we can obtain the cumulative relative frequencies for all classes. These are shown in Table 3.6 for the data of Table 3.5.

Figure 3.17   *Cumulative relative frequency polygon for the data of Table 3.6*

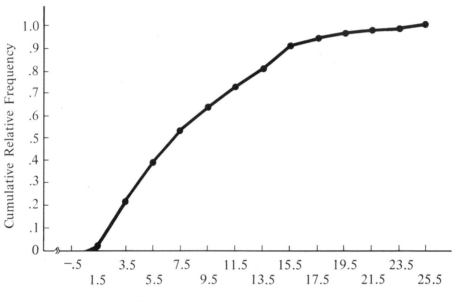

Murder Rates (real class limits shown)

The cumulative relative frequencies can be graphically displayed using a cumulative relative frequency polygon that is constructed in a manner similar to that employed for the frequency polygon of Section 3.9, with the exception that the vertical axis is labeled "Cumulative Relative Frequency" rather than "Frequency." The cumulative relative frequency associated with each class is indicated by a dot placed over the upper endpoint of the class interval with the height of the dot equal to the class cumulative relative frequency. The dots are then joined by straight lines. The cumulative relative frequency polygon for the data of Table 3.6 is presented in Figure 3.17.

31 Set up a cumulative relative frequency column for the data contained in Exercise 13 (page 66) males only.

32 Set up a cumulative relative frequency column for the data contained in Exercise 13, females only. Plot the cumulative relative frequency distribution.

33 When might a cumulative relative frequency polygon be more useful than a relative frequency polygon? Comment.

34 Set up a cumulative relative frequency column for the data contained in Exercise 16 (page 67).

35 Set up a cumulative relative frequency column for the data contained in Exercise 17 (page 68).

36 Set up a cumulative relative frequency column for the data contained in Exercise 18 (page 69).

3.11   GRAPHICAL DESCRIPTION FOR QUANTITATIVE DATA: TIME GRAPH

The final technique for graphical portrayal of quantitative data is a time graph, a graphic representation of a time series (Chapter 2). There are two types of time graphs: arithmetic and semilogarithmic. The only difference between them involves the scale of the vertical axis. We begin by examining the arithmetic time graph.

The arithmetic time graph is similar in appearance to the frequency polygon except that we label the horizontal axis in intervals of time and the vertical axis in units of the quantitative variable. For example, Figure 3.18 is a time graph that displays the fluctuations of the prime interest rate from 1950 to 1971. Note that we still require that the horizontal (time) axis be marked off in uniform time intervals and that entries be plotted over the midpoints of the time intervals as we did for the frequency polygon.

Sometimes it is instructive to display multiple time graphs on the same scale. In examining the fluctuations of the prime interest rate, it would also be useful to examine the trends in 3-month treasury bills. These two rates are displayed simultaneously in Figure 3.19.

Figure 3.18   *Time graph of the prime interest rate.* [SOURCE: *Finance Facts*, published by National Consumer Finance Association, May 1972; by permission]

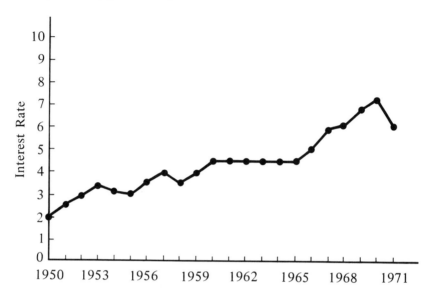

It is not always possible to maintain an unbroken vertical scale starting at zero as was done in Figures 3.18 and 3.19 without making the time graph unwieldy and out of proportion. In these cases we form a *break* in the vertical axis. Such a break is illustrated in the time graph for an index of consumer attitudes and inclinations to buy (Figure 3.20).

The arithmetic time graph we have just discussed provides a means for graphically displaying absolute change in a quantitative variable over time. In contrast, the semilogarithmic time graph enables us to represent relative change in a quantitative variable over time. The only difference between the two graphs is the vertical axis. Recall that we labeled the vertical axis of the arithmetic time graph in units of the quantitative variable. Now, although we shall label the vertical axis in units of the quantitative variable, distances between labeled numbers on the axis will be altered to enable us to readily identify a variable with a constant rate of change.

What do we mean when we say a variable has a constant rate of change? Any variable that has the same *relative* change from time period to time period has a constant rate of change. For example, if total sales for a particular company start at 1 million dollars and increase in successive years to 2, 4, 8, 16, and 32 million dollars, the company would have a constant rate of increase in sales

Figure 3.19 *Money market rates.* [SOURCE: *Finance Facts*, published by National Consumer Finance Association, May 1972; by permission]

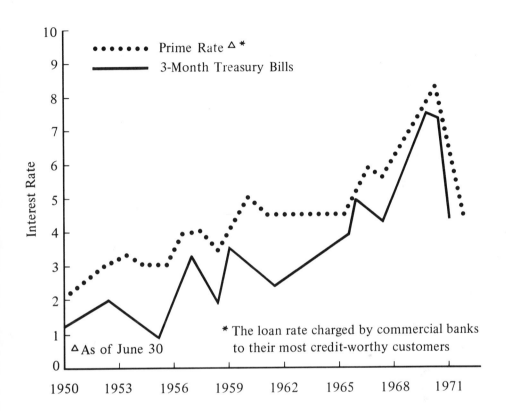

since each year's sales is double that of the previous year. The vertical axis of a semilogarithmic scale is scaled so that the distance between 1 and 2, 2 and 4, 4 and 8, 8 and 16, and 16 and 32 is identical. Similarly, any two sets of numbers that have the same ratio have identical distances between them. The distance between 5 and 15 is identical to that between 1 and 3 or 20 and 60.

Although one could draw the appropriate vertical axis for a semilogarithmic time graph, it is convenient to use prepared semilogarithmic graph paper. This paper comes marked with two or more cycles and can be identified by the identical sets of horizontal markings. Sheets of two-cycle and three-cycle semilogarithmic graph paper are displayed in Figure 3.21(a) and (b), respectively. The number of cycles required for the graph paper depends on the range of values for the quantitative variable. The cycles run in the

Figure 3.20   *Index of consumer sentiment (attitudes and inclinations to buy).* [SOURCE: *Finance Facts*, published by National Consumer Finance Association, May 1972; by permission]

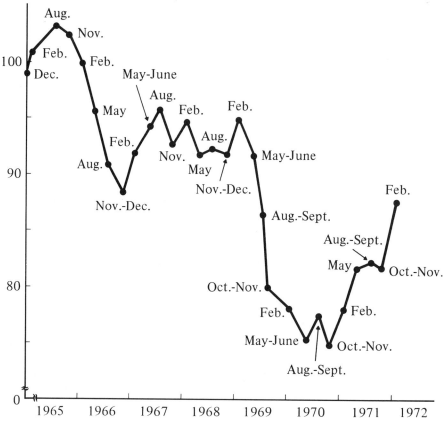

following sequence:

.01 to .1	100 to 1000
.1 to 1	1000 to 10,000
1 to 10	10,000 to 100,000
10 to 100	etc.

If the variable lies in the interval from 13 to 88, we would need one cycle, 10 to 100. If the variable ranged from 105 to 10,800, we would use three-cycle graph paper (100 to 1000, 1000 to 10,000, and 10,000 to 100,000).

Example 3.1   *Enrollments for kindergarten, elementary school, high school, college, and the population of the United States are displayed in Table 3.7 for the years 1930 to 1970. Plot the kindergarten data on semilogarithmic time graph.*

Figure 3.21   *(a) Two-cycle semilogarithmic paper; (b) three-cycle semilogarithmic paper*

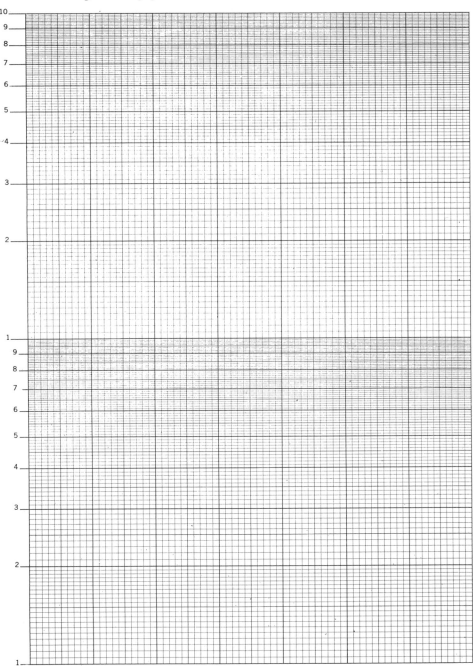

(a)

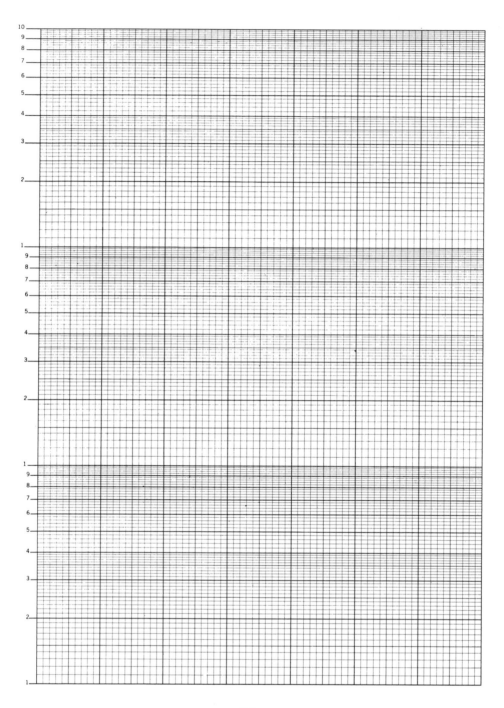

(b)

Table 3.7   *Kindergarten Through College Enrollment and the U.S. Population,* 1930–1970 (*in units of* 1000)

	1930	1940	1950	1960	1970
Kindergarten	786	661	1,175	2,293	2,821
Elementary school	22,953	20,466	21,032	30,119	34,290
High school	4,812	7,130	6,453	9,600	14,518
College	1,101	1,494	2,659	3,216	7,136
U.S. population*	123,077	132,457	151,868	179,976	203,810

* Total excludes the military.

SOURCE: U.S. Bureau of the Census, *Statistical Abstract of the United States: 1972* (Washington, D.C.), pp. 5, 105.

Solution   *Since all the numbers are in units of* 1000, *we can use the data as given in Table 3.7. The kindergarten enrollments are between* 661 *and* 2821, *which requires two-cycle semilogarithmic paper with cycles of* 100 *to* 1000 *and* 1000 *to* 10,000. *The kindergarten data have been graphed in Figure 3.22.*

Example 3.2   *Use the data of Table 3.7 to plot all four sets of data corresponding to enrollments for kindergarten, elementary school, high school, and college on a semilog time graph.*

Solution   *The enrollments for all four populations lie in the interval* 661 *to* 34,290, *so we need to use three-cycle semilog graph paper (with cycles of* 100 *to* 1000, *and* 1000 *to* 10,000, *and* 10,000 *to* 100,000). *These data have been graphed in Figure 3.23.*

We mentioned previously that the vertical axis of the semilogarithmic time graph was constructed so that it would be easy to detect a variable that has a constant rate of change over time. Indeed, this is so. In fact, anytime the trend in the semilog time graph is a straight line, the rate of change is constant over time. Typical examples of increasing, decreasing, and constant trends over time are given in Figure 3.24.

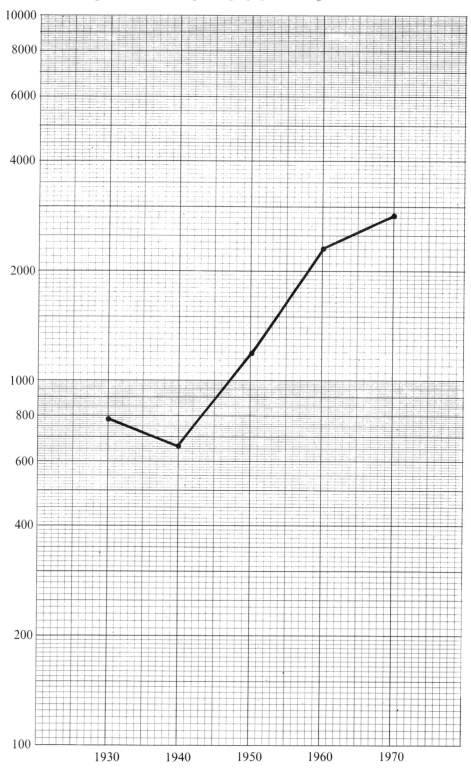

Figure 3.22 *Semilog time graph for kindergarten enrollments*

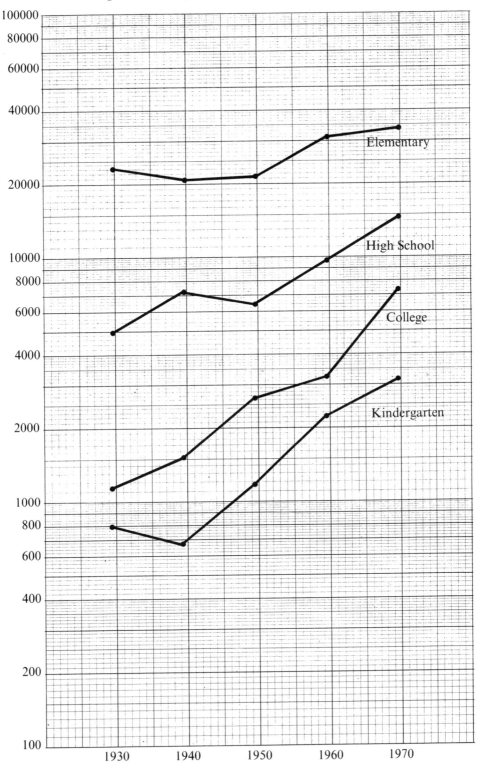

Figure 3.23 *Time graph of kindergarten, elementary school, high school, and college enrollments, 1930–1970*

Figure 3.24 *Typical semilogarithmic time graph patterns:* (*a*) *variable is increasing at an increasing rate;* (*b*) *variable is increasing at a decreasing rate;* (*c*) *variable is increasing at a constant rate;* (*d*) *variable is decreasing at a decreasing rate;* (*e*) *variable is decreasing at an increasing rate;* (*f*) *variable is decreasing at a constant rate*

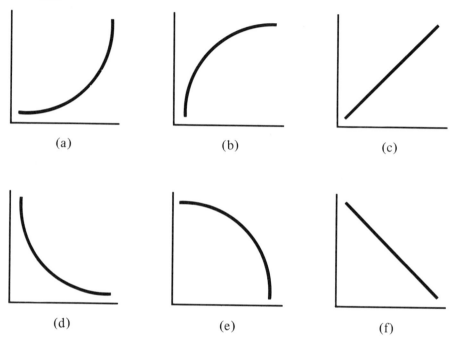

(a)   (b)   (c)

(d)   (e)   (f)

EXERCISES

37   Plot the data in the table on an arithmetic time graph.

Civilian per Capita Consumption of Meats in the United States by Pound (1940–1970)

Year	Consumption
1940	142.4
1950	144.6
1960	160.9
1970	186.3

SOURCE: U.S. Bureau of the Census, *Statistical Abstract of the United States: 1972,* Washington, D.C., p. 84.

38 Plot the data in the table first on arithmetic graph paper and then on semilog graph paper. Comment on the two graphs.

Admissions Rate to Mental Hospitals per 10,000 *U.S. Population* (1940–1970)

Year	Rate
1940	14
1950	20
1960	23
1970	33

SOURCE: U.S. Bureau of the Census, *Statistical Abstract of the United States: 1972*, Washington, D.C., p. 74.

39 Plot on semilog paper the two sets of data given. What trends are suggested, if any?

Incidence Rates of Encephalitis and Measles per 1,000,000 *U.S. Population* (1940–1970)

Year	Encephalitis Rate	Measles Rate
1945	5.6	1039.5
1950	7.5	2095.8
1955	13.1	3345.7
1960	13.0	2444.8
1965	13.9	1347.9
1970	9.5	231.0

SOURCE: U.S. Bureau of the Census, *Statistical Abstract of the United States: 1972*, Washington, D.C., p. 80.

40 Plot on arithmetic graph paper the two sets of data given. Comment on the two patterns.

Number of Physicians and Nurses per 100,000 *U.S. Population* (1950–1970)

Year	Physicians	Nurses
1950	149	249
1955	150	259
1960	148	282
1965	153	319
1970	171	345

SOURCE: U.S. Bureau of the Census, *Statistical Abstract of the United States: 1972*, Washington, D.C., p. 70.

J

3.12   DIFFERENT SHAPES FOR
FREQUENCY DISTRIBUTIONS

n Sections 3.6 through 3.9 we studied procedures for organizing quantitative data and presented the frequency distribution as a graphical technique for data description. At this time we would like to describe briefly typical shapes one might encounter in these histograms.

The most common shape for a frequency distribution is that of the bell-shaped (or normal) curve pictured in Figure 3.25(a). Although we shall

Figure 3.25   *Typical shapes for frequency distributions: (a) bell-shaped curve; (b)* J-shaped curve; *(c)* U-shaped curve

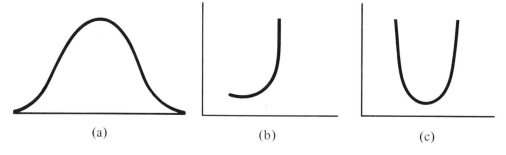

(a)   (b)   (c)

discuss the normal curve in greater detail in Chapters 4 and 5, it is important to note that the normal curve is *the* most important frequency distribution that we will encounter in this text. This is because so many populations of real data, collected in many areas of the social and physical sciences, possess nearly normal frequency distributions. For example, social science variables, such as examination test scores, average hourly wage paid to construction workers by county, Social Security average payments to retirees by area of residence, and length of patient confinement in hospitals, all possess frequency distributions that are approximately bell-shaped.

Two other curves, the J-shaped and U-shaped curves, are indicated in Figure 3.25(b) and (c), respectively. Although less common than the bell-shaped normal curve, they are not rare. Almost any behavior that is governed by an enforced social norm would have a J-shaped frequency distribution. Typical examples might include the number of people (frequency) who commit incestuous acts, kill people, or run stoplights on a particular day.

Examples of variables with a U-shaped distribution include the number of people who utilize physician services $X = 0, 1, 2, \cdots$ times per year. This curve will vary as a function of social class and age of the people.

Similarly, looking only at the totality of patients seeking a medical service, you obtain a U-shaped curve if you plot frequency (number of patients) as a function of age of the patients.

3.13 GRAPHICAL DISTORTIONS

ictures provide an excellent way to distort the truth. You have seen those tobacco ads that feature a lovely seductress who creates in many of us an almost inhuman urge to dash out and buy a pack of cigarettes. Mail-order catalog sketches of products are frequently more attractive than the real thing, but we usually take this type of "lying" for granted, and we submit to these minifrauds with much less distress than perhaps we should. Statistical pictures are the histograms, frequency polygons, pie charts, and bar graphs of this chapter. These types of drawings or displays of numerical results are more difficult to combine with sketches of lovely females and are hence secure from the most common form of graphic distortion. Instead, one can shrink or stretch axes of the various figures to imply the desired results, based on our intuitive understanding that shallow and steep slopes are associated with small and large increases, respectively.

For example, suppose that the number of near-fatal collisions between aircraft per month at a major airport is recorded as 13, 14, 14, 15, and 15 for the period January through May. If you want this growth to appear small (perhaps you represent the Civil Aeronautics Administration), you show the results using the frequency polygon of Figure 3.26. The growth is apparent,

Figure 3.26 *Number of near-collisions per month*

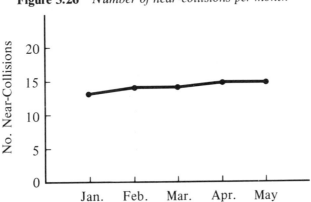

but it does not appear to be very great. If you want the growth to appear large (perhaps you belong to the Citizens' Safety Group), look at the graph of

Figure 3.27 *Number of near-collisions per month*

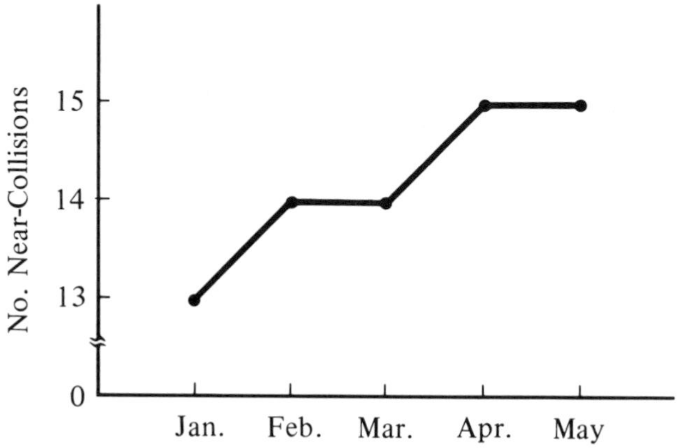

the same data in Figure 3.27. The vertical axis is stretched and does not include zero. Note the impression of a substantial rise indicated by the steeper slope.

Another way to achieve the same effect—to decrease or increase a slope—is to stretch or shrink the horizontal axis. Of course, you are sometimes limited in the amount of shrink or stretch you can apply and still achieve a picture that appears reasonable to the viewer. For example, you could not shrink or stretch the horizontal axes of Figures 3.26 and 3.27 very much because of the limited number of data points ($n = 5$).

Shrinking or stretching axes to increase the slopes in bar graphs, histograms, frequency polygons, or other figures usually catches the hasty reader unaware; the distortions are apparent only if you look closely at the axes. The important point, however, is that increases or decreases in responses are judged large or small depending on the arbitrary importance to the observer of the change, not on the slopes shown in graphic representations.

3.14 SUMMARY

We study how to describe a set of measurements for two reasons. First, one of the uses of statistics in the social sciences is to make inferences about a population based on information contained in a sample. Since populations are sets of measurements, existing or imaginary, we need some method for talking about the population or, equivalently, for describing a set of measurements. The second reason is the need for the condensation and description of large quantities of social science data that are collected annually by various government agencies or even every 10 years by the U.S. Bureau of the Census.

Data can be described either graphically or numerically. Graphical techniques such as the pie chart, the bar chart, histograms, and frequency polygons are presented in this chapter. Numerical descriptive measures are discussed in Chapter 4.

Note the role of data description in statistical inference. It is not possible to make an inference about anything unless you are able to describe the object of your interest. Graphical descriptions of sample data are similar to corresponding figures for the population from which the sample has been drawn, but this type of inference possesses limitations that will be made apparent in Chapter 4. The sole purpose of this chapter is data description.

QUESTIONS AND PROBLEMS

41 A researcher asked a group of high school dropouts their reasons for leaving school. He coded their responses as given in the table.

Reason	Percent
Dissatisfied with school	28
Needed income to help support family	25
Joined the armed forces	17
Arrested and placed in a detention home	12
Do not know	9
Gave no reason	5
Other	4

Evaluate the organization of responses based upon the principles of exclusiveness and inclusiveness.

42 Construct a pie chart for the 1970 estimated membership of the principal world religions in North America. See Figure 3.4 (page 51). (*Suggestion:* Combine Eastern Orthodox and Others.)

43 Using the data in Table 3.3 (page 47), construct a bar graph.

44 Find an example of a statistical map used in the social science literature.

45 Use the data in Table 3.4 (page 60) to construct a frequency distribution where the interval width is 3; that is, $i = 3$.

46   Compute the relative frequencies for the data contained in Exercise 45.

47   Construct a frequency histogram for the data contained in Exercise 45.

48   Construct a frequency polygon for the data contained in Exercise 45.

49   Plot the time series for man-hours needed to produce 1 net ton of raw steel for both the United States and Japan (individually) using the data in Table 2.13 (page 32).

50   The vertical axis of a time series may be an unbroken line that includes 0. In other instances, we include only a portion of the vertical axis (omitting 0). Under what conditions might you select one of these two alternatives?

51   Use the data contained in Exercise 7 (page 57) to prepare a statistical map showing the percentage distribution, irrespective of country of origin, by state, of foreign stock in the United States. Set up your own appropriate unit for grouping percentages.

52   Can we construct a bar graph using the data contained in Exercise 39 (page 89)?

53   Construct a cumulative relative frequency polygon for the data in the table.

Degree of Professionalism Among 219 Nurses*

Index Score	f
111–113	1
108–110	0
105–107	3
102–104	2
99–101	10
96–98	7
93–95	9
90–92	9
87–89	21
84–86	36
81–83	31
78–80	25
75–77	28
72–74	14
69–71	6
66–68	7
63–65	2
60–62	1
Total	212

* Seven nurses were eliminated because they did not respond.

SOURCE: Kimball P. Marshall, "A Study of Professionalism Among 219 Nurses in the Southeast," unpublished paper.

54 Construct a relative frequency polygon for the data in the table.

Index of Presence of Depressive Symptoms in South County

Index Score	f	Percent
60–64	1	.1
55–59	2	.1
50–54	6	.4
45–49	9	.5
40–44	27	1.6
35–39	44	2.7
30–34	81	4.9
25–29	146	8.9
20–24	214	13.0
15–19	330	20.1
10–14	352	21.4
5–9	350	21.3
0–4	83	5.0
Total	1645	100.0

SOURCE: George J. Warheit, *Southern Mental Health Needs and Services Project*, NIMH 15900–05, unpublished data; by permission.

55 Construct a cumulative relative frequency polygon for the data in Exercise 54.

REFERENCES

Anderson, T. R., and M. Zelditch. *A Basic Course in Statistics*, 2nd ed. New York: Holt, Rinehart and Winston, Inc., 1968. Chapters 2, 3, and 4.

Blalock, H. M. *Social Statistics*, 2nd ed. New York: McGraw-Hill Book Company, 1972. Chapter 4.

Champion, D. J. *Basic Statistics for Social Research*. Scranton, Pa.: Chandler Publishing Company, 1970. Chapter 2.

Mueller, J. H., K. F. Schuessler, and H. L. Costner. *Statistical Reasoning in Sociology*, 2nd ed. Boston: Houghton Mifflin Company, 1970. Chapters 3 and 4.

Palumbo, D. J. *Statistics in Political and Behavioral Science*. New York: Appleton-Century-Crofts, 1969. Chapter 2.

Weiss, R. S. *Statistics in Social Research*. New York: John Wiley & Sons, Inc., 1968. Chapter 5.

4

NUMERICAL TECHNIQUES FOR DESCRIBING DATA FROM A SINGLE SAMPLE

n

umerical descriptive measures are commonly used to convey a mental image of physical objects or phenomena. Many students understand the expression "the 100 in 9.2" and have no difficulty in creating a mental picture of a runner dashing 100 yards in 9.2 seconds. So it is with statistics. Although satisfied by the ability of a frequency distribution to describe a set of measurements, we seek one, two, or more numbers, called *numerical descriptive measures,* that will create a mental picture of the frequency distribution for a set of data. We have two good reasons for desiring numerical as well as graphical techniques for data description.

First, we frequently wish to discuss with other people sets of measurements, populations, and large sets of sociological, economic, or census data, and it is inconvenient to carry frequency histograms about in one's pocket. It would be much easier if we could conjure a picture of the frequency distribution in the minds of our listeners using one or two descriptive numbers. Second, the frequency distribution is an excellent method for characterizing a population, but it possesses severe limitations when used to make inferences. The irregular frequency histogram of the sample will be similar to the corresponding distribution for the population, but how similar? How do we measure the goodness of our inference? How do you measure the degree of dissimilarity between two irregular figures?

The sample frequency histogram can be used to make an inference concerning the shape of the population frequency distribution, but there is no satisfactory method of saying how good the inference is. In contrast, numerical descriptive measures of the population can be *estimated* using the sample measurements, and one can say, with a measured degree of confidence, how close the estimate will lie to the population descriptive measure. Since population numerical descriptive measures will frequently be the target of our inferences (that is, we shall estimate or make decisions about them), we shall give them a special name.

Definition 4.1

Numerical descriptive measures of a population are called parameters.

The two most important types of parameters are those that locate the center and describe the spread of the distribution. They are called measures of *central tendency* and *variation,* respectively. Although perhaps inconceivable, we will show that two numbers, one locating the center of a distribution and one the spread, do provide a very good description of the frequency distribution for a set of measurements. As you might suspect, we shall frequently use a descriptive measure of the sample to estimate the value of the corresponding parameter of the population.

Measures of central tendency, their definitions, interpretations, and applications will be presented in Sections 4.2 through 4.6. Measures of variability and, more important, their calculation and interpretation occupy the remainder of the chapter. Chapter 4 will provide the final touches to the description of a set of measurements—the first step in our study of statistical inference. We shall use these descriptive measures in later chapters to phrase inferences about populations based on sample measurements.

4.2 A MEASURE OF CENTRAL TENDENCY: THE MODE

The first measure of central tendency is the *mode* of a distribution of measurements. (Keep in mind that scores and values are types of measurements.)

Definition 4.2

The *mode* of a set of measurements is defined to be the measurement that occurs with greatest frequency.

The mode is the least common of the three measures of central tendency considered in this text, but it is very useful in business planning for identifying products or product sizes in greatest demand. A shirt or dress manufacturer would be interested in the sizes most frequently purchased. Frequent reference is also made to the mode of a set of measurements in advertising campaigns. We often hear that.Brand W is preferred by housewives to any other laundry detergent, or more doctors smoke Lungs cigarettes than any other brand. Sociological surveys sometimes refer to the mode of a set of measurements, such as the most frequently observed number of children per family for a suburban demographic study. The mode is employed where it is important to locate the measurement that occurs most frequently in a set.

Example 4.1 *A research team consisting of an anthropologist and a sociologist analyzed the family structure of a large Hutterite community. One task involved counting the number of children in each family. The data given in Table 4.1 represent a portion of the total number of Hutterite families. Determine the modal number of children.*

Table 4.1 *Number of Children in 25 Hutterite Families*

7	10	8	11	9
9	9	8	9	8
9	9	9	8	9
8	8	9	10	11
10	7	10	9	7

Solution *First, let us arrange the measurements in an array, ranging from the smallest to the largest:*

7, 7, 7, 8, 8, 8, 8, 8, 8, 9, 9, 9, 9, 9, 9, 9, 9, 9, 9, 10, 10, 10, 10, 11, 11

It is clear from these data that the modal number of children per Hutterite family is 9.

Sometimes a set of measurements has more than one mode. We can label these sets of measurements (or correspondingly the frequency distributions of these sets of measurements) as bimodal, trimodal, and so on. An extension of this situation occurs when all observations appear the same number of times. In this case the mode gives no information in locating the center of the distribution and we say that the frequency distribution possesses no mode.

Identification of the mode for the data of Example 4.1 was quite easy because we were dealing with the actual measurements. However, when one tries to compute the mode for grouped data, some difficulties arise. Consider the following example. Three samples of 30 standard metropolitan statistical areas (SMSA) were drawn to obtain information on murder rates. The murder rate for each of the SMSA was recorded in Table 3.4. Recall that we organized the data to obtain a single frequency distribution of these measurements (Table 4.2).

Table 4.2 *Relative Frequency Distribution for Three Samples of 30 SMSA's*

Class Interval		Class Frequency	Relative Frequency
Real Limits	Apparent Limits*		
23.5–25.5	24–25	1	.011
21.5–23.5	22–23	1	.011
19.5–21.5	20–21	1	.011
17.5–19.5	18–19	2	.022
15.5–17.5	16–17	4	.044
13.5–15.5	14–15	9	.100
11.5–13.5	12–13	7	.078
9.5–11.5	10–11	8	.089
7.5–9.5	8–9	9	.100
5.5–7.5	6–7	13	.144
3.5–5.5	4–5	15	.167
1.5–3.5	2–3	18	.200
−0.5–1.5	0–1	2	.022
Total		90	

* Class intervals for social science data are usually presented with apparent limits, as shown. The real limits would be 23.5–25.5, 21.5–23.5, etc.

There are two problems that one would face in trying to compute the mode for the data in Table 4.2. First we must locate the modal frequency and second, we must determine a value for the mode corresponding to that frequency. We shall discuss two procedures for computing the mode for grouped data. The first is called the crude mode. As the name suggests, it is the simpler of the two procedures.

Definition 4.3

The crude mode for a set of grouped measurements is defined to be the midpoint of the class interval with the highest frequency.

Example 4.2 *Compute the crude mode for the data in Table 4.2.*

Solution *The class with highest frequency is the second class, which has real class limits 1.5 to 3.5. The midpoint of the class is 2.5; that is,*

$$crude\ mode = 2.5$$

It should be noted that the midpoint of a class interval is the same whether we use real or apparent limits. In our example the midpoint of the apparent limits, 2 to 3, is also 2.5.

Unfortunately the procedure that is simpler to compute is not always better. The crude mode is a case in point. If the class width varies, it can be shown that the crude mode will also vary. This instability is rather unsettling since it means that the mode of a set of grouped measurements depends on the way in which the data were grouped.

A modification of the crude mode, called the refined mode, is more stable, that is, it is less dependent on the class width or class interval location, but it requires more calculation.

Suppose, for example, that the frequency distribution appears as shown in Figure 4.1(a). Noting that the frequencies of the classes adjacent to the modal class are equal, we would be inclined to locate the mode at the center of the modal class interval. In contrast, if there is a difference between the frequencies of the adjacent classes, as shown in Figure 4.1(b) and (c), it would appear more reasonable to move the mode toward one end of the modal class interval in the direction of the larger adjacent class frequency. Thus we would move the mode toward the upper end of the modal class for the distribution in Figure 4.1(b) and toward the lower end for (c). A shift correction of the type described above is accomplished using the refined mode. Before proceeding with a formal definition for the refined mode, we need some new notation. Let

L = lower real limit of the interval with the highest frequency
D_1 = difference between the highest frequency and the frequency of the adjacent interval below
D_2 = difference between the highest frequency and the frequency of the adjacent interval above*
i = interval width
M_0 = refined mode

* This assumes that class intervals are listed on the page with high variates at the top and low variates at the bottom. If class intervals are listed with the high variates at the bottom of the page and low variates at the top, the roles of D_1 and D_2 are reversed.

Figure 4.1 *Histograms for three different sets of data*

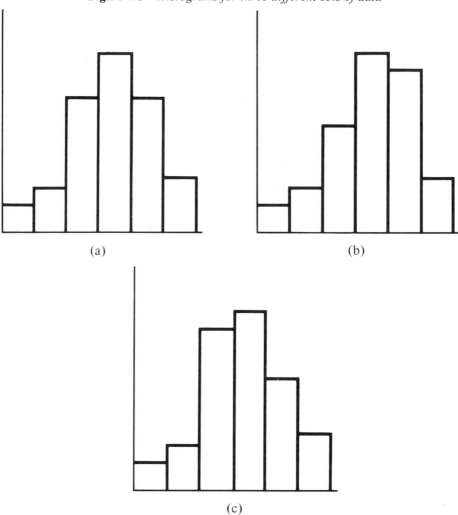

Then the refined mode is defined as follows:

Definition 4.4

The refined mode is defined by the formula

$$M_0 = L + i\left(\frac{D_1}{D_1 + D_2}\right)$$

Example 4.3 *Compute the refined mode using the data in Table 4.2.*

Solution   *We must first determine L, D_1, D_2, and i. The interval with the highest frequency has apparent limits 2–3. Therefore,*

$$L = 1.5$$

The frequencies for the lower and upper adjacent intervals are 2 and 15, respectively. Therefore,

$$D_1 = 18 - 2 = 16$$
$$D_2 = 18 - 15 = 3$$

Similarly, it is readily apparent that the interval width is i = 2. Therefore,

$$M_0 = L + i\left(\frac{D_1}{D_1 + D_2}\right) = 1.5 + 2\left(\frac{16}{16 + 3}\right)$$
$$= 3.18$$

Thus the refined mode, the value of X that occurs with greatest frequency, is computed to be 3.18.

EXERCISES

1   The following represent a sample of 10 measurements on the time (in minutes) required for two-person teams to resove a complex interpersonal argument:

63	45
56	45
32	45
48	39
48	41

Determine the mode.

2   The following represent a sample of 10 measurements on a social sensitivity scale (the higher the score, the greater the sensitivity to the needs of others):

7	0
4	3
4	8
5	2
6	10

Determine the mode.

3 Determine the crude mode for the data shown in Exercise 54, Chapter 3 (page 95). Repeat for the refined mode.

4 Determine the crude mode for the data shown in Exercise 53, Chapter 3 (page 94). Repeat for the refined mode.

4.3 A MEASURE OF CENTRAL TENDENCY: THE MEDIAN

he median is a second measure of central tendency. It is computed in the same way for either a sample or a population. The following definitions refer to ungrouped data.

Definition 4.5

The median for an odd number of measurements is the middle measurement when the measurements are arranged in order of magnitude.

Definition 4.6

The median for an even number of measurements is the average of the two middle observations when the measurements are arranged in order of magnitude.

Note that the median is a number, chosen so that half of the measurements lie below it, half above. Thus if the median wage of a government employee is $3.75 per hour, it means that 50 percent of all government employees make less than $3.75 per hour, 50 percent more. The median is a very popular measure of central tendency used to describe social science data. Newspaper reports and magazines frequently refer to the median wage increase won by unions, the median age of persons receiving Social Security benefits, the median income of families in the United States, and the gap between the median income for men and for women.

Example 4.4 *Find the median of the 7 test scores 95, 86, 78, 90, 62, 73, and 89.*

Solution *We must first arrange the scores in order of magnitude:*

$$
\begin{array}{ccc}
62 & 86 & 90 \\
73 & 89 & 95 \\
78 & &
\end{array}
$$

Since we have an odd number (7) of measurements, the median is then the middle score; that is,

$$median = 86$$

Example 4.5 *Suppose that 3 more students out of a class of 30 took the achievement test of Example 4.4 and scored 73, 75, and 91, respectively. Determine the median for the combined 10 test scores.*

Solution *Since we have an even number of observations, the sample median is the average of the two middle scores when the scores are arranged in numerical order. Arranging the scores in order of magnitude we have*

$$
\begin{array}{cc}
62 & 86 \\
73 & 89 \\
73 & 90 \\
75 & 91 \\
78 & 95
\end{array}
$$

The two midpoint test scores are 78 and 86; hence the median is given by

$$median = \frac{78 + 86}{2} = 82$$

As with the mode, we have to adjust our procedure for computing the median for grouped data. This is because the exact values of the measurements are lost when they have been grouped. Hence we may know that the "middle" observation occurs in a particular class but may not know exactly where to locate the median within the interval.

If almost half of the observations have been counted before you come to the cell containing the median, you would be inclined to locate the median toward the lower portion of the median interval. On the other hand, if the classes up to and including the median class contain exactly 50 percent of the measurements, you would be inclined to move the median to the upper end of the class interval. The following procedure makes this type of adjustment for grouped data. Let

M_d = median
L = lower real limit of the class interval that includes the median
n = total frequency
cf_b = sum of the frequencies (cumulated frequencies) for all class intervals *below* the interval that includes the median
f_m = frequency of the class interval that includes the median
i = interval width

Definition 4.7

The median for grouped data *can be computed as follows:*

$$M_d = L + \frac{i}{f_m}(50\% \text{ of } n - cf_b)$$

Example 4.6 *Compute the median for the murder-rate data of Table 4.2.*

Solution *We must first determine L, n, cf_b, f_m, and i. Recall that we sampled 90 SMSA's; so n = 90 and the interval width was i = 2. To determine the interval that contains the median we begin to sum the relative frequencies from the lowest interval until the cumulated relative frequency passes .50. Backtracking we then determine the interval whose relative frequency makes the cumulative relative frequency more than .50. This interval contains the median. For our data the frequencies, cumulative frequencies, and cumulative relative frequencies are as listed in Table 4.3.*

Note that the cumulative relative frequency for intervals 1 through 3 from the bottom is .389; for 1 through 4 it is .533. Interval 4 contains the median and

$$L = 5.5$$

Similarly, the frequency for class interval 4 is f_m = 13 and the cumulative frequency for intervals below interval 4 is cf_b = 35. The median for the

Table 4.3 *Frequencies for Murder-Rate Data of Table 4.2*

Class Interval	Frequency	Cumulative Frequency	Cumulative Relative Frequency*
24–25	1	90	1.000
22–23	1	89	.989
20–21	1	88	.978
18–19	2	87	.967
16–17	4	85	.944
14–15	9	81	.900
12–13	7	72	.800
10–11	8	65	.722
8–9	9	57	.633
6–7	13	48	.533
4–5	15	35	.389
2–3	18	20	.222
0–1	2	2	.022

*Note that these cumulative relative frequencies differ slightly from those in Table 3.6 because they were calculated directly from the cumulative frequencies. Consequently, they are more accurate because they do not contain the rounding errors of Table 3.6.

grouped data is then

$$M_d = L + \frac{i}{f_m}\left(50\,\%\ \text{of}\ n\ -\ cf_b\right) = 5.5 + \frac{2}{13}(45 - 35)$$

$$= 7.04$$

EXERCISES

5 Determine the median for the data in Exercise 1 (page 103).

6 Determine the median for the data in Exercise 2 (page 103).

7 Determine the median for the data in Exercise 54, Chapter 3 (page 95).

8 Determine the median for the data shown in Exercise 53, Chapter 3 (page 94).

T 4.4   A MEASURE OF CENTRAL
           TENDENCY: THE MEAN

he most widely used measure of central tendency is the arithmetic mean of a set of measurements.

Definition 4.8

An *arithmetic mean* *is the sum of a set of measurements divided by the number of measurements in the set.*

The arithmetic mean, often called the average, is employed extensively in all fields of science and business. Thus we commonly observe phrases such as the mean income for persons living in a ghetto area, the mean tensile strength of a cable, the mean velocity of the first stage of a missile, the mean increase in the cost of living index over the past 6 months, and the mean closing price of a group of stocks (such as the Dow Jones average of 30 industrials).

Example 4.7 *Seven students were given a reading achievement test. Find the mean for the seven test scores:*

$$
\begin{array}{ccc}
95 & 90 & 73 \\
86 & 62 & 89 \\
78 & &
\end{array}
$$

Solution

$$
mean = \frac{sum\ of\ test\ scores}{n}
$$

$$
= \frac{95 + 86 + 78 + 90 + 62 + 73 + 89}{7} = 81.86
$$

The mean for both the sample and the population are defined in the same way since both are sets of measurements, but we shall use separate symbols for each. Although we shall seldom actually calculate the population mean, we will estimate or make decisions about it based on the sample mean. Thus it is important to draw a distinction between the two quantities. We shall use the symbol $\bar{X}$ (X bar) to denote the mean of a sample and μ (the Greek letter mu) to denote the mean of a population.

$\overline{X}$ is the *sample mean*

μ is the *population mean*

It is convenient to introduce some notations that we shall use in the computational formulas encountered in this and later chapters. First, let the letter X represent the measurement or value we are observing. If we refer specifically to a sample, X will represent any measurement in the set. It may also be convenient sometimes to use a subscript to denote a particular measurement in the set. If we consider the seven measurements from Example 4.7,

$$95 \quad 90 \quad 73$$
$$86 \quad 62 \quad 89$$
$$78$$

we could let X_1 denote the first observation. Thus $X_1 = 95$. In the same manner we could let $X_2 = 86$, $X_3 = 78, \ldots, X_7 = 89$.

To indicate a sum we shall use the Greek symbol Σ (sigma). Thus ΣX would indicate the sum of the measurements that we denoted by the symbol, X. In Example 4.7

$$\Sigma X = X_1 + X_2 + X_3 + X_4 + X_5 + X_6 + X_7$$

$$= 95 + 86 + 78 + \cdots + 89 = 573$$

If we have a sample of n measurements that we denote by $X_1 \ X_2, \ldots, X_n$, the sample mean is given by

$$\overline{X} = \frac{\Sigma X}{n} = \frac{X_1 + X_2 + \cdots + X_n}{n}$$

Example 4.8 *Compute the mean murder rate for the original sample of 90 cities (see Table 3.4).*

Solution *The sample of 90 murder rates can be labeled from $X_1, X_2, \ldots, X_{90}$. Hence we obtain*

$$\overline{X} = \frac{X_1 + X_2 + \cdots + X_{90}}{90}$$

$$= \frac{20 + 22 + \cdots + 4}{90} = \frac{742}{90} = 8.24$$

The formula for computing the mean of a set of measurements is slightly altered when working with grouped data. Since we cannot reconstruct the actual measurements prior to grouping, we represent all values in a given class interval by the midpoint of the interval. If we set X equal to the midpoint of a class interval, then the product, fX, will denote the sum of all measurements in that interval. For example, if $X = 5$ is a class midpoint and the class contains $f = 10$ measurements, then we would approximate the class sum as 10 times 5 and denote the sum of the measurements in the class as $fX = 10(5) = 50$. Similarly, $\Sigma\ fX$ will represent the sum of all measurements accumulated over all classes.

Definition 4.9

Let X be the midpoint of a class interval with class frequency f. Then the mean for grouped data *is computed using*

$$\overline{X} = \frac{\Sigma fX}{n}$$

Example 4.9 *Compute the sample mean from the grouped data of Table 4.2.*

Solution *The appropriate class intervals, midpoints, and frequencies are listed in Table 4.4.*

Using the formula listed in Definition 4.9, we have

$$\overline{X} = \frac{\Sigma fX}{n} = \frac{747.0}{90} = 8.30$$

Note that had we grouped our data into intervals three units wide, the midpoints would have been whole numbers, thus avoiding decimals in the table.

Note that the mean computed for the grouped data (Example 4.9) differs from that computed from the ungrouped data (Example 4.8). This

Table 4.4 *Data for Example 4.9*

Class Interval	Midpoint X	Class Frequency f	Class Sums fX
24–25	24.5	1	24.5
22–23	22.5	1	22.5
20–21	20.5	1	20.5
18–19	18.5	2	37.0
16–17	16.5	4	66.0
14–15	14.5	9	130.5
12–13	12.5	7	87.5
10–11	10.5	8	84.0
8–9	8.5	9	76.5
6–7	6.5	13	84.5
4–5	4.5	15	67.5
2–3	2.5	18	45.0
0–1	0.5	2	1.0
Total		90	747.0

difference is a consequence of the method in which the data were grouped. The sample mean for grouped data represents an approximation to the sample mean for the actual observations.

Having discussed three measures of central tendency (the mode, median, and mean), one might wonder which quantity to use in describing a set of measurements. The answer depends upon the application. The mean is the most widely used measure of central tendency for statistical inference, particularly in estimation and testing hypotheses about social phenomena. As noted earlier, the mode is used when we wish to know the most frequently observed value of X. It tends to be a measure of "popularity" and provides a measure of central tendency for qualitative data where the mean and median are not applicable. In some situations the median provides more descriptive information about the center of a distribution. For example, suppose that we are interested in describing the distribution of ages for bridegrooms in a particular city over a given year. It would be easy to sample the marriage licenses on file with a Justice of the Peace. We would undoubtedly find many grooms who were in their late teens or early twenties, but some grooms would be much older. The median, which reflects the midpoint of the sample, would therefore be a better indicator of the ages of bridegrooms because the mean age would be inflated by the much older grooms.

Figure 4.2 *Relationships among the mean (μ), median (M_d), and mode (M_0): (a) bell-shaped curve; (b) skewed to the left; (c) skewed to the right*

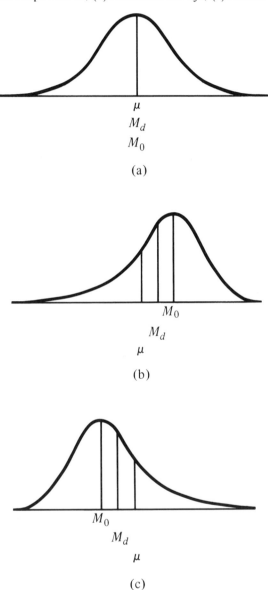

In a normal bell-shaped curve, the values of the mean, mode, and median are identical [see Figure 4.2(a)]. If the distribution is skewed to the left (negative skew), the mean, μ, is the smallest value of the three, while the mode is the largest value [see Figure 4.2(b)]. When the distribution is skewed to the right (positive skew), μ is the largest value of the three, the mode is the smallest. In each case the median is the more central value, and it is for this reason that the median is a widely used measure of location of the center of a distribution.

4.5 CODING TO REDUCE COMPUTATIONAL LABOR FOR THE SAMPLE MEAN

*D*ata are frequently coded to simplify the calculation of $\overline{X}$. It is easier to calculate the mean of the set of measurements $(-.1, .2, .1, 0, .2)$ than for $(99.9, 100.2, 100.1, 100.0, 100.2)$. The first set was obtained by subtracting 100 from each measurement of the second set. Similarly, we might wish to simplify a set of measurements by multiplying or dividing by a constant. It is certainly easier to work with the set $(3, 1, 4, 6, 4, 2)$ than with $(.003, .001, .004, .006, .004, .002)$. The first set was obtained by multiplying each element of the second set by 1000.

Data are coded by performing one or both of two operations. One may subtract (or add) a constant m to each measurement, multiply (or divide) each measurement by a constant k, or do both. The objective of coding is to obtain data for which we can more easily determine the mean. Then we wish to use these data to find $\overline{X}$, the mean of the original data.

Let $X_1, X_2, \ldots, X_n$ be the original measurements with mean $\overline{X}$. Similarly, let $\overline{X}_c$ be the mean of the coded data. How are $\overline{X}$ and $\overline{X}_c$ related for the two operations of coding? Two theorems answer this question.

First consider subtracting a constant from each measurement.

Theorem 4.1

Let $X_1, X_2, \ldots, X_n$ be n measurements with mean $\overline{X}$. If we subtract a constant m from each of the measurements, the mean for the coded measurements will be

$$\overline{X}_c = \overline{X} - m$$

The implication of Theorem 4.1 is that the difference in means for the uncoded and coded data, $\overline{X}$ and $\overline{X}_c$, will equal the quantity m. One can compute $\bar{X}_c$ for the coded data and obtain the uncoded mean as

$$\bar{X} = \bar{X}_c + m$$

We shall demonstrate this theorem with an example.

Example 4.10 *Find $\bar{X}$ for the sample measurements 99.9, 100.2, 100.1, 100.0, and 100.2 by subtracting 100 from each observation.*

Solution *Subtracting $m = 100$ from each of the measurements, we obtain the coded data $-.1, .2, .1, 0,$ and $.2$. Then*

$$\bar{X}_c = \frac{-.1 + .2 + .1 + 0 + .2}{5} = \frac{.4}{5} = .08$$

By Theorem 4.1

$$\bar{X} = \bar{X}_c + m = .08 + 100 = 100.08$$

To show that this convenient method of computing $\bar{X}$ (using smaller coded numbers) gives the same result as that computed using the standard method, note that the sum of the original uncoded measurements is

$$\Sigma X = 99.9 + 100.2 + 100.1 + 100.0 + 100.2 = 500.4$$

Then

$$\bar{X} = \frac{\Sigma X}{n} = \frac{500.4}{5} = 100.08$$

We see that this is the same answer obtained using Theorem 4.1.

The second method of coding is multiplying each measurement by a constant k. The following theorem relates the coded and uncoded means.

Theorem 4.2

Let $X_1, X_2, \ldots, X_n$ be n measurements with mean $\bar{X}$. If we multiply each measurement by a constant k, the mean for the coded data will be

$$\bar{X}_c = k\bar{X}$$

The effect of multiplying each observation by a constant k on the relationship between $\bar{X}$ and $\bar{X}_c$ was perhaps predictable. The coded mean $\bar{X}_c$ is k times as large as $\bar{X}$.

Example 4.11   *Calculate $\bar{X}$ for the sample .003, .001, .004, .006, .004, and .002 by multiplying each observation by 1000.*

Solution   *The coded data are 3, 1, 4, 6, 4, and 2. Then*

$$\bar{X}_c = \frac{3 + 1 + 4 + 6 + 4 + 2}{6} = \frac{20}{6} = 3.67$$

Hence, by Theorem 4.2,

$$\bar{X} = \frac{\bar{X}_c}{k} = \frac{3.67}{1000} = .00367$$

Again, to demonstrate that Theorem 4.2 works, note that $\bar{X}$ computed for the original coded measurements is

$$\bar{X} = \frac{\Sigma X}{n} = \frac{.003 + .001 + .004 + .006 + .004 + .002}{6}$$

$$= \frac{.020}{6} = .00367$$

the same answer as that obtained using the coded data.

Using the data found in Table 4.2, we shall show how to reduce the labor in computing the mean by using coded measurements. First, we shall look at the class intervals and their corresponding frequencies. Based on inspection alone, we shall select the class interval that we *think* is likely to contain the mean. The midpoint of that class interval, defined as m, is equal to 6.5 for this example. (Note that the selection of m is not critical; just pick a class interval near the middle of the distribution.) Then we shall subtract m from each midpoint value X. We can further reduce the difficulty of the computation by dividing each measurement by a suitable constant. It is convenient to use the interval width i. Then compute each coded value of the class interval midpoints as

$$X_c = \frac{X - m}{i}$$

Applying Theorems 4.1 and 4.2 and letting k (which appears in the formula, Theorem 4.2) equal the interval width i, the mean value of X, $\overline{X}$, is

$$\overline{X} = i\overline{X}_c + m$$

Example 4.12   *Use the coding technique just described to compute $\overline{X}$, the mean for the grouped data of Table 4.2.*

Table 4.5   *Table for the Computation of $\overline{X}_c$ and $\overline{X}$ for the Data of Table 4.2*

Class Interval	Midpoint (X)	f	$X - m$	$X_c = \dfrac{X - m}{i}$	fX_c
24–25	24.5	1	18	9	9
22–23	22.5	1	16	8	8
20–21	20.5	1	14	7	7
18–19	18.5	2	12	6	12
16–17	16.5	4	10	5	20
14–15	14.5	9	8	4	36
12–13	12.5	7	6	3	21
10–11	10.5	8	4	2	16
8–9	8.5	9	2	1	9
6–7	6.5	13	0	0	0
4–5	4.5	15	-2	-1	-15
2–3	2.5	18	-4	-2	-36
0–1	0.5	2	-6	-3	-6
Total		90			81

Solution   *Compute $X - m$ as shown in column 4, Table 4.5, and then divide each of these numbers by the class width, $i = 2$, to obtain the coded measurements listed in column 5. Finally, calculate the products of fX_c, shown in column 6, by multiplying corresponding numbers in columns 3 and 5. Now we are ready to calculate $\overline{X}_c$, the average of the numbers in column 6. Thus*

$$\overline{X}_c = \frac{\Sigma fX_c}{n} = \frac{81}{90} = .9$$

We can then find the mean for the uncoded measurements, $\overline{X}$, as follows:

$$\overline{X} = i\overline{X}_c + m$$
$$= 2(.9) + 6.5 = 8.3$$

Note that this result is identical to that obtained for the same data in Example 4.8. Once you get accustomed to this procedure, it can greatly reduce the labor in computing $\overline{X}$.

Having considered the major measures of central tendency in previous sections, we shall present a more general measure of location.

EXERCISES

9 Determine the mean for the data in Exercise 1 (page 103). Use the uncoded formula.

10 Determine the mean for the data in Exercise 2 (page 10?). Use the uncoded formula.

11 Determine the mean for the data in Exercise 17, Chapter 3 (page 68). Use the uncoded formula.

12 Repeat Exercise 11 using the coded formula with m equal to the midpoint of the interval 20–21.

13 Repeat Exercise 12 using the coded formula with m equal to the midpoint of the interval 18–19. Are your answers to Exercises 11, 12, and 13 the same? If not, check your calculations.

14 Determine the mean for the data in Exercise 18, Chapter 3 (page 69). Use the coded formula with m equal to the midpoint of the interval 16–19.

4.6   A MEASURE OF LOCATION:
PERCENTILES

ercentiles determine the fraction or percentage of measurements (scores) above or below specified values and hence can be used to locate an observation in relation to the remaining ones.

Definition 4.10

Let $X_1, X_2, \ldots, X_n$ be a set of n measurements arranged in order of magnitude. The Pth percentile is the value of X such that P percent of the measurements are less than that value of X and $(100 - P)$ percent are greater.

For example, the 80th percentile of a large set of measurements on a variable X is the value X such that 80 percent of the values or scores fall below it and 20 percent lie above it (see Figure 4.3). Note that the 50th percentile of a set of measurements is the median and that all measurements of central tendency are measures of location because they locate the center of a distribution of measurements.

Figure 4.3 *80th percentile*

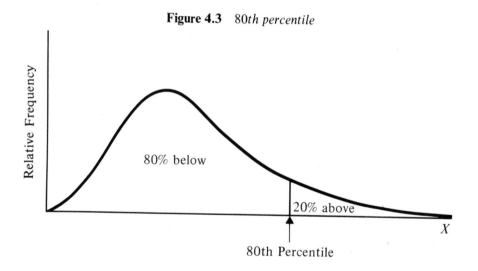

Percentiles are frequently used to describe the results of achievement tests and the final ranking of a person in comparison to the rest of the people taking an examination. Such phrases as "John S. scored at the 90th percentile on the Scholastic Aptitude Test this year" or "Susie J. scored at the 85th percentile on a national speed-reading examination for fourth graders" locate a person in relation to the many others who took the same examination.

Special names are given to the 25th and 75th percentiles of a set of measurements.

Definition 4.11

The lower quartile of a large set of data is defined to be the 25th percentile. Twenty-five percent of the measurements fall below the lower quartile and 75 percent above it (see Figure 4.4).

Definition 4.12

The upper quartile of a large set of measurements is defined to be the 75th percentile. Seventy-five percent of the measurements will lie below the upper quartile and 25 percent above it. See Figure 4.4.

Figure 4.4 *Lower quartile, median, and upper quartile*

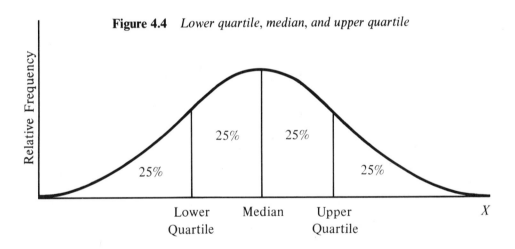

Having learned how to compute the median, percentiles are easy to compute. After all, the median is the 50th percentile.

P = percentile

L = lower real limit of the class interval that includes the percentile we wish to find

n = total frequency

cf_b = sum of the frequencies (cumulated frequencies) for all class intervals *below* the interval that includes the percentile of interest

f_p = frequency of the class interval that includes the percentile of interest

In working with percentiles, we are giving the rank of an individual and attempting to determine the individual's score. If we asked, "What is the value of the murder rate which is so low that only 40 percent of the cities have lower murder rates?", the answer would be

$$P_{40} = L + \frac{i}{f_p}(40\% \text{ of } n - cf_b)$$

$$= 5.5 + \frac{2}{13}(36 - 35)$$

$$= 5.65$$

EXERCISES

15 Use the data from Exercise 18, Chapter 3 (page 69) to compute the 75th percentile.

16 Use the data from Exercise 15 to compute the 50th percentile and compare that value with both the crude mode and the mean.

17 Use the data from Exercise 17, Chapter 3 (page 68) to determine the 25th percentile.

18 Use the data from Exercise 17 to determine the 50th percentile and compare that value with both the crude mode and the mean.

4.7 DATA VARIATION

The three measures of central tendency, the mean, median, and mode, locate the center of a distribution of data, but they tell you nothing about the spread or variation of the measurements. The importance of data variation is exemplified by a story that a friend of ours, a law professor, always tells at cocktail parties. Gathering a group of listeners about him and casting a mischievous glance in our direction, he asks: "Have you ever heard the story about the statistician who could not swim and drowned in a river with an average depth of 3 feet?" Although we admit to some discomfort every time we hear the joke, it does stress the importance of describing data variation. The mean (or any other measure of central tendency) only tells part of the story—only partially describes a distribution of measurements.

The law professor's joke illustrates the importance of describing data variation or spread. A machine manufacturing size 9 shoes, "on the average," would not be considered satisfactory if the actual sizes varied from $8\frac{3}{8}$ to $9\frac{1}{2}$. A machine producing 1-inch nails, "on the average," would not be very satisfactory if the actual lengths varied from $\frac{1}{2}$ to $1\frac{1}{2}$ inches. Who would sort them? Indeed, variation of product quality is probably of far greater importance to a manufacturer than corresponding measures of central tendency.

Keep in mind that the objective of numerical description is to obtain a set of measures, one or more, that will create a mental reconstruction of the frequency distribution of the data we wish to describe. A measure of central tendency only locates the center of the distribution. The above examples amply illustrate the need for numerical measures of data variation or spread.

Six numerical descriptive measures of data variation are described in this chapter, the index of qualitative variation, the range, the interquartile range, the average deviation, the variance, and the standard deviation. The percentiles of Section 4.6 constitute a seventh measurement because they provide information on data variability as well as location. Each measure has certain advantages and disadvantages, and some tend to be favored over others in the various fields of science. We shall begin by describing the index of qualitative variation.

4.8 A MEASURE OF VARIABILITY FOR
QUALITATIVE DATA: THE INDEX
OF QUALITATIVE VARIATION

We noted previously that a qualitative variable has observations that vary in quality from trial to trial. Just how variable are these observations? Since we cannot compute actual numerical differences between observations, we count the number of observations that differ in quality. We illustrate a common method for counting the number of observations that differ through an example. A sample of 20 faculty members was selected to determine the sex and age of pre-school-age children eligible for a university-supported nursery school for three-year-olds. Each of the faculty members was asked to state the sex of his eligible children. Typical results are listed in Table 4.6.

In Table 4.6(a) all 20 of the three-year-olds were boys; hence there is no variability in the qualitative variable, sex, and the number of observations

Table 4.6 *Sex Classification of Three-Year-Olds*

(a)			(b)	
Sex	Frequency		Sex	Frequency
Male	20		Male	19
Female	0		Female	1

(c)			(d)	
Sex	Frequency		Sex	Frequency
Male	18		Male	10
Female	2		Female	10

that differ in quality is zero. In contrast, the results in Table 4.6(b) indicate that 19 were males and 1 was a female. Since each of the 19 boys is of a different sex than the 1 girl, there are 19 observations that differ in quality. Similarly, for the outcome of 18 boys and 2 girls, each of the 18 boys differs in sex from each of the 2 girls; hence there are $18 \times 2 = 36$ observations that differ in sex. Utilizing the same reasoning, if the outcome were as listed in Table 4.6(d), 10 boys and 10 girls, each of the 10 boys differs in sex from each of the 10 girls, making $10 \times 10 = 100$ observations that differ in the qualitative variable, sex.

The counting scheme that we have just illustrated is by no means restricted to a qualitative variable at two levels. For example, if the previous survey were conducted to determine the sex and age of pre-school children eligible for a university nursery school for three- and four-year-olds, typical results might appear as in Table 4.7.

Table 4.7 *Sex and Age of Children Eligible for a Pre-School Nursery*

(a)			(b)	
Classification	Frequency		Classification	Frequency
Female, 3	10		Female, 3	4
Female, 4	5		Female, 4	2
Male, 3	5		Male, 3	7
Male, 4	0		Male, 4	7

Each of the 10 three-year-old girls in Table 4.7(a) is of a different sex–age classification than each of 5 four-year-old girls. There are thus $10 \times 5 = 50$ observations that differ. In the same way, the 10 three-year-old girls are of a different age–sex classification than the 5 three-year-old boys, which makes $10 \times 5 = 50$ more observations that differ in the age–sex classification. Finally, the 5 four-year-old girls and the 5 three-year-old boys make $5 \times 5 = 25$ more observations of a different sex–age classification. According to our counting scheme, the total number of observations in Table 4.7(a) that differ in their sex and age classification is then

$$50 + 50 + 25 = 125$$

In general, the total number of differences for a set of observations is determined by first multiplying each frequency times every other frequency (this would produce one product for every pair of frequencies) and then summing all these products.

Example 4.13 *Determine the total number of differences for the data listed in Table 4.7(b).*

Solution *The frequencies for the classifications are given as 4, 2, 7, and 7, respectively. Hence to find the total number of differences given we first multiply each frequency times the other classification frequencies as follows:*

$$4 \times 2 = 8$$
$$4 \times 7 = 28$$
$$4 \times 7 = 28$$
$$2 \times 7 = 14$$
$$2 \times 7 = 14$$
$$7 \times 7 = 49$$

Note now that each frequency is multiplied times every other frequency. The total number of observed differences is then the sum of all these products:

$$8 + 28 + 28 + 14 + 14 + 49 = 141$$

The total number of observed differences is not to be confused with the maximum number of possible differences. The latter quantity is the largest possible number that the total of observed differences can obtain. Although proof is omitted, this maximum number is given in the following theorem:

Theorem 4.3

The maximum number of possible differences for n observations on a qualitative variable with l levels is

$$\frac{n^2(l-1)}{2l}$$

We can now express the total number of observed differences as a percentage of the maximum possible number of differences. This quantity is called the index of qualitative variation (IQV).

Definition 4.13

The index of qualitative variation is the total number of observed differences expressed as a percentage of the maximum possible number of differences:

$$IQV = \frac{\text{total number of observed differences}}{\text{maximum number of possible differences}} \times 100$$

Example 4.14   *Compute the IQV for the data of Table 4.7(b).*

Solution   *Recall that in Example 4.13 we computed the total number of observed differences to be 141. The maximum number of possible differences can be found from Theorem 4.3 with n = 20 and l, the number of classifications, equal to 4:*

$$\text{maximum number of possible differences} = \frac{(20)^2(3)}{2(4)}$$

$$= 150$$

Hence the index of qualitative variation is

$$IQV = \frac{141}{150} \times 100 = 94\%$$

If an IQV equals 0 percent, there is perfect homogeneity in our population; if the IQV equals 100 percent, there is perfect heterogeneity in our population with respect to the qualitative characteristics being considered.

The IQV is not the only measure of data variation for qualitative variables, but it is the most widely accepted measure in the social sciences and will be used exclusively in this text. We omit further discussion of variation in qualitative variables and proceed to the important topic of measures of variation for quantitative variables.

EXERCISES

19 For a group of 4 Catholics, 3 Jews, and 9 Protestants, calculate the IQV.

20 For the 4 racial–ethnic groups shown in Exercise 6, Chapter 3 (page 56), calculate the IQV for males.

21 For the 4 racial–ethnic groups shown in Exercise 6, Chapter 3 (page 56), calculate the IQV for females.

22 If a group of students is composed of 3 men and 7 women; 1 Catholic, 1 Jew, and 8 Protestants; and 4 whites and 6 blacks, is it possible to represent this diversity by means of a single IQV? Explain your answer.

4.9 A MEASURE OF VARIABILITY
FOR QUANTITATIVE DATA:
THE RANGE

The simplest measure of data variation is the range.

Definition 4.14

The range of a set of measurements is defined to be the difference between the largest and smallest measurements.*

---

* Most social scientists would compute the range for ungrouped data using the difference between the upper true limit of the largest measurement and the lower true limit of the smallest measurement. Note that we use "true" limits when referring to ungrouped data and "real" limits for grouped data.

> **Definition 4.15**
>
> *If data are grouped in classes, the* range *is defined to be the
> difference between the upper real limit of the highest class and the
> lower real limit of the lowest class.*

The range is employed extensively as a measure of variability in
summaries of data that are made available to the general public. We might
read that the range of salaries for psychologists with the rank of assistant
professors is $4000, the range in temperature in Miami throughout the year
is 50°F, and the range in personal property taxes for a given state is $600. It is
also widely used to describe variation in the quality of an industrial product
when small samples are selected periodically from an operating production
line. For small samples the range is about as good as any other measure of
variation.

Example 4.15   *Give the range for the 90 measurements on murder
rates, Table 4.2.*

Solution   *The range for grouped data is the difference between the
upper real limit of the highest class and lower real limit of the lower
class. Hence the range* $= 25.5 - (-.5) = 25.5 + .5 = 26.0$.

Figure 4.5   *Two distributions with the same range*

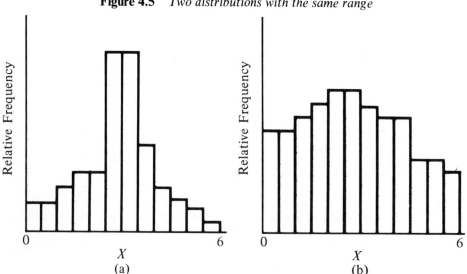

Although simple to define and calculate, the range is not always satisfactory as a measure of variability. The two distributions of measurements shown in Figure 4.5(a) and (b) have the same range (6), but it is readily apparent that the data for the two distributions differ greatly in variation. The data for Figure 4.5(a) are much less variable than that for part (b). Figure 4.5(a) has most of its measurements very close to the mean; part (b) has the measurements spread evenly throughout the range.

A sample range is extremely variable because any change in the extreme observations alters its value. This instability is reduced by considering certain intermediate ranges. One in particular is the interquartile range.

Definition 4.16

The interquartile range (IR) of a set of measurements is defined to be the difference between the upper and lower quartiles (see Figure 4.6).

Figure 4.6   *Interquartile range*

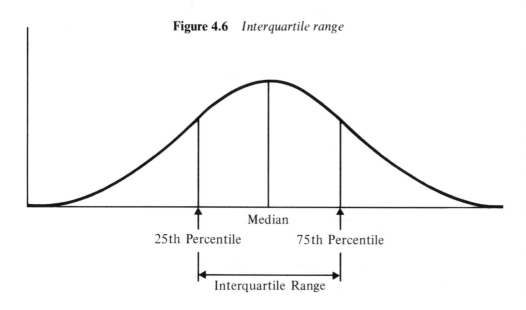

Although specification of the median, the interquartile range, and the range of a set of measurements provides a fairly good description of a set of measurements, it still requires three measures to create a mental image of the frequency distribution. Since a single number would be easier to interpret, we seek a more sensitive measure to describe the variation of a set of measurements.

23 Using the data in Exercise 13, Chapter 3 (page 66), calculate the interquartile range for the age of males.

24 Repeat Exercise 23 for the interquartile range of the female ages.

4.10   A MEASURE OF VARIABILITY FOR QUANTITATIVE DATA: THE AVERAGE DEVIATION

The shortcoming of the range as a measure of variation is that it is insensitive to a pileup of data near the center of the distribution (see Figure 4.5). To overcome this, we use the deviations or distances of the data from their center. Thus a less variable set will contain measurements, most of which are located near the center of the distribution, as shown in Figure 4.5(a). A more variable or dispersed set of measurements is shown in Figure 4.5(b).

To illustrate, suppose that we have the set of five measurements $X_1 = 8$, $X_2 = 7$, $X_3 = 6$, $X_4 = 3$, and $X_5 = 1$. These are shown in a dot diagram, Figure 4.7. (Dot diagrams are used to depict very small sets of

Figure 4.7   *Dot diagram*

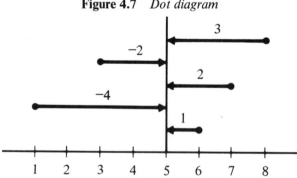

measurements.) Each measurement is located by a dot above the horizontal axis of the diagram.

We shall use the mean

$$\bar{X} = \frac{\Sigma X}{n} = \frac{25}{5} = 5.0$$

to locate the center of the set and construct horizontal lines on Figure 4.7 to represent the distances (deviations) of the measurements from their mean. Surely the larger the deviations, the greater will be the variation of the set of measurements. The deviations of the measurements are computed using the formula $(X - \bar{X})$. The deviation of X_1 from the mean is

$$(X_1 - \bar{X}) = 8 - 5 = 3$$

The five measurements and their deviations from the mean are shown in Table 4.8 (and in Figure 4.7).

Table 4.8 *Deviation of the Five Measurements from Their Mean*

Measurement X	Deviation $(X - \bar{X})$
8	3
7	2
6	1
3	−2
1	−4
$\Sigma X = 25$	$\Sigma(X - \bar{X}) = 0$

$$\bar{X} = \frac{\Sigma X}{n} = \frac{25}{5} = 5$$

Many different measures of variation could be constructed using the deviations of the measurements from their mean. A first thought would be to use their arithmetic average, but this will always equal zero as it does in Table 4.8. This is because the negative and positive deviations always balance one another so that their sum equals zero. A second possibility would be to ignore the minus sign and compute the average of the absolute values of the deviations.

Two different measures of variation are based on the average of the absolute values of deviations, one based on deviations from the mean, the other from the median. Both quantities clearly measure the variation in a set of data, but they are difficult to interpret. We discuss them because they occasionally appear in the literature.

Definition 4.17

The average deviation of a set of n measurements from their mean is defined to be the average of the absolute values of the deviations,

$$AD_{\bar{x}} = \frac{\Sigma|X - \bar{X}|}{n}$$

Example 4.16 *Use the data in Table 4.8 to compute the average deviation of the measurements about their mean.*

Solution *The absolute deviations* $|X - \bar{X}|$ *are 3, 2, 1, 2, and 4. Substituting these measurements into the formula presented in Definition 4.17 we have*

$$AD_{\bar{x}} = \frac{\Sigma|X - \bar{X}|}{n}$$

$$= \frac{3 + 2 + 1 + 2 + 4}{5} = 2.4$$

As we said previously, for symmetrical frequency distributions, the mean and median are identical; however, when a frequency distribution "tails off" sharply to one side or the other, the frequency distribution is said to be *skewed*. For such cases the mean and median would not be identical, leading some social scientists to prefer to measure data variation using the deviations of the measurements from their median.

Definition 4.18

The average deviation of a set of n measurements from their median is defined to be the average of the absolute values of the deviations:

$$AD_{M_d} = \frac{\Sigma|X - M_d|}{n}$$

Example 4.17 *A fertility ratio for each of the 11 SMSA's was obtained from the 1970 census of the population. These data are presented in Table 4.9. Note that the fertility ratio is defined to be the number of children under 5 per 1000 women between the ages of 15 and 49.*

Table 4.9 *Fertility Ratios by SMSA*

SMSA	Fertility Ratio
Altoona, Pa.	342
Bakersfield, Calif.	366
Buffalo, N.Y.	349
Cedar Rapids, Iowa	390
Columbia, S.C.	334
El Paso, Tex.	401
Gainesville. Fla.	300
Hartford, Conn.	347
Honolulu, Hawaii	362
Wichita, Kans.	351
Pueblo, Colo.	345

Use these data to determine the average deviation of the measurements from their median.

Solution *Before we can compute AD_{M_d}, we must first determine the median fertility ratio. Perhaps the easiest way to do that is to arrange the measurements in order of magnitude.*

$$
\begin{aligned}
&300\\
&334\\
&342\\
&345\\
&347\\
&349 \leftarrow median\\
&351\\
&362\\
&366\\
&390\\
&401
\end{aligned}
$$

Using $M_d = 349$, we compute the deviations and absolute deviations in Table 4.10.

Table 4.10 *Calculations for Example 4.17*

Measurement X	Deviation $X - M_d$	Absolute Deviation $\lvert X - M_d \rvert$
300	-49	49
334	-15	15
342	-7	7
345	-4	4
347	-2	2
349	0	0
351	2	2
362	13	13
366	17	17
390	41	41
401	52	52
		$\Sigma\lvert X - M_d \rvert = 202$

Using Definition 4.18,

$$AD_{M_d} = \frac{\Sigma\lvert X - M_d \rvert}{11} = \frac{202}{11} = 18.36$$

Social researchers almost never employ the average deviation—either from the mean or median—unless they compare it to its own origin. For example, if we compute the average deviation from the mean, we compare that value to the mean and multiply by 100. The end product is the "coefficient of variation for the average deviation"

$$CV_{\bar{X}} = \frac{AD_{\bar{X}}}{\bar{X}} \text{ times } 100$$

If we were dealing with the average deviation from the median, the formula becomes

$$CV_{M_d} = \frac{AD_{M_d}}{M_d} \text{ times } 100$$

The coefficient of variation then expresses the average deviation as a percentage of the value of a measure of central tendency. Let us assume, for example, that in one instance we have a sample with a mean of 100 and an average deviation of 5 and in a second sample a mean of 25 and an average deviation of 5. The $CV_{\bar{x}}$ for the first sample is 5 percent; the $CV_{\bar{x}}$ for the second sample is 20 percent. To most people, the first sample is less variable than the second, but not all scientists agree with that interpretation. Fortunately, statisticians have developed other measures of variability which are much more meaningful and easier to interpret.

EXERCISES

25 Use the data from Exercise 2 (page 103) to calculate the average deviation from the mean.

26 Use the data from Exercise 2 to calculate the average deviation from the median.

27 Determine the coefficient of variation for the average deviation from the mean using the data of Exercise 25.

28 Determine the coefficient of variation for the average deviation from the median, using the data of Exercise 26.

4.11 A MEASURE OF VARIABILITY FOR QUANTITATIVE DATA: THE VARIANCE

more easily interpreted function of the deviations than the average deviation involves the sum of the squared deviations of the measurements from their mean. Recall that the deviation of a score X from the sample mean $\bar{X}$ was expressed as $(X - \bar{X})$. The square of a deviation could then be represented as $(X - \bar{X})^2$ and the sum of the squared deviations can be written as

$$\Sigma (X - \bar{X})^2$$

Definition 4.19

The variance of a set of n measurements is defined to be the sum of the squared deviations of the measurements about their mean, divided by n − 1.

The variance is used most frequently as a measure of variability in technical and professional journals. Very seldom do we read of the variance of a set of measurements when data are summarized for public consumption. We shall soon show that the variance can be used to obtain a very practical and easily understood measure of data variation.

Although we would calculate the variance of any set of interval or ratio data in the same way, it becomes convenient to draw a distinction between the variance of a set of sample measurements and the variance of a population. We shall use the symbol s^2 to represent the sample variance. Thus

$$s^2 = \frac{\Sigma (X - \overline{X})^2}{n - 1}$$

The corresponding population variance will be denoted by σ^2 (σ is the Greek letter sigma).

s^2 **is the** *sample variance*

σ^2 **is the** *population variance*

Example 4.18   *Calculate the variance for the five measurements 8, 7, 6, 3, and 1 given in Table 4.8 (page 129).*

Solution   *From Table 4.8 we obtain the deviations, $(X - \overline{X})$. Hence the squared deviations are as shown in Table 4.11.*

Table 4.11   *Calculations*
for Example 4.18

$(X - \bar{X})^2$
9
4
1
4
16

The sample variance is

$$s^2 = \frac{\Sigma (X - \bar{X})^2}{n - 1} = \frac{9 + 4 + 1 + 4 + 16}{5 - 1} = \frac{34}{4} = 8.5$$

We can make a simple modification of our variance formula to apply to grouped data. Recall that in computing the mean for grouped data (Section 4.4) we used the interval midpoint to represent each observation in the interval. We use this same procedure in calculating the variance for grouped data.

Definition 4.20

If X represents the midpoint of a class interval with frequency f and $\bar{X}$ denotes the mean for the grouped data, the variance of a set of n grouped measurements is defined to be

$$\frac{\Sigma f(X - \bar{X})^2}{n - 1}$$

Computational Formulas for s^2, Grouped and Ungrouped Data

$$Ungrouped: \quad s^2 = \frac{1}{n - 1} \left[\Sigma X^2 - \frac{(\Sigma X)^2}{n} \right]$$

$$Grouped: \quad s^2 = \frac{1}{n - 1} \left[\Sigma fX^2 - \frac{(\Sigma fX)^2}{n} \right]$$

The computational formulas for s^2 are particularly useful when working with large sets of data by hand or with an electronic desk calculator. Most calculators allow one to accumulate the sums and sums of squares required in the computational formulas.

Examination of Definitions 4.19 and 4.20 indicates that the larger the variation or spread in a set of measurements, the larger will be the variance. We can compare variances of sets of measurements to compare variability, but, like an average deviation, it is difficult to interpret the variance for a single set of measurements. What can we say about the spread of a set of measurements with a variance of 2.0? We shall try to answer this question and provide you with a measure of variability useful not only for comparison purposes but also for describing a single set of measurements.

Definition 4.21

The standard deviation *of a set of measurements is defined to be the positive square root of the variance.*

The sample standard deviation will be denoted by s and the corresponding population standard deviation by the symbol σ.

---

s is the sample standard deviation

σ is the population standard deviation

As an aid to remembering the notation, we merely note that the letter s is the first letter of standard in standard deviation.

Example 4.19 *Calculate the sample variance and standard deviation for the data of Table 4.8 using the shortcut formula for s^2.*

Solution *It is convenient to make a table of calculations when using the shortcut formula for s^2.*

Table 4.12 *Calculations for* s^2

X	X^2
8	64
7	49
6	36
3	9
1	1
Total 25	159

The totals for the X and X^2 columns of Table 4.12, ΣX and ΣX^2, are needed in the formula for s^2. For $n = 5$,

$$s^2 = \frac{1}{n-1}\left[\Sigma X^2 - \frac{(\Sigma X)^2}{n}\right]$$

$$= \frac{1}{4}\left[159 - \frac{(25)^2}{5}\right]$$

$$= \frac{1}{4}\left(159 - 125\right) = \frac{34}{4} = 8.5$$

Note that this is identical to the result for Example 4.18, where we used the formula

$$s^2 = \frac{\Sigma (X - \bar{X})^2}{n-1}$$

The sample standard deviation is then

$$s = \sqrt{8.5} = 2.92$$

For the computational formula using grouped data, we multiply f times X and f times X^2 for each class. If we then add fX for each class and fX^2 for each class we obtain ΣfX and ΣfX^2, respectively. These quantities are used in the computation of s^2.

We now state a rule that gives practical significance to a standard deviation of a set of measurements. Particularly, it explains the spread (variation) of a set of measurements by stating the percentage of measurements that we might expect to find within certain distances from the mean. The rule applies only to mound-shaped frequency distributions (frequency histograms that pile up near the center). Because mound-shaped distributions of data occur so frequently in nature, the rule possesses wide applicability and hence is called the empirical rule.

The Empirical Rule

Given a frequency distribution that is mound-shaped (see Figure 4.8), the interval

1   $\overline{X} - s$ to $\overline{X} + s$ will contain approximately 68 percent of the measurements.

2   $\overline{X} - 2s$ to $\overline{X} + 2s$ will contain approximately 95 percent of the measurements.

3   $\overline{X} - 3s$ to $\overline{X} + 3s$ will contain approximately 99.7 percent of the measurements.

Figure 4.8   *Mound-shaped frequency distribution*

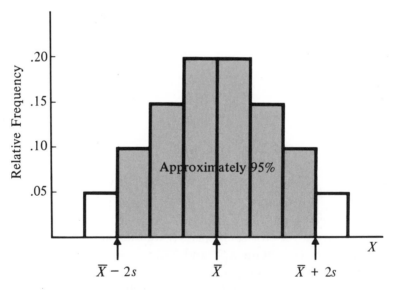

Example 4.20   *A study was conducted to determine the average length of time it takes for an item to be completed on an assembly-line operation. A sample of 50 items is timed. The mean and standard deviation (in hours) for the 50 measurements are $\overline{X} = 4.8$ and $s = .42$. If we assume that the measurements have a mound-shaped frequency histogram, describe the data using the empirical rule.*

Solution   *The empirical rule tells us that approximately 68 percent of the measurements lie in the interval $\overline{X} - s$ to $\overline{X} + s$, that is, the interval 4.38–5.22. The interval $\overline{X} - 2s$ to $\overline{X} + 2s$ (i.e., the interval 3.96 to 5.64) should contain approximately 95 percent of the measurements. Approximately 99.7 percent of the measurements should be in the interval $\overline{X} - 3s$ to $\overline{X} + 3s$ or the interval 3.54–6.06.*

We have discussed five measures of variability for quantitative data in this chapter: the range, the interquartile range, the average deviation, the variance, and the standard deviation. Although each of these measures is useful in data description, the variance and the standard deviation of a set of measurements provide us with information to

1   Compare variability between sets of measurements.

2   Interpret the variability of a single set of measurements using the empirical rule. Recall that the empirical rule only applies to data that possess mound-shaped frequency distributions.

In the remainder of this text, the standard deviation will be used almost exclusively to measure the variability of a single set of measurements.

A final comment concerning the standard deviation should be made. Some researchers like to compare the magnitude of the standard deviation of a set of measurements relative to a particular baseline, such as $\overline{X}$.

Definition 4.22

The coefficient of relative variation (CRV) for the standard deviation is defined to be the standard deviation expressed as a percentage of the mean $\overline{X}$:

$$\text{CRV} = \frac{s}{\overline{X}} \times 100$$

For the purpose of a quick comparison of two different s's, the social researcher will occasionally compute the CRV for each and then discuss the difference between them. To some, a standard deviation of 10 with $\overline{X} = 1000$ represents, relatively speaking, less variation than a standard deviation of

10 with $\overline{X} = 100$. The respective CRV's are 1 and 10. In terms of their means, the relative variation in the first sample is only 1/10 of the relative variation in the second sample.

4.12   CODING TO REDUCE COMPUTATIONAL LABOR FOR THE SAMPLE STANDARD DEVIATION

As with calculations for the mean, data are frequently coded to simplify our computations. We might subtract (add) a constant m to each measurement, or multiply (divide) each measurement by a constant k.

Subtracting a constant m has no effect on the standard deviation. However, if you code by dividing the original data by a constant k, then the standard deviation of the coded data s_c is equal to the standard deviation for the original data divided by k. Similarly, if you multiply each measurement by a constant k, the standard deviation of the coded data is equal to k times the original standard deviation (Table 4.13).

Table 4.13   *Summary of the Effect of Coding on the Sample Mean and Sample Standard Deviation*

	Sample Mean	Sample Standard Deviation
Add a constant m to each measurement	$\overline{X}_c = \overline{X} + m$	$s_c = s$
Subtract a constant m from each measurement	$\overline{X}_c = \overline{X} - m$	$s_c = s$
Multiply each measurement by a constant k	$\overline{X}_c = k\overline{X}$	$s_c = ks$
Divide each measurement by a constant k	$\overline{X}_c = \dfrac{\overline{X}}{k}$	$s_c = \dfrac{s}{k}$

Why code data? We remind you that coding reduces computational labor. It is particularly useful when doing computations by hand or by electronic calculator.

Example 4.21 *Using the data of Table 4.2 and Table 4.14, apply the coding principles just discussed to compute s^2, the sample variance for the grouped murder-rate data.*

Table 4.14 *Murder-Rate Data of Table 4.2*

Class Interval	Midpoint (X)	f	$X - m$	$X_c = \dfrac{X - m}{i}$	fX_c	fX_c^2
24–25	24.5	1	18	9	9	81
22–23	22.5	1	16	8	8	64
20–21	20.5	1	14	7	7	49
18–19	18.5	2	12	6	12	72
16–17	16.5	4	10	5	20	100
14–15	14.5	9	8	4	36	144
12–13	12.5	7	6	3	21	63
10–11	10.5	8	4	2	16	32
8–9	9.5	9	2	1	9	9
6–7	6.5	13	0	0	0	0
4–5	4.5	15	-2	-1	-15	15
2–3	2.5	18	-4	-2	-36	72
0–1	0.5	2	-6	-3	-6	18
		90			$\Sigma fX_c = 81$	$\Sigma fX_c^2 = 719$

Solution *The midpoints (X) are coded in the same way as they were for Example 4.12 with $m = 6.5$ and $k = i = 2$:*

$$X_c = \frac{X - m}{i} = \frac{X - 6.5}{2}$$

These results are listed in the fifth column of Table 4.14. Two additional columns have also been added: one for computing fX_c and the other for fX_c^2. Using the shortcut formula we have

$$s_c^2 = \frac{1}{n-1}\left[\Sigma fX_c^2 - \frac{(\Sigma fX_c)^2}{n}\right]$$

$$= \frac{1}{89}\left[719 - \frac{(81)^2}{90}\right]$$

$$= \tfrac{1}{89}(646.1) = 7.26$$

and

$$s_c = \sqrt{7.26} = 2.69$$

Applying our results on coding we know that dividing each measurement by a constant k, $s_c = s/k$. Hence for $k = i$, the interval width, we can solve for s to obtain

$$s = is_c = 2(2.69) = 5.38$$

EXERCISES

29 Calculate the standard deviation of the ungrouped data in Exercise 1 (page 103).

30 Calculate the standard deviation of the ungrouped data in Exercise 2 (page 103).

31 Calculate the standard deviation of the data on males in Exercise 13, Chapter 3 (page 66). Use the coded formula and set m equal to the midpoint of the interval 20–21.

32 Calculate the standard deviation of the data in Exercise 15, Chapter 3 (page 67). Use the coded formula and set m equal to the midpoint of the interval 52–55.

33 Calculate the standard deviation of the data in Exercise 13, Chapter 3 (page 66), females only. Use the coded formula and set m equal to the midpoint of the interval 18–19.

34 Calculate the standard deviation of the data in Exercise 16, Chapter 3 (page 67). Use the coded formula and set m equal to the midpoint of the interval 6–8.

35 Calculate the standard deviation of the data in Exercise 17, Chapter 3 (page 68). Use the coded formula and set m equal to the midpoint of the interval 20–21.

36 Calculate the standard deviation of the data in Exercise 18, Chapter 3 (page 69). Use the coded formula and set m equal to the midpoint of the interval 20–23.

4.13 A CHECK ON THE CALCULATION OF A STANDARD DEVIATION

Arithmetic mistakes or errors in reading a square-root table can easily occur when calculating a standard deviation. We present a shortcut method for approximating the standard deviation of a set of measurements that can be used as a check on our calculations. This approximation (and hence our check) will work best for mound-shaped distributions.

The *empirical rule*, which is useful in interpreting the variability of a mound-shaped distribution, states that approximately 95 percent of the measurements in a set will be within two standard deviations of their mean. Using notation for a sample, approximately 95 percent of the measurements would be in the interval $\overline{X} - 2s$ to $\overline{X} + 2s$ (see Figure 4.9).

Figure 4.9 *Illustration of the empirical rule*

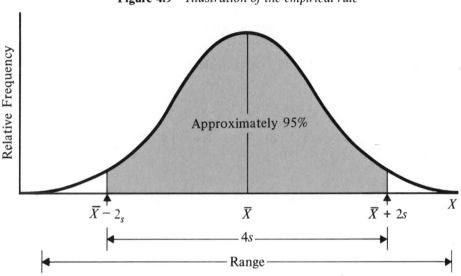

We note also from Figure 4.9 that the interval $\overline{X} - 2s$ to $\overline{X} + 2s$ has a length equal to $4s$. Since the range of the measurements is approximately $4s$, one-fourth of the range will provide an approximate value for s. This approximation can then be used to check our calculation of s:

$$approximate\ value\ of\ s = \frac{range}{4}$$

Example 4.22 *Fifteen students with similar socioeconomic backgrounds were compared on an IQ test at the end of their freshmen year. Their scores are presented in Table 4.15.*

Table 4.15   *IQ Scores*

116	118	126
129	114	130
129	122	128
132	131	125
123	134	126

Assume that these data are approximately mound-shaped and calculate s using the usual formula. Check your results.

Solution   *If we use the shortcut formula for $\Sigma (X - \bar{X})^2$ we need to compute ΣX^2 and ΣX:*

$$\Sigma X^2 = (116)^2 + (129)^2 + \cdots + (126)^2 = 236{,}873$$

$$\Sigma X = 116 + 129 + \cdots + 126 = 1883$$

Hence

$$\Sigma (X - \bar{X})^2 = \Sigma X^2 - \frac{(\Sigma X)^2}{n}$$

$$= 236{,}873 - \frac{(1883)^2}{15}$$

$$= 236{,}873 - 236{,}379.267$$

$$= 493.733$$

The sample variance s^2 is then

$$s^2 = \frac{\Sigma (X - \bar{X})^2}{n - 1} = \frac{493.733}{14}$$

$$= 35.267$$

and the standard deviation is

$$s = \sqrt{35.267} = 5.94$$

We can approximate the value of s using the range of the measurements divided by 4. From Table 4.15 we see that the range is 134.5 − 113.5 = 21. Our approximation is then

$$approximate\ value\ of\ s = \frac{range}{4} = \frac{21}{4} = 5.25$$

Although the approximate value, 5.25, differs somewhat from the actual value of s, 5.94, it still provides a check on our calculations. For example, we know that our calculation of 5.94 is reasonable. If we had forgotten to divide the sum of squares of the deviations (493.733) by $(n - 1) = 14$, our computed value of s would have been much larger (22.22). This error would have been detected by comparison with the range estimate of s. We would advise running this simple check every time you compute a standard deviation.

EXERCISES

37 If an approximate value of the standard deviation is the range divided by 4, check the standard deviation determined in Exercise 31 (page 142).

38 Repeat Exercise 37 but for the standard deviation determined in Exercise 32 (page 142).

T 4.14

he research-oriented social scientist often requires statistical tools in order to make a meaningful inference about a population based on information contained in a sample. Since making an inference implies description of the population, we seek a way to describe a set of measurements.

Graphical descriptive techniques, presented in Chapter 3, are very effective for describing a set of measurements or scores, but they are unsatisfactory for statistical inference. The difficulty is that one can say that the sample frequency histogram will be similar to the population frequency distribution, but we have no way to measure the goodness of the inference. Equivalently, we have no way to measure the degree of difference between two irregularly shaped figures.

Numerical descriptive measures numbers that create a mental image of the frequency distribution for a set of measurements. The most important of these are the mean and standard deviation, which measure the center and spread, respectively, of a frequency distribution. The standard deviation of a set of measurements is a meaningful measure of variability when interpreted using the empirical rule. Numerical descriptive measures of the population are called parameters.

Numerical descriptive measures are suitable both for description and inference making. We can use a numerical descriptive measure of the sample, say the sample mean, to estimate a parameter such as the population mean. The great advantage of the numerical descriptive measure in making inferences is that we can give a quantitative measure of the goodness of the inference. In particular, we shall show that the sample mean will lie within a specified distance of the population mean with a predetermined probability.

Now that we know how to phrase an inference, that is, how to talk about a population of interest to us, we turn to the problem of making the inference. As suggested in an earlier discussion, all types of inferences based on partial information, such as sample data, are subject to a degree of uncertainty. If we decide that the population possesses a certain characteristic, we may be correct, but there is always some element of doubt or uncertainty lurking in our minds. We shall find that probability, which is a measure of uncertainty, plays a major role both in making an inference and in measuring how good it is.

QUESTIONS AND PROBLEMS

39   The following measures represent the ratio of male to female executives in 10 corporations.

12.67	11.85
12.15	17.74
18.14	14.65
15.73	16.91
13.35	10.28

(a) Round the data to the nearest tenth.
(b) Compute the mode for the 10 scores of part (a).
(c) Compute the median for the 10 scores of part (a).
(d) Compute the mean for these 10 scores.
(e) Compute the average deviation from the mean.
(f) Compute the average deviation from the median.

40   What is the range for the sex ratios in Exercise 39?

41   Use the data in Table 4.2 (page 100) to compute the 30th percentile.

42   Randomly select 10 cities from the North in Table 3.4 (page 60) and 10 from the South. Given these two samples of murder rates, compute the average deviation from the median for each sample and then the coefficient of variation.

43 The following represent a sample of 10 measurements on the degree of urbaniza-
tion of 10 localities:

63	45
56	45
32	96
56	57
48	71

(a) Compute the range.
(b) Calculate the mean and standard deviation.
(c) Use the range approximation of Section 4.13 to check your calculation
of s.

44 The mean contribution per person for a college alumni fund drive was $100.50,
with a standard deviation of $38.23. What is the contribution of the person who
is 1 standard deviation below the mean? One standard deviation above the
mean?

45 Seven different families have the following number of children: 2, 1, 1, 0, 4, 5,
and 2. Find the mean and standard deviation. Check your calculation of s using
the range approximation of Section 4.13.

46 Using the data in Exercise 39, estimate the standard deviation. (*Hint*: See Section
4.13.) Compute s and compare with your approximation.

47 From the following scores on the percentage of family income allocated to rent,
a researcher calculated s to be .263. On what grounds might we doubt his
accuracy? What is the correct value (nearest hundredth)?

17.2	17.1	17.0	17.1	16.9
17.0	17.1	17.0	17.3	17.2
17.1	17.0	17.1	16.9	17.0
17.1	17.3	17.2	17.4	17.1

48 One hundred and twenty-eight individuals were given a test for social awareness.
The scores produced a mean and a standard deviation equal to 92 and 10,
respectively, based on a scale of 100. Would you expect the distribution of
these 128 scores (where a score cannot exceed 100) to be bell-shaped? Why?

49 One hundred fundamentalist ministers were administered a 40-item test concerning knowledge about causes of mental illness. Compute the mean and standard deviation for the scores in the table.

Class Interval	f	X_c	fX_c	fx_c^2
34–35	4	10		
32–33	0	9		
30–31	2	8		
28–29	6	7		
26–27	3	6		
24–25	5	5		
22–23	6	4		
20–21	6	3		
18–19	7	2		
16–17	9	1		
14–15	22	0		
12–13	10	-1		
10–11	8	-2		
8–9	6	-3		
6–7	6	-4		

50   Compute the refined mode for the data in Exercise 49.

51   What is the median for the data of Exercise 49?

52   Fifteen students enrolled in a course called Statistics for the Social Sciences were administered a pencil-and-paper IQ test. The results are as follows:

10	7	14	19	17
17	16	16	16	20
15	14	12	15	8

Determine the mean, median, and mode. Compare your answers.

53   Determine the standard deviation for ungrouped data (Exercise 52).

54   Check your computation of the standard deviation in Exercise 53 by comparing it with the range estimate of the standard deviation.

55   The levels of educational attainment (in years) of 15 heads of households living in an upper-middle-class residential area are as follows:

12	7	14	21	19
16	16	18	16	20
15	14	12	15	8

Determine the mean, median, and mode. Compare your answers.

56   Determine the standard deviation for ungrouped data (Exercise 55).

57   Check your computation of the standard deviation in Exercise 56 by comparing it with the range estimate of the standard deviation.

REFERENCES

Anderson, T. R., and M. Zelditch. *A Basic Course in Statistics*, 2nd ed. New York: Holt, Rinehart and Winston, Inc., 1968. Chapter 5.

Blalock, H. M. *Social Statistics*, 2nd ed. New York: McGraw-Hill Book Company, 1972. Chapters 5 and 6.

Champion, D. J. *Basic Statistics for Social Research*. Scranton, Pa.: Chandler Publishing Company, 1970. Chapters 3 and 4.

Mueller, J. H., K. F. Schuessler, and H. L. Costner. *Statistical Reasoning in Sociology*, 2nd ed. Boston: Houghton Mifflin Company, 1970. Chapters 5 and 6.

Palumbo, D. J. *Statistics in Political and Behavioral Science*. New York: Appleton-Century-Crofts, 1969. Chapters 2 and 3.

Weiss, R. S. *Statistics in Social Research*. New York: John Wiley & Sons, Inc., 1968. Chapters 6 and 7.

5

PROBABILITY, INFERENCES, AND SAMPLING DISTRIBUTIONS IN THE SOCIAL SCIENCES

*W*e stated in Chapter 1 that the social scientist uses inferential statistics to make an inference about a population based on information contained in a sample. Because populations are sets of measurements, we need a way to phrase an inference about a set of numbers. Graphical and numerical descriptive techniques were presented in Chapters 3 and 4, respectively. Now let us examine the mechanism employed in making inferences. This is best illustrated by means of an example.

Jones, a candidate for Congress, publicly announces that his forthcoming election is a guaranteed success and, in particular, he forecasts victory by a substantial margin in a precinct of a certain city. Somewhat dubious about his claims, we randomly select 20 names from the voter registration list of the precinct, call on each voter, and inquire who they will vote for in the upcoming election. Not one of the 20 states that he will vote for Jones—all favor his opponent. What do you conclude about Jones's claim to victory in the sampled precinct?

The answer to the preceding question requires an inference about the population of 1s and 0s that corresponds to the set of voters in the sampled precinct who favor or do not favor Jones. A 1 denotes a voter favoring Jones, a 0 corresponds to a voter favoring his opponent. In particular, we wish to infer whether the fraction of 1s in the population is greater than 1/2. If it is, Jones will be the victor. Noting the results of the sample, that all 20 of 20 voters favor Jones's opponent, we would conclude that Jones will lose the precinct. Let us examine the mental process, the line of reasoning that led us to this conclusion.

We reasoned as follows. If Jones were correct in his claim of victory, at least half the voters in the precinct would favor him and somewhat near this same fraction should be observed in the sample. In contrast to these expectations, none of the voters in the sample favored Jones, a result highly contradictory to his claim. Hence we infer that the fraction of voters in the population (the precinct) favoring Jones is less than 1/2.

Note that we concluded that Jones would not win because the sample yielded results highly contradictory to his claim. By contradictory, we do not mean that it was impossible to select at random 20 of the 20 voters favoring Jones's opponent (assuming Jones to be a winner); we mean simply that it was highly *improbable. We measured the degree of contradiction in terms of the probability of the observed sample.*

To get a better view of the role that probability plays in making this inference, suppose that the sample produced 9 in favor of Jones and 11 in favor of his opponent. We would not consider this result highly improbable and would probably not reject Jones's claim. How about 7 in favor and 13 against—or 5 in favor and 15 against? Where do you draw the line? At what point do you decide that the result of the observed sample is so improbable, assuming Jones to be the winner, that you disagree with his claim? To answer this question you must know how to find the probability of acquiring a particular sample. Probability is the mechanism for making inferences.

We would like to make it very clear to the reader that our line of reasoning is perfectly consistent with the logic he employs in everyday decision making. A trial jury convicts an indicted man if the prosecution's circumstantial evidence is highly contradictory to the man's claim of innocence. Because of an occasional misinterpretation of evidence, the presentation of only partial evidence, or an occasional lying witness, the juror is always faced with some degree of uncertainty and probability enters the picture. The difference between the statistical and the lay approach to decision making is that the statistican does not rely on his own intuition to determine the probability of observing the evidence, assuming the victim to be innocent. Intuitions about probability can be very much in error. Rather, he applies a precise evaluation of the probability of the sample before reaching a decision.

Since probability is the mechanism used to make inferences, we might ask: What is probability? We have employed the term in the previous discussion in the manner that it is used in everyday conversation. Let us examine this more closely.

5.2   WHAT IS PROBABILITY?

The motivation for the use of statistical techniques by social scientists is that social scientists deal with the observations of social phenomena which generate outcomes that cannot be predicted in advance. We cannot say in advance that a specific underdeveloped nation will become a developed nation by 1985. Indeed, the status of any society is likely to bob about, depending upon the balance of payments, the introduction of modern medicine, and the decline of death and birth rates. This kind of uncertainty creates a need for inferences in the social sciences, especially for the appropriate statistical tools for making inferences.

Definition 5.1

An experiment is the process by which an observation (or measurement) is obtained.

Note that the observation need not be numerical. Typical examples of experiments are:

1 Recording a voter's preference.
2 Observing a consumer's preference for a particular type of product.
3 Observing the number of blacks moving from an urban to a rural area during a specified period of time.
4 Examining a teenager to measure his political views.
5 Observing the number of homicides in a city precinct during a period of time.
6 Observing the number of divorces in a geographical area during a specified period of time.

We obtain a population of measurements when an experiment is repeated many times. For instance, we might be interested in the percentage of registered voters in a particular precinct that exercised the right to vote in the previous election. Questioning a voter to determine whether or not he voted in the previous election represents a single experiment; repetition of the experiment for all registered voters during this period would generate the entire population. A sample represents the results of a small group of experiments selected from the population.

A particular observation of a social science phenomenon can result in many different outcomes, some of which are more likely than others.

Definition 5.2

The probability of a particular outcome is a measure of one's belief in its occurrence when the experiment is conducted a single time.

Numerous attempts have been made to give a more precise statement of Definition 5.2. For some purposes we may view the probability of an outcome as the fraction of times the outcome will occur in a long series of observations. This is called the relative frequency concept of probability. Observations that are not repeatable are not so easy to interpret. An entre- preneur, Sam A., gambles $5,000,000 on a new business venture. He will never be able to repeat exactly the same observation, but he is vitally interested in the probability of success. The surgical operation to remove John T.'s spleen is also an observation that can never again be repeated as far as John is concerned, and John is certainly desirous of a high probability of success. Nonrepeatable observations require a subjective assignment of probability to various outcomes by a person or persons familiar with the mechanism under study. An experienced entrepreneur might be able to give a very satisfactory measure of his belief that Sam A.'s venture will be a success. Similarly, an experienced surgeon will be able to attach a fairly reasonable probability to the success of John's operation.

The theory of probability is a branch of mathematics concerned with the calculation of the probabilities of experimental outcomes. Of course, the outcomes of interest to us are the measurements we observe in a sample. Basic to the theory is the assignment of probabilities to a set of experimental outcomes called simple events. How one acquires the probabilities of these events is irrelevant as long as the quantities so acquired give fairly accurate measures of the likelihood of their occurrence when the experiment is conduc- ted once. The relative frequency concept of probability seems most reasonable to the authors because it can be checked by the observation of repeated experiments. However, we leave it to the reader to select an interpretation of probability that is philosophically satisfying to him.

The mathematical theory of probability provides a precise analytical method for finding the probability of an event, assuming that you agree with and satisfy the axioms upon which the theory is based.

A second method for finding the probability of an event is available to those who believe in the relative frequency approach to probability. Repeat the experiment a large number of times, say N, count the number of times that an event A is observed, say n, and calculate the *approximate* probability of A as

$$P(A) \approx \frac{n}{N}$$

We say "approximate" because we think of $P(A)$ as the relative frequency of the occurrence of event A over a very large number of repetitions of the experiment. Since our resources and patience are limited, the simulation of the experiment will be repeated only a finite number of times. If the experiment cannot actually be repeated, we can often simulate the process using a high-speed electronic computer.

We have noted two ways to find the probability of an observed event, an analytical approach using the mathematical theory of probability and an empirical method. We omit a discussion of the theory of probability and direct the interested reader to the references at the end of the chapter. Instead, we shall employ the earthier and more easily understood empirical method to find probabilities relevant to social science data that we shall use in making inferences.

In conclusion we note the difference between probability and statistics. The probabilist assumes that the population is known and wishes to infer the nature of a sample. He assumes a coin to be balanced (that is, 50 percent of the hypothetical population of coin tosses are heads, 50 percent tails) and calculates the probability of two heads in two successive tosses of the coin. In contrast, the statistician observes a sample and reasons backward to the unknown population. He would observe the results of a number of tosses of a coin and then attempt to infer whether the coin was balanced. As indicated in Jones' election survey (Section 5.1), we used probability as the mechanism for making an inference.

5.3 FINDING THE PROBABILITY OF AN EVENT: AN EMPIRICAL APPROACH

The empirical method for finding the probability of an event, called empirical sampling, relies on the repetition of an experiment (actual repetitions or simulations) a very large number of times. If the experiment is conducted N times of which n outcomes result in event A, then the probability of A is approximately

$$P(A) \approx \frac{n}{N}$$

Finding $P(A)$ by repeating the experiment a "large" number of times might seem easier than employing a mathematical theory of probability to obtain the same result, but it can be more difficult than it appears. Some experiments are very complex, and their repetition involves a sizable expenditure of time and money. Fortunately, many complex experiments can be simulated on high-speed electronic computers in short periods of time at very small costs.

To illustrate the empirical method for finding the probability of an event, consider an experiment consisting of tossing two coins, a penny and a dime, and observing the upturned faces. Denote four events as follows:

(TT):   Observe tails for both the penny and dime.
(TH):   Observe a tail for the penny and a head for the dime.
(HT):   Observe a head for the penny and a tail for the dime.
(HH):   Observe heads for both the penny and dime.

The penny and dime were tossed 2000 times with the results shown in Table 5.1. You will see that empirical sampling yields approximate probabilities that are in agreement with our intuition. That is, we might expect these outcomes to be equally likely and any one to occur with probability equal to .25.

Table 5.1   *Results of* $N = 2000$ *Tossings of a Penny and a Dime*

Outcome	Frequency	Relative Frequency
TT	474	.237
TH	502	.251
HT	496	.248
HH	528	.264

Suppose that we wish to find the probability of tossing two coins and observing exactly one head. Calling this event A, then

$$P(A) \approx \frac{502 + 496}{2000} = .499$$

This is very close to the actual probability that can be shown to equal .500.

Whether one finds the probability of observed events empirically or analytically depends on one's preference and the nature of the experiment. Some probabilities are easier to find by use of the theory of probability, others are easier to find by use of repeated sampling.

$\mathcal{T}$ 5.4   CONDITIONAL PROBABILITY
AND INDEPENDENCE

wo events are often related in such a way that the probability of occurrence of the first depends on whether the second has or has not occurred. For instance, suppose that a labor committee of two is to be chosen from a group of three women and four men. Let *A* be the event that the second person chosen is a woman. If each person has an equal chance (namely 1/7) of being selected on the committee, then, without knowledge of the sex of the first person drawn, we would intuitively state that the probability of selecting a woman on the second draw is 3/7. The probability of event *A*, denoted by $P(A)$, is $P(A) = 3/7$.

Suppose now that *B* is the event that the first person chosen is a woman. What is the probability that event *A* occurs given that we know that event *B* has occurred; that is, what is the probability that the second person chosen is a woman when we know that the first person chosen was also a woman? If the first person chosen were a woman, only two of the remaining six people are women. If we assume that each person has an equal chance (namely 1/6) of being the second person selected, the probability of drawing a woman is 2/6. This probability is called the conditional probability of event *A* given that event *B* has occurred and is denoted as $P(A|B)$. For our example, $P(A|B) = 2/6 = 1/3$.

Sometimes the occurrence of an event *B* has no effect on whether an event *A* occurs or not. In such a situation, event *A* would not depend on event *B*.

Definition 5.3

Two events A and B are independent *if P(A), the probability of event A, equals P(A|B), the conditional probability of A given that B has occurred; that is, A and B are independent if*

$$P(A) = P(A|B)$$

Otherwise they are dependent.

Example 5.1   *Determine whether event A, the second person selected for the labor committee is a woman, and event B, the first person selected for the committee is a woman, are independent.*

Solution   *We previously showed that $P(A) = 3/7$. Similarly, we reasoned that the probability of A occurring given that B has occurred, $P(A|B)$, is $2/6$. Since $P(A)$ does not equal $P(A|B)$, events A and B are dependent.*

Many social science studies result in the classification of people (or behavior) according to one or more systems of classification. For example, we might wish to classify people according to sex and income. One interesting question associated with this type of data is to determine whether the distribution of people in income categories is independent of sex. Or, in general, whether one variable is independent of another variable. We shall answer questions of this type in Chapter 9.

5.5   VARIABLES: DISCRETE AND CONTINUOUS

In Section 2.2 we distinguished between quantitative and qualitative variables. We shall now designate certain quantitative variables as random variables.

We have defined an experiment to be the process of obtaining an observation. Most experiments of interest result in numerical observations or measurements. If a quantitative variable measured (or observed) in an experiment is denoted by the symbol X, we are interested in the values that X can assume in an experiment. These values or outcomes are called numerical events. The number of students in a class of 50 who earn an A in their elementary statistics course is a numerical event. The percentage of registered voters who cast ballots in a given election is also a numerical event. The quantitative variable X is called a random variable because the value that X assumes in a given experiment is a chance or random event.

Random variables are classified as one of two types.

Definition 5.4

A discrete random variable is one that can assume a countable number of values.

Examples of discrete random variables would be:

1. The number of homicides per 100,000 individuals during a particular year.
2. The number of accidents per year at an intersection.
3. The number of voters in a sample favoring Jones.

Note that it is possible to count the number of values that each of these random variables can assume.

Definition 5.5

A *continuous random variable* is one that can assume an infinitely large number of values in a line interval.

For example, the daily maximum temperature in Rochester, New York, can assume any of the infinitely large number of values on the real line. It could be 89.6, 89.799, 89.7611114, etc. Typical continuous random variables are temperature, pressure, height, weight, and distance.

The distinction between discrete and continuous random variables is pertinent when one is seeking the probabilities associated with specific values of a random variable. The need for the distinction will be apparent when probability distributions are presented in Sections 5.6 and 5.8.

EXERCISES

1. Are social science data subject (amenable) to the notions of probability? Explain.

2. What is meant by the "relative frequency concept of probability"?

3. Describe the empirical approach to probability.

4. Conduct a class experiment in which each student flips two coins and records the number of heads face up. Repeat this experiment 10 times and combine the class results in a table similar to Table 5.1. Compare the two tables and comment upon the differences in relative frequencies.

5.6 PROBABILITY DISTRIBUTIONS
FOR DISCRETE RANDOM
VARIABLES

s previously stated, we need to know the probability of sampled observations in order to make an inference about the population from which the sample was drawn. Thus we need to know the probability associated with each value of the random variable X, because repeated observations of X yield the measurements in a sample. Viewed as relative frequencies, these probabilities generate a distribution of theoretical frequencies called the probability distribution of X. Probability distributions differ for discrete and continuous random variables but the interpretation is essentially the same.

The probability distribution $P(X)$ for a discrete random variable X displays the probability associated with each value of X. It can be presented as a table, graph, or formula. To illustrate, consider the two-coin tossing experiment of Section 5.3 and let X equal the number of heads observed. Then X can take the value 0, 1, or 2. From the data of Table 5.1, we would observe the approximate probabilities for each of these values of X given in Table 5.2. You will note that the relative frequencies are very close to the

Table 5.2 *Empirical Sampling Results for X, the Number of Heads in* 2000 *Tosses of Two Coins*

X	Frequency	Relative Frequency
0	474	.237
1	998	.499
2	528	.264

actual probabilities, which can be shown to be .25, .50, and .25. If we had employed 2,000,000 tosses of the coins instead of 2000, the relative frequencies would be indistinguishable from the actual probabilities. Thus the probability distribution for X, the number of heads, in the coin-tossing experiment is as shown in Table 5.3. It is presented graphically as a probability histogram in Figure 5.1.

Table 5.3 *Probability Distribution for the Number of Heads When Two Coins Are Tossed*

X	P(X)
0	1/4
1	1/2
2	1/4

Figure 5.1 *Probability distribution for the number of heads when two coins are tossed*

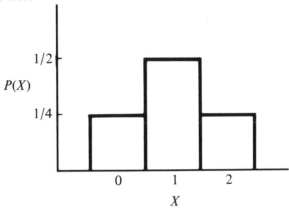

The probability distribution for this simple discrete random variable illustrates three important properties of discrete random variables.

Important Properties of Discrete Random Variables

1 The probability associated with every value of X must lie between 0 and 1.
2 The sum of the probabilities for all values of X must equal 1.
3 The probabilities for a discrete random variable are additive. Hence the probability that $X = 1$ or 2 is equal to $P(1) + P(2)$.

The relevance of the probability distribution to statistical inference can be illustrated by reconsidering the deduction based on the poll of the 20 precinct voters concerning Jones' prospects for election (Section 5.1). We concluded that Jones would lose the precinct because it seemed intuitively improbable that all 20 would favor his opponent if Jones were truly a winner. Although satisfied with our inference, we noted that less conclusive results would require more than an intuitive assessment of the probability of the observed sample. These probabilities can be obtained as follows.

If Jones possessed at least 50 percent of the votes (let us say exactly 50 percent), then drawing a single voter at random from the population of responses for all voters in the precinct would be equivalent to flipping a balanced coin. Letting a head denote a preference for Jones and a tail denote a preference for his opponent, the selection of 20 voters would be analogous to flipping 20 coins. We would be interested in X, the number of heads (those favoring Jones), in the sample. You will note that this experiment is just a simple extension of the two-coin-tossing experiment of Section 5.3.

The probability distribution for X could be obtained using empirical sampling. Many repetitions of the 20 coin tosses would result in a probability distribution for X, the number of voters favoring Jones (number of heads),

Figure 5.2   *Probability distribution for X, the number of voters who favor Jones in a sample of n = 20*

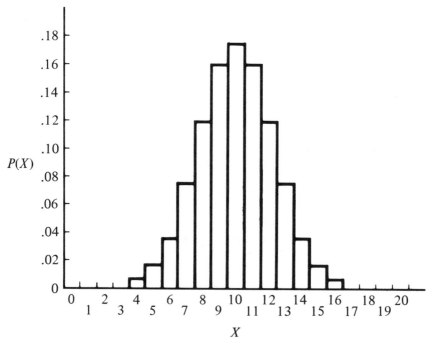

X

Number of Voters Favoring Jones

as shown in Figure 5.2. (*Note:* The probability that $X = 0, 1, 2, 3, 17, 18, 19,$ and 20 is very small and does not appear in the figure.)

Observe that $X = 10$ occurs with the largest probability, which is what we would expect if 50 percent of the voters favor Jones. We also note that it is highly improbable that we would draw a sample with as few as $X = 3$ or, for that matter, even $X = 5$. The improbable result that none in a sample favor Jones is highly contradictory to the assumption that he is a winner. We infer that the population of voter results contains less than 50 percent favoring him and therefore that he will be a loser.

The example illustrates the important role of probability in making inferences. We have used the probability distribution of a discrete random variable to make an inference about a very real population that is of interest both to Jones and to the general public.

5.7 A USEFUL DISCRETE RANDOM VARIABLE: THE BINOMIAL

One of the most useful discrete random variables is associated with the coin-tossing experiment described in Sections 5.3 and 5.6. Actually numerous coin-tossing experiments are conducted daily in the social sciences, physical sciences, and industry.

To illustrate, consider a taste-test survey conducted to determine which of two flavors (lime or grape) should be marketed in a new soft drink. A panel of consumers is selected and the preference of each member recorded. Determining a single panel member's preference is, in many respects, similar to tossing an unbalanced coin (a coin for which the probability of tossing a head is not equal to 1/2). A member's choice may be either the lime flavor (a "head") or the grape flavor (a "tail").

Similar experiments are conducted in other areas of industry, in the social sciences, and in education. The politician seeking reelection is interested in the fraction of voters supporting him in his district, the teacher is interested in the fraction of students who pass his course, and the research physician is interested in the fraction of white mice who survive a specified dose of a new drug.

Although dissimilar in some respects, the experiments described all exhibit, to a reasonable degree of approximation, the characteristics of a *binomial experiment*.

Definition 5.6

A binomial experiment is one that possesses the following properties:

1   The experiment consists of *n* identical trials.
2   Each trial results in one of two outcomes. We shall label one outcome a success, *S*, and the other a failure, *F*.
3   The probability of success on a single trial is equal to *p* and remains from trial to trial. The probability of failure on a single trial is $q = 1 - p$ and remains constant from trial to trial.
4   The trials are independent; that is, the outcome of one trial does not influence the outcome of any other trial.
5   We are interested in *X*, the number of successes observed during the *n* trials.

The foundation for the binomial experiment was laid by Jacob Bernoulli (1654–1705), who was the first of a series of gifted mathematicians to come from the Bernoulli family of Antwerp. Bernoulli's contributions to statistics include the formulation of the *Bernoulli trial,* which is the basis for the binomial experiment. Each of the *n* trials in a binomial experiment is called a Bernoulli trial.

It should be noted that very few real-life situations will satisfy perfectly the requirements stated in Definition 5.6, but for many the lack of agreement is so moderate that the binomial experiment still provides a very good model for reality.

Having defined the binomial experiment and suggested several practical applications, we now examine the probability distribution for the binomial random variable *X*, the number of successes observed in *n* trials.

The general formula for the probability of observing successes, *P(X)*, in *n* identical trials is given next.

Binomial Probability Distribution

$$P(X) = \frac{n! p^X q^{n-X}}{X!(n-X)!}$$

where

> n = *number of trials*
>
> p = *probability of success on a single trial*
>
> $q = 1 - p$

$$X = \text{number of successes in } n \text{ trials}$$

$$n! = n(n - 1)(n - 2) \cdots (3)(2)(1)$$

[*Note: The formula for P(X) is frequently written as*
$P(X) = C_X^n p^X q^{n-X}$, *where* $C_X^n = n!/X!(n - X)!$.]

The notation $n!$ (referred to as *n-factorial*) is used to indicate the product

$$n! = n(n - 1)(n - 2) \cdots (3)(2)(1)$$

For $n = 3$,

$$n! = 3! = (3)(3 - 1)(3 - 2) = (3)(2)(1) = 6$$

Similarly, for $n = 4$,

$$4! = (4)(3)(2)(1) = 24$$

We also note that $0!$ is always set equal to 1.

Now let us use the probability distribution to reason from a known binomial population to the results of a sample. Consider the following example.

Example 5.2 *A labor union's examining board for the selection of apprentices has a record for admitting 70 percent of all applicants who satisfy a set of basic requirements. Five members of a minority group recently came before the board and four out of the five were rejected. First let us reason from the population to the sample and find the probability that one or less would be accepted if p is really .7. We shall then reason in reverse and use our intuition to deduce whether the board applied a lower probability of acceptance when reviewing the five members of the minority group.*

Solution *Let X be the number selected out of the five examined. The probability of the event that $X = 0$ or 1 is, from Section 5.6,*

$$P(X = 0 \text{ or } 1) = P(0) + P(1)$$

Since we regard "acceptance" as a success, $p = .7$. Using $n = 5$ and substituting into the formula

$$P(X) = \frac{n!}{X!(n - X)!} p^X q^{n-X}$$

we have

$$P(0) = \frac{5!}{0!(5-0)!}(.7)^0(.3)^5$$

$$= \frac{5!}{(1)5!}(.7)^0(.3)^5 = .002$$

and

$$P(1) = \frac{5!}{1!(5-1)!}(.7)^1(.3)^4$$

$$= \frac{(5)(4)(3)(2)(1)}{(1)(4)(3)(2)(1)}(.7)(.3)^4 = .028$$

Then

$$P(X = 0 \text{ or } 1) = P(0) + P(1)$$

$$= .002 + .028 = .030$$

This indicates that if the examinations are independent and the committee is accepting each applicant with probability equal to .7, the probability that only one or less of the five is selected is 3 chances in 100. We would regard this improbable occurrence ($X \leqslant 1$) as a rare event.

Now let us reason from the sample to the population. Having observed the rare event $X \leqslant 1$, we could draw one of two conclusions. Either we have been unlucky and observed an unlikely event, or the board is currently admitting with a probability less than .7. We would be inclined to accept the latter conclusion. Note we do not conclude that the board is necessarily biased against the minority group. A lower probability of acceptance (which appears to exist) may also have been caused by over-crowded conditions in this particular labor speciality.

Now consider a second example.

Example 5.3 *A drug company advertises a new drug that is an effective treatment for a disease but produces an undesirable side effect on 10 percent of the patients treated. A doctor is contemplating using the drug on three patients. What is the probability that two or more will experience side effects? If this event occurs, what might we conclude concerning the drug company's advertisement?*

Solution *This example satisfies the five characteristics of a binomial experiment with n = 3 and p, the probability of observing a side effect, equal to .1. The probability of observing two or more patients with side* *effects is equal to*

$$P(2) + P(3)$$

Using the general formula

$$P(X) = \frac{n!}{X!(n - X)!} p^X q^{n-X}$$

and substituting n = 3, p = .10, we have

$$P(2) = \frac{3!}{2!(1)!} (.1)^2 (1 - .1)^1$$

$$= \frac{(3)(2)(1)(.1)^2(.9)}{(2)(1)(1)} = .027$$

and

$$P(3) = \frac{3!}{3!(0)!} (.1)^3 (1 - .1)^0$$

$$= (.1)^3 = .001$$

The desired probability is then

$$P(2) + P(3) = .027 + .001 = .028$$

Thus two or more in a sample of 3 could experience side effects approximately 3 percent of the time. Since this is an improbable event and we observed two or more with side effects, we would conclude that there is enough evidence to indicate that the drug firm's advertisement is false.

We have discussed probability distributions for discrete random variables and have given an example of a very useful discrete random variable, the binomial. We proceed now with probability distributions for continuous random variables.

EXERCISES

5   Describe, in your own words, the binomial experiment. Compare your answer with Definition 5.6.

6   If $n = 5$, what is $n!$? $0!$?

7   Consider the following class experiment: Toss three coins and observe the number of heads, X. Let each student repeat the experiment 10 times, combine

the class results, and construct a relative frequency table for X. Note that these frequencies give approximations to the actual probabilities that $X = 0, 1, 2,$ and 3. (*Note*: The exact values of these probabilities can be shown to be $\frac{1}{8}, \frac{3}{8}, \frac{3}{8},$ and $\frac{1}{8},$ respectively.)

8   A community is evenly divided between black families and white families. If a random sample of 10 families is selected to be interviewed concerning the advisability of rezoning a school district, what is the probability that all families are white? What is the probability that all 10 are black families? Comment. If all 10 families selected were either all white or all black, would you suspect a racial bias in the selection of the sample? (*Note*: Your decision should be made on the basis of probability.)

9   Refer to Exercise 8. Suppose that the sample contained only 5 families and that 4 of the 5 were white. What is the probability of selecting 4 or more whites in a random sample of 5? Does the sample suggest racial bias in the selection? (Base your decision on the calculated probability.)

5.8   PROBABILITY DISTRIBUTIONS
FOR CONTINUOUS RANDOM
VARIABLES

ou will recall that a continuous random variable is one that can assume values associated with the infinitely large number of points contained in a line interval. Without elaboration, we flatly state that it is impossible to assign a small amount of probability to each value of X (as was done for a discrete random variable) and retain the property that the probabilities sum to 1. We overcome this difficulty by reverting to the concept of the relative frequency histogram, Chapter 3, where we talked about the probability of X falling in a given interval. Recall that the relative frequency histogram for a population containing a large number of measurements will almost be a smooth curve because the number of class intervals can be made large and the width of the intervals can be decreased. Thus we envision a smooth frequency curve that provides a model for the population frequency distribution generated by repeated observation of a continuous random variable. This will be similar to that shown in Figure 5.3(a).

Recall that the histogram relative frequencies were proportional to areas over the class intervals and these areas possess a probabilistic interpretation. That is, if a measurement is randomly selected from the set, the probability that it will fall in an interval is proportional to the histogram area above

Figure 5.3 *Probability distribution for a continuous random variable*

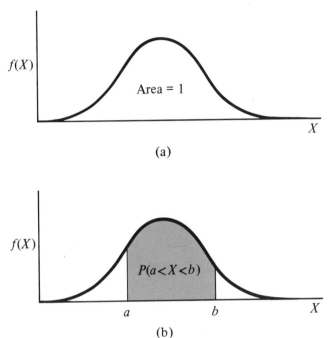

(a)

(b)

the interval. If we let the total area under the histogram equal 1, then areas over intervals are exactly equal to the corresponding probabilities. Since a probability distribution is a model for a population relative frequency histogram, we shall want the total area under the probability curve to equal 1.

The graph for the probability distribution for a continuous random variable is shown in Figure 5.3(a). The ordinate (height of the curve) for a given value of X is denoted by the symbol $f(X)$.

The probability that a continuous random variable falls in an interval, say between two points a and b, follows directly from the probabilistic interpretation given to the area over an interval for the relative frequency histogram (Section 3.8). Thus the probability that X falls between two points a and b is equal to the area under the curve over the interval a to b as shown in Figure 5.3(b). This probability is written $P(a < X < b)$.

We have curves of many shapes that can be used to model population frequency distributions for measurements associated with continuous random variables. Fortunately, the areas under these curves are all tabulated and ready for use. Thus if we know that student examination scores possess a particular probability distribution, as in Figure 5.4, and areas under the curve have been tabulated, we could find the probability that a particular student will score in excess of 80 percent by looking up the tabulated area, as shaded in Figure 5.4.

You probably wonder how we know which shape to use for the probability distribution in a given situation. The selection is quite critical if we wish to be probabilists and reason from the population to the sample, that is, if we wish to talk about the probability of X falling in some interval. Then we need information about the behavior of the random variable in previous sampling.

Figure 5.4 *Hypothetical probability distribution for student examination scores*

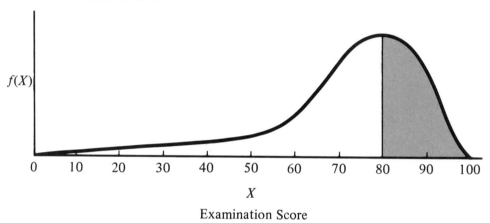

Examination Score

The relative frequency histogram of the sample would be very useful in making the selection. However, you recall that our objective as statisticians *is not* to reason from the population to the sample but to travel in reverse. Thus we wish to estimate or make decisions about the parameters of a population from information contained in a sample. As you will see subsequently, the specific shape chosen for the population frequency distribution (or, equivalently, the probability distribution for the observed random variable) will often have little effect on the probability statements associated with our inferences. Thus we can relax in the knowledge that the selection of the exact shape for the probability distribution for a continuous random variable is not crucial.

We shall find that most data collected on continuous random variables in nature possess mound-shaped frequency distributions and many of these are nearly bell-shaped. A continuous random variable that provides a good model for these types of data is called a normally distributed random variable. The normal random variable also plays a very important role in statistical inference. We shall study its bell-shaped probability distribution in detail in the next section.

5.9 A USEFUL CONTINUOUS RANDOM VARIABLE: THE NORMAL PROBABILITY DISTRIBUTION

ou recall that probability is the mechanism used to make inferences about populations based on information contained in a sample. Thus we learned the meaning of "probability" earlier in the chapter, and found that probabilists commence with a known population and deduce the outcome of a sample. Statisticians reason in reverse. Knowing the details of a sample and the probability of its occurrence for various populations, the statistician deduces the nature of the population. A prerequisite for making the inference is knowledge of the probability distribution for the random variable measured in the experiment.

With this background, it would seem natural to present the properties of many different random variables with probability distributions that could be used to approximate distributions of data observed in nature. Fortunately, this task can be greatly simplified because a large number of experimental populations satisfy the properties required by the empirical rule. That is, they often possess mound-shaped frequency distributions that approximate the bell shape of the normal curve. Therefore, the probabilities associated with many experimental outcomes can be evaluated using the empirical rule (Chapter 4). (Indeed, the probabilities in the empirical rule were obtained by calculating areas under the normal curve.) To summarize, the probability distribution for the normal random variable is an excellent model for the frequency distributions of many random variables observed in nature. The

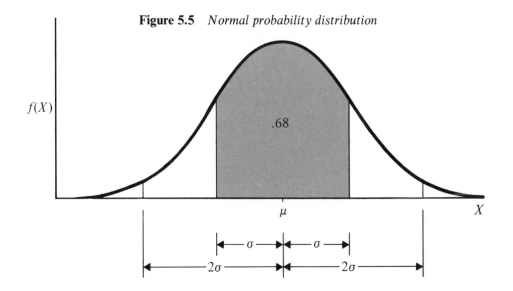

Figure 5.5 *Normal probability distribution*

bell-shaped normal probability distribution evolved in the work of the famous mathematician Johann Karl Friedrich Gauss (1777–1855) and is sometimes called a Gaussian distribution or a Gaussian curve.

It is not necessary to write a mathematical expression to describe the normal probability distribution.* Instead, we shall portray it graphically and will identify some of its important properties. These properties are:

1 A normal probability distribution is bell-shaped (see Figure 5.5). The area within one standard deviation of the mean is approximately .68, within two standard deviations, .95.
2 A normal probability distribution is symmetrical about its mean, μ (see Figure 5.5).

Figure 5.6 *Three normal probability distributions with different means and standard deviations*

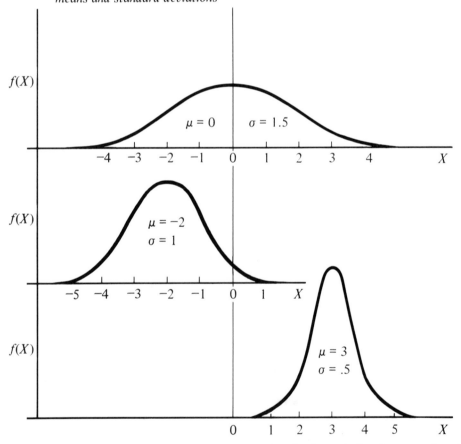

*The references at the end of the chapter give the formula for the normal probability distribution.

3   The location and spread of the normal distribution will depend
on the values of μ and σ. For example, three normal
distributions with different means and standard deviations are
shown in Figure 5.6.

The usefulness of the normal probability distribution is due to the
frequent occurrence of near-normal frequency distributions of data.

5.10   THE NORMAL APPROXIMATION TO THE BINOMIAL DISTRIBUTION

The binomial probability distribution for the number of
successes X will be approximately normal when the number of trials n is
large. The normal distribution that provides the best approximation to the
binomial probability distribution is one with the mean and the standard
deviation computed according to the following formulas:

Mean and Standard Deviation of the Binomial Random Variable

$$\mu = np$$

$$\sigma = \sqrt{npq}$$

where p is the probability of success in a given trial and $q = 1 - p$.

The binomial probability distribution for the number of voters favoring
Jones (Figure 5.2) is an excellent example of the application of the bell-shaped
curve. The number of voters X favoring Jones in the sample of 20 can be
viewed as the sum (or total) of the individual "yes" votes. We would expect
the probability distribution for X to be approximately normal.* The probability distribution for X (Figure 5.2) is shown in Figure 5.7 with a normal curve
superimposed. Note how well the normal distribution approximates the
actual probability distribution for X.

* One of the most fundamental theorems in the theory of statistics, the central
limit theorem, implies that the binomial probability distribution can be approximated
by the normal curve for large samples. Actually, the approximation is quite good for
n as small as 10 if p is near .5.

Figure 5.7   *Probability distribution for X, the number of voters favoring Jones in a sample of n = 20, and the approximating normal curve*

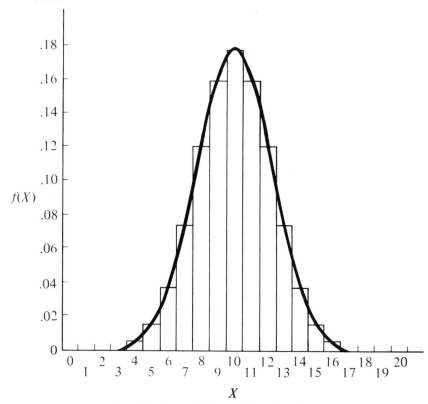

X

Number of Voters Favoring Jones

The approximating normal curve to the probability distribution for Jones' survey, Figure 5.7, was based on $p = .5$ (assuming that 50 percent of the voters favor Jones) and $n = 20$. Then

$$\mu = np = 20(.5) = 10$$
$$\sigma = \sqrt{npq} = \sqrt{20(.5)(.5)} = 2.24$$

We have not explained how to draw the approximating normal curve, but this is unimportant. The relevant point to note is that we would *not expect* X to fall more than two standard deviations [$2\sigma = 2(2.24) = 4.48$] away from the mean, $\mu = 10$, because the probability of this occurrence is quite small. From the empirical rule, the probability that X will lie more than two standard deviations away from μ is approximately .05. This probability is shown in Figure 5.8 as the shaded area under the approximating normal curve. We would regard a probability of .05 as relatively small.

Figure 5.8   *Approximating normal curve for Jones's survey*

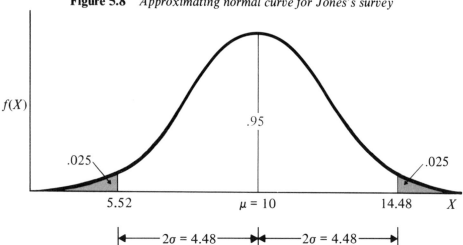

Example 5.4   *In a national survey to determine the fraction of voters favoring the Republican presidential candidate, a sample of 1000 voters was polled to determine their preference. Let X denote the number of persons favoring the Republican in the 1000 sampled. Assume that 50 percent of the registered voters favor him and find the mean and standard deviation of the random variable X. Use this information to describe the variability of X in repeated sampling.*

Solution   *This election survey satisfies the properties of a binomial experiment (n = 1000, p = .5). The mean and standard deviation of the binomial random variable X are*

$$\mu = np = 1000(.5) = 500$$

and

$$\sigma = \sqrt{npq} = \sqrt{1000(.5)(.5)}$$
$$= \sqrt{250} = 15.81$$

respectively. By the empirical rule, 95 percent of the values of X in repeated sampling will fall in the interval $\mu - 2\sigma$ to $\mu + 2\sigma$; that is,

$$[500 - 2(15.81)] \ to \ [500 + 2(15.81)] \quad or \quad 468.38 \ to \ 531.62$$

The reader will readily see how inference could enter this example. Suppose that X assumed an improbable value; that is, suppose that we drew a sample of 1,000 persons and observed 468 or fewer who favored the Republican candidate. Since the probability of this occurring is approximately .025, we could infer either that we observed a rare event (assuming 50 percent of the voting public favors the Republican), or that less than 50 percent of the voters favor him. We would be inclined to choose the latter.

Suppose that 480 voters favor the Republican candidate of Example 5.4. Is this an improbable result if we assume that 50 percent or more of the voters favor him? To answer this question, we must extend the empirical rule and find the probability that X will lie within any number of standard deviations of μ. To find these probabilities, we need a table of areas under the normal curve.

EXERCISES

10   List three important properties of a normal probability distribution.

11   What is the mean of the binomial distribution? The standard deviation?

12   Assume that 50 percent of the residents of a city are female. Suppose that of 1000 workers employed by a large contractor, only 48 are women. If the workers were randomly selected from the city, what is the mean number of females you would expect to be employed by the contractor? What is the standard deviation?

13   Refer to Exercise 12. Observing that only 48 of the employees are women, does this suggest bias on the employer's part (assuming that sampling was random)? What is the fallacy in this argument?

14   Previous experience indicates that 1 of every 100 major crimes in a large city is murder. A new crime-prevention program has been initiated and in effect for a year. During the year, the city experiences 3000 major crimes. From past experience, what is the average number of murders that you would expect? Standard deviation?

15   Refer to Exercise 14. Suppose that the observed number of murders in the city was 10. Assuming past experience, is this a highly improbable event? Would you conclude that the new program has been effective?

5.11   TABULATED AREAS UNDER
THE NORMAL PROBABILITY
DISTRIBUTION

ou will recall that we need to know the probability of a sample
in order to make inferences about a population. Hence we need to be able to
calculate the probability that a random variable with a normal probability
distribution will fall in a specified interval. For example, if X is normally
distributed with mean μ and standard deviation σ, we might need to know the
probability that X lies in an interval between two points a and b. Recall that
probabilities for continuous random variables correspond to areas under a
smooth curve. The shaded portion of Figure 5.9 represents the area under the
curve between points a and b and corresponds to the probability that X will
fall within the interval.

Figure 5.9   *Area under a normal curve*

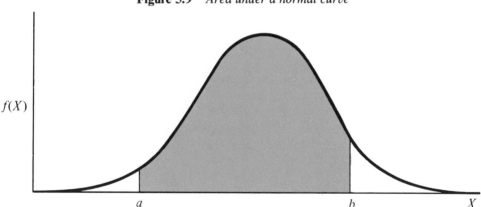

The desired area can be obtained using a table of areas under the
normal curve. Unfortunately there are infinitely many normal distributions
determined by supplying various values to the parameters μ and σ. It would
be an impossible task to provide tables of area for every normal distribution,
but we have an easy way to circumvent this difficulty.

Table 1 of the Appendix gives areas under the normal curve between
the mean and a measurement z standard deviations to the right of μ. The
tabulated area is shown in Figure 5.10.

Areas to the left of the mean need not be tabulated because the normal
curve is symmetric about the mean. The area between the mean and a point

Figure 5.10   *Tabulated areas under the normal curve*

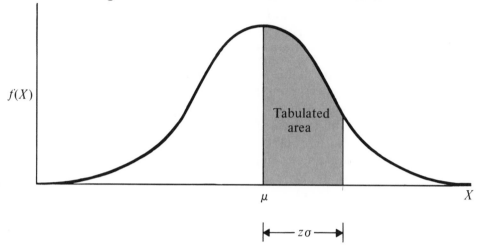

Figure 5.11   *Tabulated area corresponding to z = 2*

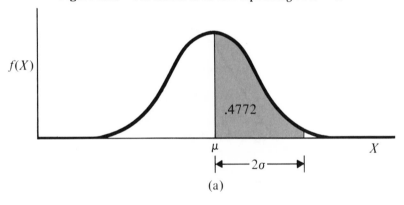

(a)

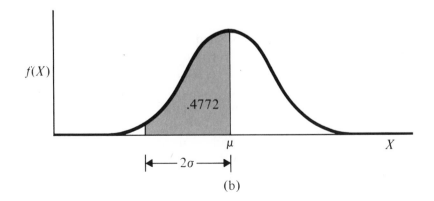

(b)

2σ to the right is the same as the area between the mean and a similar point 2σ to the left.

The number of standard deviations, z, is given to the nearest tenth in the left-hand column of Table 1. Adjustments to take z to the nearest one-hundredth are given in the top row of the table. Entries in the table are the areas corresponding to particular values of z. For example, the area between the mean and a point $z = 2$ standard deviations to the right of the mean is shown in the second column of the table opposite $z = 2.0$. This area, shaded in Figure 5.11(a), is .4772. The area between the mean and a point two standard deviations to the left of the mean, shown in Figure 5.11(b), is also .4772. Then the area *within* two standard deviations of the mean is $2(.4772) = .9544$. This explains the origin of the "approximately 95 percent" in the empirical rule.

Similarly, the area one standard deviation to the right of the mean (i.e., $z = 1$) is .3413. The area *within* one standard deviation of the mean is .6826, or approximately 68 percent, as stated in the empirical rule. This area is shown in Figure 5.12.

Figure 5.12

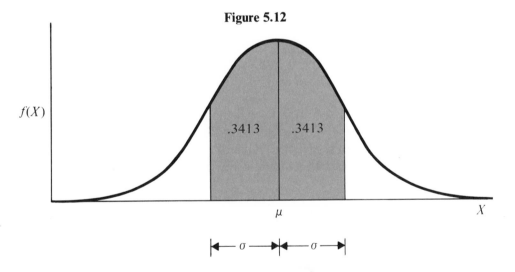

Suppose that we wish to find the area corresponding to $z = 1.64$. Proceed down the left column to the row $z = 1.6$ and across the top of the page to the .04 column. The intersection of the $z = 1.6$ row with the .04 column gives the desired area, .4495. This area is shown in Figure 5.13.

To determine how many standard deviations a measurement X lies from the mean, μ, we first determine the distance between X and μ. Recall that this distance can be represented as

$$\text{distance} = X - \mu$$

Figure 5.13 *Area corresponding to z = 1.64*

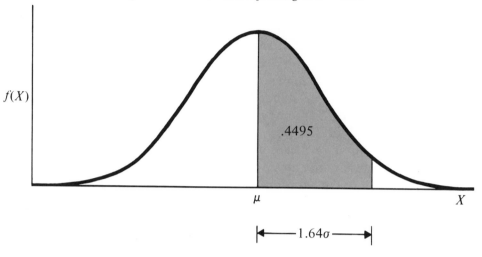

The distance between X and μ can then be converted into a number of standard deviations by dividing by σ, the standard deviation of X. Hence

$$X \qquad z = \frac{\text{distance}}{\text{standard deviation}} = \frac{X - \mu}{\sigma}$$

The probability distribution for z, which has $\mu = 0$ and $\sigma = 1$, is often called the standard normal distribution (see Figure 5.14). The area under the curve between $z = 0$ and a specified value of z, say z_0, has been tabulated in Table 1 of the Appendix and is shown in Figure 5.14.

Figure 5.14 *Standard normal distribution*

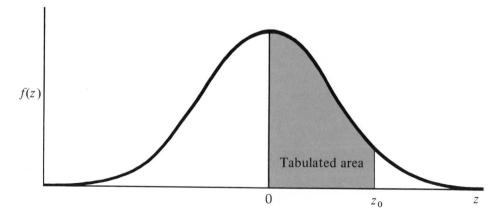

To calculate the area under a normal curve between the mean μ and a specified value X to the right of the mean we first determine z, the number of standard deviations X lies from μ, using

$$z = \frac{X - \mu}{\sigma}$$

We then refer to Table 1 and obtain the entry corresponding to the calculated value of z. This entry is the desired area (probability) under the curve between μ and the specified value of X. Note that z gives the distance between a measurement X and its mean expressed in units of the standard deviation and is therefore a standardized distance. Social scientists frequently refer to a computed value of z as a z score.

We shall illustrate the use of the table of normal-curve areas with a simple example to assist our pedagogy and then will proceed to more practical applications.

Example 5.5   *Suppose that X is a normally distributed random variable with a mean $\mu = 8$ and standard deviation $\sigma = 2$. Find the probability that X lies in the interval from 8 to 11; that is, what fraction of the total area lies under the curve between 8 and 11? (See the shaded portion of Figure 5.15.)*

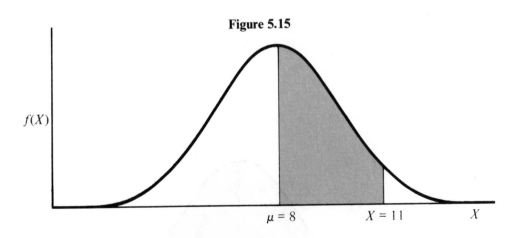

Figure 5.15

$f(X)$

$\mu = 8$ $X = 11$ X

Solution   *To determine the desired area we compute how many standard deviations $X = 11$ lies from the mean $\mu = 8$.*

$$z = \frac{X - \mu}{\sigma} = \frac{11 - 8}{2} = 1.50$$

*The corresponding area can then be determined from the entry in Table
1 opposite z = 1.50. We see that the desired area is .4332 (see Figure
5.16). Therefore, the probability that X lies between 8 and 11 is equal to
.4332.*

Figure 5.16   *Area between μ = 8 and X = 11*

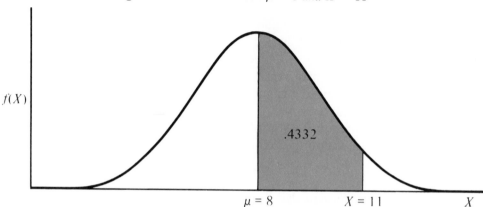

Example 5.6   *The quantitative portion of a nationally administered
achievement test is scaled so that the mean score is 500 and the standard
deviation is 100. If we assume the distribution of scores is normal (bell-
shaped), what fraction of the students throughout the country should
score between 500 and 682? What fraction should score between 340 and
682?*

Figure 5.17   *Area between X = 340 and X = 682*

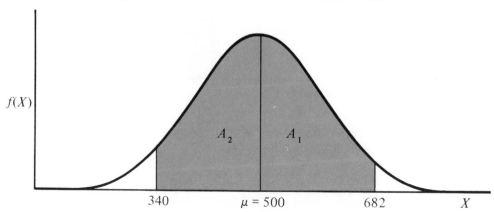

Solution   *Consider Figure 5.17. To answer the first question we need to compute the area A_1 between $\mu = 500$ and $X = 682$:*

$$z = \frac{X - \mu}{\sigma} = \frac{682 - 500}{100} = 1.82$$

The tabulated area for this value of z is $A_1 = .4656$. Thus we expect 46.56 percent of the students to score between 500 and 682. The area between 340 and 682 is equal to the sum of A_1 and A_2 in Figure 5.17. To find A_2 we compute the number of standard deviations that $X = 340$ lies from $\mu = 500$. Hence

$$z = \frac{X - \mu}{\sigma} = \frac{340 - 500}{100} = -1.60$$

(Negative values of z indicate a point to the left of the mean.) The appropriate area found by ignoring the negative sign is then .4452. We would expect $A_1 + A_2 = .4656 + .4452 = .9108$ of the students to score between 340 and 682 on the examination.

Example 5.7   *An anthropology instructor has found that the length of time for a student to complete his favorite final examination possesses a normal distribution with $\mu = 2.2$ hours with a standard deviation $\sigma = .25$. What is the probability that a randomly selected student will complete the examination in less than 1.50 hours? On a particular examination, a mediocre student, Sam J., not only received an A but completed the examination in $1\frac{1}{2}$ hours. Would you suspect Sam of an unusually fine preparation for the examination or perhaps an unknown and unseen supporting hand?*

Figure 5.18   *Probability that an anthropology student requires less than 1.50 hours*

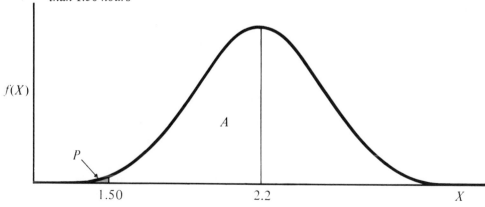

Solution   *A sketch of the desired area P is shown in Figure 5.18. First compute*

$$z = \frac{X - \mu}{\sigma} = \frac{1.50 - 2.2}{.25} = -2.8$$

This area, A, corresponding to z = 2.8, is .4974. The area on one side of the mean is .5, so the desired area is

$$P = .5 - A = .0026$$

This probability is quite small, which means that Sam J.'s spectacular feat is indeed a rare event or else he has taken exceptional steps to cope with the examination. Although rare events do occur sometimes, we would be inclined to accept the latter explanation.

In this section we have computed areas under a normal curve. To do this we converted the distance between a measurement X and the mean μ into a number of standard deviations (denoted by the symbol z) and then referred to Table 1. These areas are equal to the probabilities that measurements will fall in particular intervals.

EXERCISES

16   Using Table 1 of the Appendix, specify the area under the normal curve over the intervals identified by the following z scores:

(a)   0 to 1.00   (b)   0 to 2.00   (c)   0 to $+1.43$
(d) $+.73$ to $+.98$ (e) $-.37$ to $+.37$ (f) -1.43 to 0

17   Repeat Exercise 16 for
(a) -2.58 to -1.00 (b) $-.61$ to $+1.78$
(c) $+1.00$ to $+2.00$ (d) $+2.00$ to $+3.00$

18   Repeat Exercise 16 for
(a) -1.56 to $+1.36$ (b) $+1.65$ to $+2.00$
(c) -1.04 to $+1.27$ (d) $-.39$ to $+1.83$

19   Find the value of z such that 25 percent of all the scores in the distribution are to the left of z.

20   A normally distributed random variable X possesses a mean and a standard deviation equal to 8 and 3, respectively. Find the z score that corresponds to $X = 8.5$.

21 Refer to Exercise 20. Find the z score corresponding to $X = 6.7$.

22 Refer to Exercise 20. Find the probability that X lies within the interval 7.4–8.3.

23 Annual incomes for intracity social workers throughout the country are assumed to be normally distributed with $\mu = \$8500$ and $\sigma = \$1600$. What fraction of social workers receive an income greater than $\$10,000$? Less than $\$5500$?

5.12 PARAMETERS OF POPULATIONS

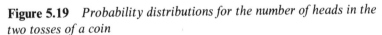

e shall assume that the probability distribution for a random variable is a good model for the population relative frequency histogram of real data that would be generated by repeated experimentation. In fact, let us assume that the two are identical. Then the mean of the probability distribution, called the *expected value* of the random variable, will equal μ, the mean of the population. Also, its standard deviation will equal σ.

For example, refer again to the probability distribution for X, the number of heads in two tosses of a coin, shown in Figure 5.19. From Figure 5.19 you can see that the center of the distribution is at $X = 1$ and, in fact, it is easy to show that $\mu = 1$. Similarly, it can be shown that $\sigma = .5$. Both of these values can be verified by calculating the mean and standard deviations of values of X generated by repeated sampling. The mean and standard deviation of any probability distribution could be acquired by repeated

Figure 5.19 *Probability distributions for the number of heads in the two tosses of a coin*

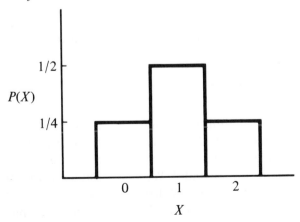

sampling, but we will give you the necessary formulas for all probability distributions presented in this text.

You will recall that numerical descriptive measures of populations are called parameters. Thus μ and σ are parameters of a population, as are the median, percentiles, and other numerical descriptive measures defined in Chapter 4.

Since the objective of inferential statistics is inference and we aim at description of the population based on information contained in a sample, it is reasonable that we would wish to make inferences about μ, σ, σ^2, and other parameters of the population. In fact, it is quite logical that we would use the sample mean $\overline{X}$ to estimate μ and would use other sample numerical descriptive measures to estimate their population counterparts.

5.13   RANDOM SAMPLING

We have talked about finding the probability distribution of a random variable by observing the frequency of occurrence of values of X in many repetitions of the experiment. Frequently the experiment will involve the random selection of one (or more) measurements from a population. Although we have a layman's concept of the term "random," we shall give more precise meaning to the term random sampling.

Definition 5.7

A random sample of a single measurement from a population of N is one in which each of the N measurements has an equal probability of selection.

The notion of random sampling can be extended to the selection of more than one measurement, say n measurements. For purposes of illustration, we shall take a population that contains a very small number of measurements, say $N = 4$, which are 1, 4, 3, and 2. We shall assume further that we wish to select a random sample of $n = 2$ from the $N = 4$. How many different (distinct) samples could one select? The six distinct samples are listed in

Table 5.4   *Samples of Four Measurements*

Sample	Measurements in Sample
1	1, 2
2	1, 3
3	1, 4
4	2, 3
5	2, 4
6	3, 4

Table 5.4. A *random sample* of two measurements from the population of four would be one that would give each of the six different samples the same probability of selection.

Definition 5.8

A random sample of n measurements from a population is one that gives every different sample of size n from the population an equal probability of selection.

It is unlikely that one could ever draw a sample with all probabilities of selection exactly equal, but random sampling can be achieved (for all practical purposes) by ensuring that we have a thorough mix in our sample. Card shuffling, although never perfect, is an attempt to distribute a random sample of cards to each player. If populations are generated by physically conducting an experiment and making a measurement, such as tossing two coins, the experiment could be repeated indefinitely and would generate a conceptual population. Each toss should be generated under the same experimental conditions so that the observed sample will, for all practical purposes, satisfy our definition of random sampling. Good mixing is the key.

The simplest and most reliable way to select a random sample of n elements from a large population is to employ a table of random numbers such as that shown in Table 8 of the Appendix. Random-number tables are constructed so that integers occur randomly and with equal frequency. For example, suppose that the population contains 1000 elements. Number the elements in sequence from 1 to 1000. Then turn to a table of random numbers such as the excerpt shown in Table 5.5.

Select *n* of the random numbers in order. The population elements to be included in the random sample will be given by the first three digits of the random numbers (unless the first four digits are 1000). If *n* = 5, we would include elements numbered 104, 223, 241, 421, and 150. So as not to use the same sequence of random numbers over and over, the experimenter should select different starting points in Table 8 to begin the selection of random numbers for different samples.

Table 5.5 *Portion of a Table of Random Numbers*

10480	15011	01536
22368	46573	25595
24130	48360	22527
42167	93093	06243

What is the importance of random sampling? The answer is that we must know how the sample was selected so that we can evaluate its probability. The probabilities of samples selected in a random manner are known and thus enable us to make valid inferences about the population from which the sample was drawn. Non-random-sampling procedures have unknown probabilities associated with the samples and are useless for purposes of inference.

EXERCISES

24 What is the difference between a parameter and a statistic (a numerative descriptive measure of the sample)? Give examples of three parameters.

25 Define what we mean by a random sample. Is it possible to draw a truly random sample? Comment.

26 Suppose that you wish to draw a random sample of *n* = 20 households from a city that contains 75,000 households. Further, suppose that the households have been identified and numbered. Use Table 8 of the Appendix to identify the households to be included in a random sample.

5.14   SAMPLING DISTRIBUTIONS

hat happens when we draw a random sample from a popula-
tion and compute the sample mean (or median)? How close does this computed
statistic lie to the corresponding population parameter? Similar questions
might be asked concerning other statistics (for example, the range, percentiles,
and the index of qualitative variation) when they are computed from sample
data. To answer this question, we employ the same reasoning we did in
analyzing Jones's prospects for election in Section 5.1. That is, we need to
know the probability distribution (called a sampling distribution) for the
sample mean. We shall use the empirical sampling approach of Section 5.3.
That is, we shall obtain an approximation to this distribution for a given
sample size by repeating the experiment (drawing a sample) over and over,
computing each time the value of the sample statistic. The resulting frequency
histogram for the sample statistic gives an approximation to its probability
distribution.

To return to the sample mean, let us examine its behavior when we
sample from a known population. Although it is unrealistic to work with a
small population (since most populations of interest in the social sciences are
large), for purposes of illustration we will take the 90 murder rates of Table
3.4 as our population and randomly draw 50 samples of five measurements.
For example, numbering the cities from 1 to 90 and using the table of random
numbers (Table 8) we could proceed down column 1 and use the first two

Table 5.6   *Results of Sample* 1

City Number	City	Murder Rate
10	Dallas, Tex.	18
22	Nashville, Tenn.	13
24	Orlando, Fla.	11
42	Kalamazoo, Mich.	4
37	Cleveland, Ohio	14
	Total	60

digits of each five-digit random number. We obtain 10, 22, 24, 42, and 37, and
would select the murder rates from cities possessing these numbers. The
result of the first of the 50 samples is shown in Table 5.6. The sample mean

is

$$\overline{X} = \frac{\Sigma\, X}{n} = \frac{60}{5} = 12$$

We repeated this procedure 49 more times to acquire 49 new samples. Their sample means are listed in Table 5.7.

Table 5.7   *Means of 50 Samples of Size 5 Selected from the 90 Murder Rates of Table 3.4*

12.0	8.8	11.4	5.0	8.8
7.0	12.4	15.0	8.6	8.6
14.2	7.4	10.2	8.4	7.2
11.2	5.4	7.8	9.6	10.2
9.2	8.8	6.2	4.4	9.0
10.0	6.4	6.0	10.4	8.4
9.0	7.0	11.0	11.4	8.6
9.8	8.2	8.2	7.4	5.4
7.4	10.0	13.2	7.4	8.4
7.0	7.8	7.6	4.2	9.6

To see how the sample means relate to the population mean ($\mu = 8.24$) for the 90 murder rates of Table 3.4, we construct a histogram showing the distribution of the 50 sample means of Table 5.7. This distribution is shown in Figure 5.20. Note how the sample means cluster about the population mean $\mu = 8.24$. Let us relate this phenomenon to a practical situation wherein we wish to estimate the population mean murder rate based on a single sample of five metropolitan areas. Since most of the sample means in Figure 5.20 fall within a distance of 4 murders (per 100,000 inhabitants) of the population mean ($\mu = 8.24$), we would expect our single sample mean to lie within the same distance of μ. This tells us something about the accuracy acquired in estimating the mean murder rate based on a sample of only five measurements. Of course, in a practical situation we would not have the results of a sampling experiment available, but we shall see in Chapter 6 that the sampling distributions (frequency histograms for sample statistics) are well known. Using knowledge of these distributions, we can say, with a given probability, how close a single sample statistic will lie to the corresponding population parameter.

You may be dismayed to note how little accuracy one acquires in estimating a population mean murder rate based on a sample of five metropolitan areas. In Chapter 6 you will find that this accuracy increases if you select a larger sample.

Figure 5.20 *Frequency histogram of sample means, n = 5*

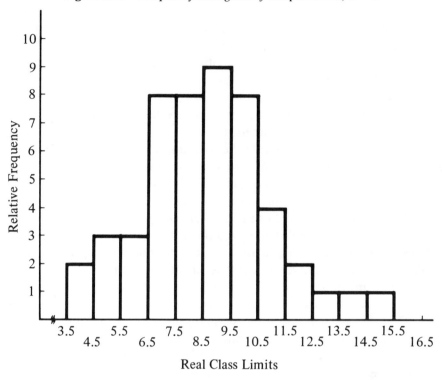

5.15 **SUMMARY**

ecall that social scientists frequently wish to make an inference about a population based on information contained in a sample. Anyone can use sample data to estimate the fraction of voters favoring candidate Jones, but the real question is how far will their estimate lie from the true fraction? How much faith can we place in their inference-making ability? In a nutshell, what is the probability that their estimate will differ from the actual value by 5 percent? To answer this question, we need to know the probabilities associated with values of the sample fraction favoring Jones; that is, we need to know its probability distribution. This probability is at the heart of the science of inference making.

Random variables can be of two types, discrete and continuous. Their probability distributions can be derived mathematically using the theory of probability or acquired empirically by repeating the experiment a large number of times (called empirical sampling) and constructing a frequency histogram of the results. Each probability distribution is a model for the population relative frequency histogram, and each can be described by a mean, standard deviation, or any of the other numerical descriptive measures discussed in Chapter 4.

Numerical descriptive measures of the population are called parameters. We may attempt to describe the population by estimating or making decisions about population parameters.

In Chapter 5 we demonstrate the role that probability plays in making inferences and explain in general terms how one can find the probabilities associated with the values of a random variable that appear in a sample. The binomial and normal random variables occupy prominent positions in statistical methodology. The binomial random variable crops up in the conduct of many social science sample surveys as the number of persons in the sample who favor a particular issue, favor a particular political candidate, or, in general, possess a specific attribute. The normal random variable assumes a position of prime importance in the social sciences because so many random variables of interest possess frequency distributions that are approximately normal. Even the binomial random variable possesses a distribution that is approximately normal for large sample sizes. Consequently, we often can make approximate probability statements about observed sample results by using the areas (probabilities) under a normal probability distribution. We now apply these results in Chapters 6 and 7. Specifically, we shall consider ways to estimate parameters of populations and judge how far away from the parameter our estimate might be. Or we might wish to hypothesize something about the population parameters and use the sample data to reach a decision concerning the believability of the hypothesis. These concepts, based on single samples from populations, are developed in Chapters 6 and 7. Methods for the comparison of parameters from two or more populations are considered in subsequent chapters.

QUESTIONS AND PROBLEMS

27 What is probability? Briefly describe its utility in the social sciences.

28 What are the differences between the probability distributions for discrete and continuous random variables?

29 If a population contains five scores of alienation, how many distinctly different samples of two scores can be selected? (*Hint*: See Section 5.13.)

30 Discuss the logic behind the empirical approach to probability. Set up an empirical sampling procedure of your own to ascertain the probability of a particular event.

31 What is the binomial experiment? Give three examples of a binomial experiment of interest to the social scientist.

32 Prenursing students throughout the country were administered an examination to determine their knowledge concerning personal hygiene. If we assume that the population of scores was normally distributed with a mean of 50 and standard deviation of 3.5, approximately what percentage of the population would we expect to fall
(a) in the interval 47–55?
(b) in the interval 44.8–48.9?
(c) below a score of 58?

33 Define z in terms of X, a normally distributed variable with a mean μ and standard deviation σ.

34 Using Table 1 of the Appendix, calculate the area under the normal curve between
(a) $z = 0$ and $z = 1.5$.
(b) $z = 0$ and $z = 1.8$.

35 Repeat Exercise 34 for
(a) $z = 0$ and $z = 2.5$.
(b) $z = -1.5$ and $z = 0$.

36 Repeat Exercise 34 for
(a) $z = -.8$ and $z = 0$.
(b) $z = -.8$ and $z = .8$.

37 Repeat Exercise 34 for
(a) $z = -1.96$ and $z = 1.96$.
(b) $z = -2.58$ and $z = 2.58$.

38 Repeat Exercise 34 for
(a) $z = -.12$ and $z = 1.8$.
(b) $z = 1.65$ and $z = 2.0$.

39 Find the value of z such that 30 percent of the area lies to its right. (*Note*: This is the 70th percentile of the standard normal distribution.)

40 Find the value of z such that 5 percent of the area lies to its right.

41 Find the value of z such that 2.5 percent of the area lies to its right.

42 A normally distributed random variable X possesses a mean and standard deviation equal to 7 and 2, respectively. Find the z value corresponding to $X = 6$.

43 Refer to Exercise 42. Find the value of z corresponding to $X = 8.5$.

44 Refer to Exercise 42. Find the probability that X lies in the interval 6–8.5.

45 One week prior to an important vote, a survey of members of a large all-male student organization indicated that 40 percent were in favor of changing their constitution to allow women members. Two days prior to the vote, the campus student government suggested that it would withhold operating funds for the organization if it did not vote to allow women members. That same day 50 members were polled at random to ascertain their opinion on the issue and 30 said they would vote for the constitutional change. Assuming that 40 percent of the entire membership still favor women members, what is the mean and standard deviation of the number favoring a constitutional change in a random sample of 50 members? (Assume that the student organization is large enough so that the sampling satisfies the assumptions required of a binomial experiment.)

46 Refer to Exercise 45. Based on the fact that 30 of the 50 favored a constitutional change, would you conclude that opinions seem to have changed following the student government announcement?

47 Find the z score such that exactly 28 percent of the scores are less than that value of z.

48 In a distribution of birth rates, the mean is 18.5 and the standard deviation is 4.8. Find the z score for a birth rate of 21.2. Would you be surprised to find a birth rate of 2.2? Why?

49 Thirty percent of all children in a large public school have been found to exhibit signs of malnutrition. If a random sample of 100 students is selected from the school, give the mean and standard deviation of the number of malnutritioned students in the sample.

50 From returns in previous years, it has been found that approximately 70 percent of the tax returns in a given income category are incorrectly filed. Assuming that 70 percent of the returns will be incorrect this year also, find the mean and standard deviation of the random variable X, the number of tax returns incorrectly filed in a random sample of 5000 returns. Use this information to describe the variability of X in repeated sampling.

51 If 2600 of the 5000 sampled returns (of Exercise 50) are filed incorrectly, would you anticipate approximately 70 percent of all the returns this year to be incorrectly filed? Explain.

52 Find the value of z such that 1 percent of the area lies to its right.

53 Find the value of z such that 10 percent of the area lies to its right.

54 The hourly local union wage for a particular type of construction worker has a national average of $5.30 and a standard deviation of $.63. Find the fraction of local unions for which the compensation is more than $6.00 per hour. If there were 850 such local unions, how many unions would have compensations in excess of $6.00 per hour?

55 If we can assume that the lengths of television commercials average 3 minutes per break and have a standard deviation of 1.20 minutes, what is the probability of observing a commercial break that exceeds 5 minutes?

REFERENCES

Anderson, T. R., and M. Zelditch. *A Basic Course in Statistics*, 2nd ed. New York: Holt, Rinehart and Winston, Inc., 1968. Chapter 5.

Blalock, H. M. *Social Statistics*, 2nd ed. New York: McGraw-Hill Book Company, 1972. Chapters 7, 9, and 10.

Champion, D. J. *Basic Statistics for Social Research*. Scranton, Pa.: Chandler Publishing Company, 1970. Chapter 5.

Hogg, R. V., and A. T. Craig. *Introduction to Mathematical Statistics*, 3rd ed. New York: Macmillan Publishing Co., Inc., 1970. Chapter 3.

Mendenhall, W., and L. Ott. *Understanding Statistics*. North Scituate, Mass.: Duxbury Press, 1972. Chapters 4 and 5.

Mendenhall, W., L. Ott, and R. L. Scheaffer. *Elementary Survey Sampling*. Belmont, Calif.: Wadsworth Publishing Company, Inc., 1971.

Mendenhall, W., and R. L. Scheaffer. *Mathematical Statistics with Applications*. North Scituate, Mass.: Duxbury Press, 1973. Chapters 3 and 4.

Mueller, J. H., K. F. Schuessler, and H. L. Costner. *Statistical Reasoning in Sociology*, 2nd ed. Boston: Houghton Mifflin Company, 1970. Chapters 6 and 8.

Palumbo, D. J. *Statistics in Political and Behavioral Science*. New York: Appleton-Century-Crofts, 1969. Chapter 5.

6

CONCEPTS OF ESTIMATION: THE ONE-SAMPLE CASE

6.1 INTRODUCTION

Inference, specifically decision making and prediction, is centuries old and plays a very important role in our lives. Each of us is faced daily with personal decisions and situations that require predictions concerning the future. The government is concerned with predicting the number of welfare recipients and the average amount paid per person in a given year. A social scientist may wish to estimate the mobility of a particular segment of our society. A political scientist seeks to know whether people's opinions concerning particular penal reforms are related to political party affiliation. The inferences that these individuals make should be based upon relevant facts which we call observations, or data.

In many practical situations the relevant facts are abundant, seemingly inconsistent, and in many respects overwhelming. As a result, our careful decision or prediction is often little better than anybody's outright guess. The reader need only refer to the "Market Views" section of the *Wall Street Journal* to observe the diversity of expert opinion concerning future stock market behavior. Similarly, a visual analysis of data by social scientists will often yield conflicting opinions regarding conclusions to be drawn from a sample survey. Many individuals tend to feel that their own built-in inference-making equipment is quite good, but experience suggests that most people are incapable of utilizing large amounts of data, mentally weighing each bit of relevant information, and arriving at a good inference. (The reader may test his individual inference-making equipment using the exercises in Chapters 6 and 7. Scan the data and make an inference before using the appropriate statistical procedure. Compare the results.)

Certainly, a study of inference-making systems is desirable, and this is the task of the research statistician. The social scientist makes use of the statistical results rather than relying upon intuitions. He employs statistical techniques to aid in making inferences. Although earlier we touched upon some of the notions involved in statistical inference, we shall collect our ideas at this point in an elementary presentation of some of the basic ideas involved in statistical inference.

The objective of inferential statistics is to make inferences about a population based upon information contained in a sample. Inasmuch as populations are characterized by numerical descriptive measures called parameters, statistical inference is concerned with making inferences about population parameters. Typical population parameters are the mean, the standard deviation, the area under the probability distribution above or below some value of the random variable, or the area between two values of the variable. Indeed the practical problems mentioned in the first paragraph of this section can be restated in the framework of a population with a specified parameter of interest.

Procedures for making inferences about parameters fall into one of two categories. We may make decisions (test hypotheses) concerning the value of the parameter, or we may estimate or predict the value of the parameter. Although some statisticians view estimation solely as a decision-making problem, it will be convenient for us to retain the two categories and concentrate particularly on estimation and tests of hypotheses.

Any statement about statistical inference would be incomplete without reference to a measure of goodness for an inferential procedure so that one procedure may be compared with another. More than comparing different procedures, we would like to state the goodness of a particular inference in a given situation. For example, we may want to predict not only the number of unemployed persons of a given ethnic background, but we also may want to know how accurate our prediction will be. Is it correct to within 1000, 10,000, or 1,000,000? Statistical inference in a practical situation, then, contains two elements: the inference itself and a measure of its goodness.

Before concluding this introductory discussion of inference, it would be well to dispose of a question that frequently disturbs the beginner. Which method of inference should be used; that is, should the parameter be estimated or should we test an hypothesis concerning its value? The answer to this question is dictated basically by the research question that has been posed and is in some cases a personal preference. Many substantive problems involve testing hypotheses concerning parameters; others involve making estimates. We shall employ estimation procedures in this chapter and tests of hypotheses in Chapter 7. In both chapters we shall confine our attention to inferences based on a single sample selected from a population.

$\mathcal{E}$ 6.2   TWO TYPES OF ESTIMATORS

stimators of population parameters can be classified into two categories, point estimators and interval estimators. For example, suppose that university officials are concerned about parental reaction to a newly instituted "open-dormitory" policy. In particular, the university would like to estimate the fraction of parents favoring the new policy. A random sample of 500 parents is obtained, and each parent's approval or disapproval is noted. If 40 percent of the sampled parents favor the new policy, university officials might then use this number to estimate the actual percentage of parents favoring the policy. Alternatively, they might estimate that the percentage of parents agreeing with the policy lies in an interval from 38 to 42 percent. We define these two types of estimates as follows:

Definition 6.1

A single number (or point) used to estimate a population parameter is called a point estimate.

Definition 6.2

An estimate of a population parameter formed by two numbers that determine an interval within which we expect the parameter to fall is called an interval estimate.

We shall now elaborate on the methods that generate the two types of estimates.

A point-estimation procedure utilizes information in a sample to calculate a single number or point that estimates the parameter of interest. The actual estimation is accomplished by an estimator.

Definition 6.3

An estimator is a rule that tells us how to calculate an estimate using sample information.

For example, the sample mean $\overline{X}$ is an estimator of the population mean μ. Thus

$$\overline{X} = \frac{\Sigma X}{n}$$

is a precise rule for computing an estimate based on the sample observations. Correspondingly, an interval estimator is a rule that tells us how to use the sample data to calculate two points—the upper and lower limits for an interval estimate.

How can we evaluate the goodness of an estimator; that is, how well does it estimate the parameter of interest? Take the case of the famous Swiss archer William Tell. History suggests that he was a *good* archer because he fired a single arrow that split an apple resting on his son's head. Suppose that he now wished to demonstrate his skill on you. Would you be willing to offer an apple on your head as a target for Tell (or even a watermelon)? Are you convinced that Tell was a good marksman?

Point estimation can be likened to firing an arrow at a target. The estimator corresponds to the archer–bow combination, an estimate to a single arrow fired at the target, and the parameter to the bull's-eye. Every marksman is an estimator. He draws a single sample from a population, constructs his estimate, and fires it at the parameter of interest. How can we evaluate the accuracy of the marksman (estimator)?

Putting you directly beneath Tell's target goes a long way toward explaining the concepts involved in evaluating the goodness of an estimator. It is usually not sufficient to observe it in operation for a single sample. We must observe how it behaves when it is employed over and over again.

Keep in mind that an estimate computed from the random sample measurements is itself a measurement on a random variable. That is, an estimator is a random variable with a sampling distribution. Repeated shots by the archer are shown on the target in Figure 6.1(a). Analogously, the grouping of the infinitely many shots fired at the parameter bull's-eye (call it theta, θ) is shown as a sampling distribution, Figure 6.1(b). Thus θ could represent any parameter, for example μ, σ, or p. The estimator of θ will be designated by the symbol $\hat{\theta}$ ("theta-hat"), where the "hat" indicates that we are estimating the parameter immediately beneath.

The probability distribution for an estimator such as that shown in Figure 6.1(b) can be simply described using the numerical descriptive measures discussed in Chapter 4. We shall refer frequently to the mean (or expected

Figure 6.1   *Distributions of arrows and estimates: (a) arrows grouping about target bull's-eye; (b) estimates grouping about the bull's-eye*

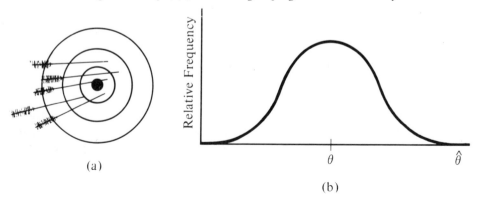

value) of an estimator and to the standard deviation of its probability distribution.

What properties should we require of a good point estimator $\hat{\theta}$; equivalently, what properties would we consider desirable for the distribution of an estimator? First, we would like the distribution of estimates to be centered about the parameter estimated; that is, we would like the mean (expected value) of the estimator to equal the parameter estimated. An "on-target" or unbiased estimator gives a grouping of estimates as shown in the probability distribution, Figure 6.2(a). A biased estimator, one that has a tendency to overestimate or underestimate a parameter, is shown in Figure 6.2(b). The biased estimator of Figure 6.2(b) is analogous to an archer who tends to fire too far to the right of the bull's-eye.

Figure 6.2 *Probability distributions for biased and unbiased estimators: (a) probability distribution for an unbiased estimator; (b) probability distribution for a biased estimator*

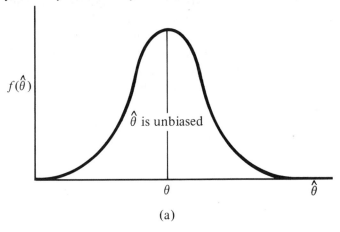

(a)

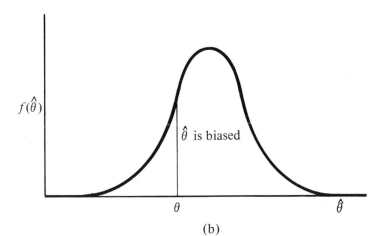

(b)

We define an unbiased estimator as follows:

Definition 6.4

An estimator, $\hat{\theta}$, of a parameter θ is said to be unbiased *if the expected value of $\hat{\theta}$ is equal to θ (i.e., the distribution of estimates obtained by using the estimator $\hat{\theta}$ is centered on the parameter θ).*

Estimators that do not satisfy this property are said to be biased.

The second property that we require of a good estimator is that the standard deviation of the distribution of estimates, denoted by $\sigma_{\hat{\theta}}$, be as small as possible. The shots at the bull's-eye in Figure 6.3(a) show a much better grouping than those shown in Figure 6.3(b). Corresponding groupings of estimates are shown as probability distributions in Figures 6.3(c) and (d). We much prefer the precision of firing indicated by Figure 6.3(a) and (c).

The positive distance between a single shot at a target and the bull's-eye is often called a firing error. Similarly, the positive distance between a single estimate and the bull's-eye parameter is called the error of estimation. Naturally we want the error of estimation to be as small as possible. All the estimators presented in this chapter are "on target" or, equivalently, unbiased estimators. How small will be the error for an estimate based on a single sample? We cannot say exactly, but we know from the empirical rule that approximately 95 percent of the estimates will lie within two standard deviations ($2\sigma_{\hat{\theta}}$) of their mean. See Figure 6.4. (If we knew for certain that the estimator possessed a normal distribution, it would be more accurate to say that estimates will lie within $1.96\sigma_{\hat{\theta}}$ of their mean 95 percent of the time.) Hence θ will be the bull's-eye and the error of estimation will be less than $2\sigma_{\hat{\theta}}$, with probability approximately equal to .95. Therefore, we call $2\sigma_{\hat{\theta}}$ a bound on the error of estimation because the error will be less than this value most of the time. We will give both the standard deviation and bound on the error of estimation for all point estimators discussed in this chapter, which will give a measure of goodness for your estimate.

The goodness of an interval estimator is evaluated in a manner similar to the method employed for point estimators. We would not want to judge the worth of an interval estimator on the basis of a single estimate. Rather we would like to observe its behavior in repeated sampling. Using empirical sampling procedures we could repeatedly draw samples of the same size and calculate an interval estimate on each occasion. This process would generate a large number of intervals (rather than points). A good interval estimator would be one that successfully enclosed the parameter of interest a large fraction of the time.

Figure 6.3   *Comparative variation*: (a) *shots closely grouped about the bull's-eye;* (b) *shots widely spread about the bull's-eye;* (c) *an estimator with a small standard deviation;* (d) *an estimator with a large standard deviation*

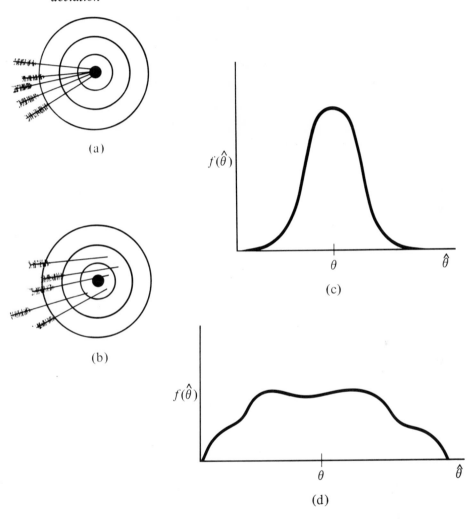

(a)

(b)

$f(\hat{\theta})$

θ $\hat{\theta}$

(c)

$f(\hat{\theta})$

θ $\hat{\theta}$

(d)

Definition 6.5

The fraction of times in repeated sampling that an interval estimator encloses the parameter estimated is called the confidence coefficient. *The estimator is called a* confidence interval.

Figure 6.4   *Probability distribution of estimates*

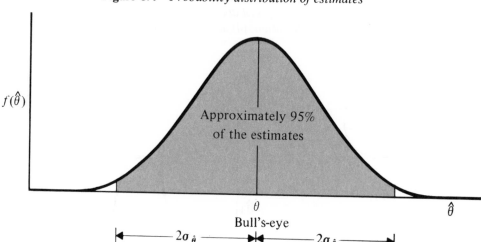

The selection of the "best" point or interval estimator in a given situation depends on numerous factors and is beyond the scope of this text. Throughout the remainder of this chapter we deal with specific problems of estimating and testing hypotheses about parameters. The general ideas presented in this section apply to all estimators. They will become more meaningful as we consider some practical examples.

6.3   ESTIMATION OF A POPULATION MEAN: A POINT ESTIMATOR

*m*any practical problems lead to the estimation of a population mean μ. For example, a psychologist may want to estimate the average time of reaction among patients given a particular stimulus, an economist may wish to estimate the mean increase in wages for various unions, and a sociologist may wish to estimate the mean family income in a ghetto area. We shall discuss both point and interval estimation of a population mean μ. In each case we shall give the appropriate estimator and provide a measure of its goodness.

Many different estimators are available for the point estimation of a population mean μ. Each would generate a probability distribution in repeated sampling and, depending on the problem, would possess certain advantages and disadvantages. For most situations the sample mean will provide the best estimate of μ.

The probability distribution of $\overline{X}$, obtained by repeated sampling of n measurements from a large population with finite mean equal to μ and standard deviation σ, possesses the following properties. Regardless of the probability distribution of the population from which the measurements are drawn,

1   The average of the distribution of sample $\overline{X}$'s is equal to μ, so $\overline{X}$ is the unbiased estimator of μ.

2   The standard deviation of the distribution of $\overline{X}$'s is equal to

$$\sigma_{\overline{X}} = \frac{\sigma}{\sqrt{n}}$$

(Note that the symbol $\sigma_{\overline{X}}$ is used to denote the standard deviation of the distribution of $\overline{X}$'s.)

Figure 6.5   *Probability distribution of $\overline{X}$'s*

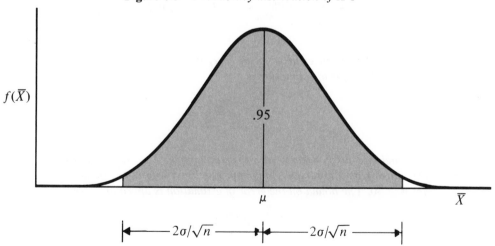

3   When n is large (30 or more), the sample mean $\overline{X}$ will be approximately normally distributed. The probability distribution for $\overline{X}$ is shown in Figure 6.5.

We summarize the point-estimation procedure of μ as follows:

Point Estimation of μ

Point Estimator of μ: $\overline{X}$.
Standard Deviation of $\overline{X}$ (sometimes called the standard error of the mean): $\sigma/\sqrt{n}$.
Bound on the Error of Estimation: $2\sigma/\sqrt{n}$.

The implication of the probability distribution for $\overline{X}$ is that 95 percent of the time an estimate $\overline{X}$ will lie within $2\sigma/\sqrt{n}$ of μ (see Figure 6.5). The error of estimation, the distance between $\overline{X}$ and μ, will be less than $2\sigma/\sqrt{n}$ with probability .95. Since the error of estimation will be less than $2\sigma/\sqrt{n}$ most of the time, we call this quantity a bound on the error of estimation. We shall illustrate the estimation procedure with an example.

Example 6.1 *In order to estimate the mean murder rate per 100,000 inhabitants in the South, a random sample of 30 SMSA's is selected and the corresponding murder rates are observed. The data are summarized below:*

$$\overline{X} = 13.36 \qquad s = 4.69$$

Obtain a point estimate of the average murder rate μ and place a bound on the error of estimation.

Solution *The point estimate of the average murder rate is 13.36. The bound on the error of estimation is*

$$2\sigma_{\overline{X}} = \frac{2\sigma}{\sqrt{n}} = \frac{2\sigma}{\sqrt{30}}$$

Although we do not know σ, we may approximate *its value using s calculated from the sample. This approximation will be reasonably good for $n \geqslant 30$. The bound on the error of estimation is approximately*

$$\frac{2s}{\sqrt{n}} = \frac{2(4.69)}{\sqrt{30}} = 1.71$$

We take this to imply that approximately 95 percent of the sample means in repeated sampling would be within two standard deviations $(2\sigma_{\overline{X}})$ of their mean. We feel quite confident that our estimate is within 1.71 of the true murder rate μ.

Three points deserve further comment. First, beginners sometimes use 2σ as a bound on the error of estimation rather than $2\sigma_{\bar{x}}$. Remember, if we want to describe the variability of the distribution of the estimator $\bar{X}$, we must refer to its probability distribution and use its standard deviation $\sigma_{\bar{x}}$. The quantity σ is the standard deviation of the population from which the sample was drawn.

Second, we sometimes have to approximate σ in the bound on the error of estimation. In Example 6.1 we used the "sample standard deviation" s to estimate σ. This approximation will be reasonably good when n is large (say 30 or more).

Third, you will note that the standard deviation of the distribution of $\bar{X}$'s depends on the standard deviation of the population σ and the sample size n. Thus

$$\sigma_{\bar{x}} = \frac{\sigma}{\sqrt{n}}$$

Although it is not obvious that the formula should take this form, it is clear that the greater the variation of the population as measured by σ, the greater will be the variation in the distribution of $\bar{X}$'s. Similarly, we would expect to

Figure 6.6 *Probability distribution for $\bar{X}$ based on samples of 5 and 25 from a normal population*

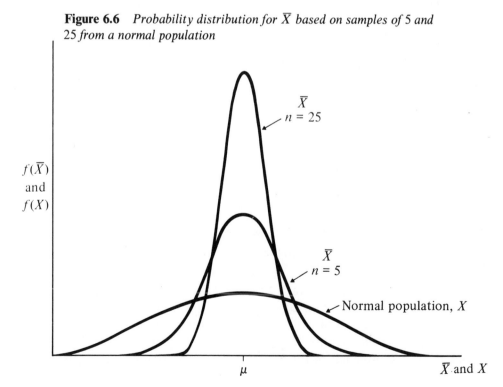

obtain more information as the sample size n is increased and hence would expect the variability of the distribution of $\bar{X}$'s to decrease. The probability distributions of $\bar{X}$ based on samples of 5 and 25 drawn from a normal distribution are superimposed on the probability distribution for a normal population in Figure 6.6.

We now consider interval estimation of a population mean μ.

EXERCISES

1   Define the terms "point estimate" and "error of estimation."

2   Sick-leave records obtained from a random sample of 100 steelworkers showed a mean number of days leave for the previous year equal to 25.6. If the sample standard deviation is 7.4, estimate the population mean number of days sick leave for steelworkers last year. Plan a bound on your error of estimation and explain its significance.

3   In a particular society, marriage counselors have found that the length of time from the first marriage to separation, for divorced couples, is 6.7 years. A random sampling of 80 divorced couples produced an average time to separation of 6.3 years with a standard deviation of 2.9 years. Estimate the mean length of time to separation and place a bound on your error of estimation.

4   A random sample of 150 terms of sentence for a particular crime (first offense) showed a mean equal to 4.2 years with a standard deviation of 2.4 years. Estimate the mean term of imprisonment awarded for the offense and place a bound on your error of estimation.

6.4   ESTIMATION OF A POPULATION MEAN: AN INTERVAL ESTIMATOR

*T*he sample mean $\bar{X}$ can be used to locate the two points needed to form an interval estimator or *confidence interval* for μ. Refer again to the probability distribution of $\bar{X}$ [shown in Figure 6.7(a)]. Note that 95 percent of the time $\bar{X}$ will lie within $2\sigma/\sqrt{n}$ of μ. Now imagine that the $\bar{X}$ computed from a single sample fell to the right of the mean as shown in Figure 6.7(b).

Figure 6.7 *Probability distribution for $\overline{X}$ showing a confidence interval for μ*

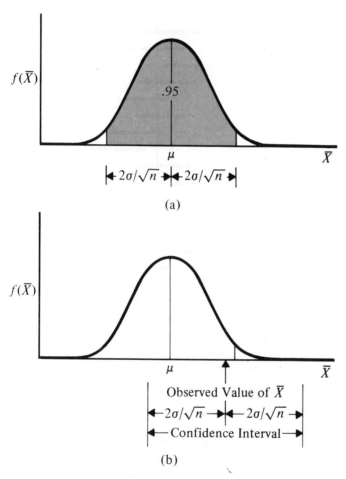

(a)

(b)

Suppose that we now measure a distance $2\sigma/\sqrt{n}$ on either side of $\overline{X}$ to form the endpoints of a confidence interval. Will the interval enclose μ? The answer is "yes." Indeed, the interval will enclose μ as long as $\overline{X}$ falls within $2\sigma/\sqrt{n}$ of μ. This will occur approximately 95 percent of the time. To be more accurate, you will recall from Chapter 5 that the area under the normal curve within 1.96 standard deviations of the mean is .95. Hence we call the interval

$$\overline{X} \pm \frac{1.96\sigma}{\sqrt{n}}$$

Figure 6.8 *Twenty different 95 percent confidence intervals for μ, each based on 30 measurements*

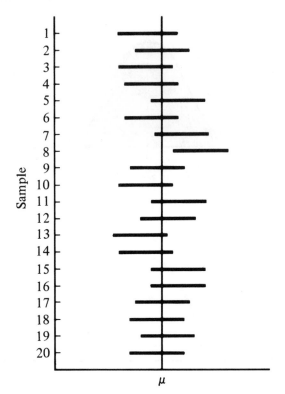

95 *percent confidence interval.* That is, intervals computed using this formula will enclose the mean μ 95 percent of the time.

To illustrate, Figure 6.8 shows twenty 95 percent confidence intervals calculated for a population mean based on samples of 30 measurements for each sample. The location and width of each confidence interval is shown as a horizontal line segment in Figure 6.8 and the location of the population mean μ is indicated by the vertical line. Note that most of the intervals cover the vertical line and hence include the population mean μ. Indeed, if we were to continue sampling, taking millions of samples, 95 percent of the computed confidence intervals would enclose μ.

Note that we shall calculate only one such interval for a given sample of n measurements, and it either will or will not cover μ. Since we know that approximately 95 percent of the intervals in repeated sampling will enclose the mean μ, we say that we are 95 percent confident that the interval we calculate does enclose μ. Because we rarely know σ and hence must approximate it with a sample standard deviation computed from a sample of size 30 or more, we call this a large-sample confidence interval.

Large-Sample 95 percent Confidence Interval for μ

$$\left(\bar{X} - \frac{1.96\sigma}{\sqrt{n}}\right) \text{ to } \left(\bar{X} + \frac{1.96\sigma}{\sqrt{n}}\right)$$

(*Note: If σ is unknown, substitute s for σ. The sample size must be 30 or more so that s will be a good approximation for σ.*)

We illustrate with an example.

Example 6.2   *The hourly wages in a particular labor union are assumed to be normally distributed. A random sample of 70 employee records is examined, and the sample mean and standard deviation are*

$$\bar{X} = \$13.00$$

and

$$s = \$2.20$$

Estimate μ using a 95 percent confidence interval.

Solution   *Since the sample size n is greater than 30, we can use the large-sample confidence interval. No prior estimate of σ is available, so we have*

$$\frac{\sigma}{\sqrt{n}} \approx \frac{s}{\sqrt{n}} = \frac{2.20}{\sqrt{70}}$$

The lower point for the confidence interval, called the lower confidence limit, is

$$\bar{X} - \frac{1.96\sigma}{\sqrt{n}}$$

which is approximately

$$\overline{X} - \frac{1.96s}{\sqrt{n}}$$

or

$$13.00 - 1.96\frac{2.20}{\sqrt{70}} = 13.00 - .515$$

$$= 12.485$$

Similarly, the upper confidence limit is approximately

$$\overline{X} + \frac{1.96s}{\sqrt{n}} = 13.00 + 1.96\frac{2.20}{\sqrt{70}}$$

$$= 13.515$$

Rounding these limits to the nearest cent, the 95 percent confidence interval for the true mean hourly wage is $12.48 to $13.52.

Now let us consider different confidence intervals. For example, if we measured one standard deviation on either side of $\overline{X}$ to form the confidence interval

$$\overline{X} \pm \frac{\sigma}{\sqrt{n}}$$

we would enclose μ approximately 68 percent of the time. This is because (from the empirical rule) $\overline{X}$ will lie within one standard deviation of μ approximately 68 percent of the time. Since the standard deviation of the distribution of $\overline{X}$'s is $\sigma/\sqrt{n}$, we have the 68 percent confidence interval shown above. This idea can be extended to obtain any desired confidence coefficient. Since 90 percent of the $\overline{X}$'s will lie within $z = 1.645$ standard deviations of μ, the interval $\overline{X} \pm 1.645\sigma/\sqrt{n}$ will enclose μ 90 percent of the time. This interval is then called a 90 percent confidence interval for μ (see Figure 6.9).

Now suppose that we wish to obtain a confidence interval with confidence coefficient $(1 - \alpha)$, where α is a number between 0 and 1. [For example, if $\alpha = .05$, $(1 - \alpha) = .95$.] Then we would want to find a value of z, call it

Figure 6.9 *Area under the normal curve between $\mu \pm 1.645\sigma/\sqrt{n}$*

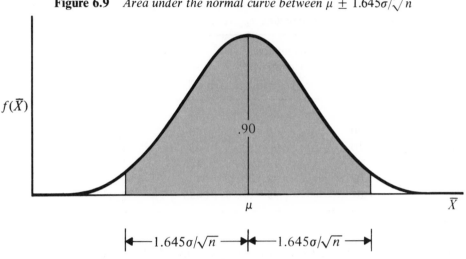

$z_{\alpha/2}$, corresponding to the area $(1 - \alpha)$, as shown in Figure 6.10. This value can be obtained from Table 1 of the Appendix by finding the value of z corresponding to an area to the right of the mean equal to $(1 - \alpha)/2$. Common large-sample confidence coefficients, with their corresponding $z_{\alpha/2}$ values, are given in Table 6.1.

We have presented point estimation and interval estimation of a population mean, Sections 6.3 and 6.4. The reader should note the fine

Figure 6.10 *Area under the normal curve equal to $(1 - \alpha)$*

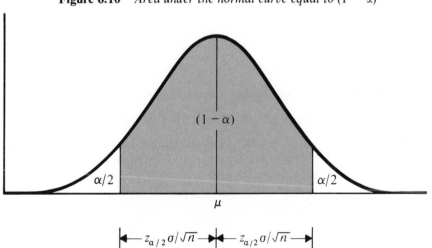

Table 6.1   *Common Large-Sample Confidence Coefficients and Their Corresponding $z_{\alpha/2}$ Values*

α	Confidence Coefficient $(1 - \alpha)$	$\dfrac{1 - \alpha}{2}$	$z_{\alpha/2}$
.32	.68	.34	1.0
.20	.80	.40	1.28
.10	.90	.45	1.645
.05	.95	.475	1.96
.02	.98	.49	2.33
.01	.99	.495	2.58

distinctions between the two estimation procedures. When we place a bound on the error of estimation for a point estimate, we are for all practical purposes constructing an approximate confidence interval. However, there is a difference. The concept "bound on error" is applied to a situation where the distribution of the estimator is unknown and where we want only a rough measure of the precision of the point estimator. Hence a two-standard-deviation bound on error is a rough bound on error and will be exceeded only approximately (very approximately) 5 percent of the time. This is because the two-standard-deviation portion of the empirical rule holds reasonably well for most distributions of point estimators. In contrast, confidence intervals are constructed for situations where the distribution of the estimator is known. This enables us to provide accurate statements of the value of the confidence coefficient.

EXERCISES

5   Distinguish between the populations associated with σ and with $\sigma/\sqrt{n}$.

6   What interpretation can we give to the interval $\bar{X} \pm 1.96\sigma/\sqrt{n}$?

7   A random sample of $n = 36$ labor contracts showed an average hourly wage increase of \$.75 with a standard deviation of \$.32. Estimate the mean hourly wage increase for all unions, using a 90 percent confidence interval. Interpret your result.

8   A random sample of $n = 40$ inner-city birth rates produced a sample average equal to 35 per thousand with a standard deviation of 6.3. Estimate the mean inner-city birth rate, using a 95 percent confidence interval.

6.5 ESTIMATION OF THE
BINOMIAL PARAMETER p:
A POINT ESTIMATOR

he point estimator of the binomial parameter p is one that we would choose intuitively. Let p be the proportion of elements in the population that are classified as successes. Then the best estimate of p would appear to be the proportion of the observed sample that are successes. If X represents the number of successes in n trials, the sample proportion of successes is

$$\hat{p} = \frac{\text{number of successes}}{\text{number of trials}} = \frac{X}{n}$$

(Note that we place a "hat" over the parameter p to denote "estimator of.")

To evaluate the goodness of this estimator, we need to know the probability distribution of $\hat{p} = X/n$. We previously noted that X possesses a mound-shaped probability distribution that approaches the normal curve as n becomes large. The probability distribution for $\hat{p}$ will possess identically the same shape as the probability distribution for X except that the mean and standard deviation will be

$$\mu = p$$

and

$$\sigma_{\hat{p}} = \sqrt{\frac{pq}{n}}$$

where

$$q = 1 - p.$$

Note that the reasoning applied in evaluating the goodness of a point estimator for p is identical to the logic employed for μ. That is, approximately 95 percent of the estimates should lie within two standard deviations $(2\sigma_{\hat{p}})$ of their mean, p. Thus the distance

$$2\sigma_{\hat{p}} = 2\sqrt{\frac{pq}{n}}$$

serves as a bound on the error of estimation. Note again that if we are sure the sample size is large enough so that $\hat{p}$ is approximately normally distributed, it would be more accurate to use $1.96\sigma_{\hat{p}}$ as a bound on the error of estimation.

Point Estimation of *p*

Point Estimator of the Binomial Parameter p: $\hat{p} = \dfrac{X}{n}$

Standard Deviation of $\hat{p}$: $\sqrt{\dfrac{pq}{n}}$

Bound on the Error of Estimation: $2\sqrt{\dfrac{pq}{n}}$

A sketch of the probability distribution for $\hat{p}$ is shown in Figure 6.11.

Figure 6.11 *Probability distribution for $\hat{p}$, an estimator of p*

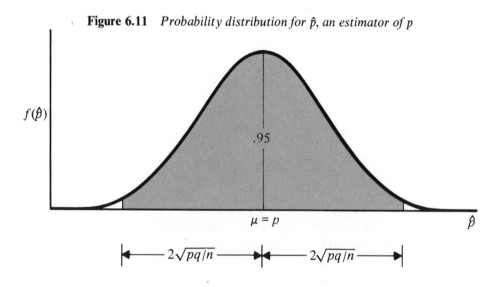

Example 6.3 *A random sample of 100 seniors from a large university was selected to estimate the fraction of graduating seniors going on to graduate school. The sample produced 15 students who plan to attend a graduate school. Estimate p, the fraction of the entire class planning to attend graduate school, and place a bound on the error of estimation.*

Solution *The point estimate of p is*

$$\hat{p} = \frac{X}{n} = \frac{15}{100} = .15$$

and the corresponding bound on the error of estimation is

$$2\sqrt{\frac{pq}{n}}$$

Note we do not know p; however, we can obtain an approximate bound on the error of estimation by substituting $\hat{p}$ for p. Provided both $n\hat{p}$ and $n\hat{q}$ are 10 or more, little error will be introduced by this substitution. Hence

$$2\sigma_{\hat{p}} = 2\sqrt{\frac{pq}{n}} \approx 2\sqrt{\frac{(.15)(.85)}{100}}$$

$$= 2(.0357) = .0714$$

We estimate that .15 of the seniors plan to attend a graduate school. The bound on the error of estimating p is .0714.

EXERCISES

9 If a sample of 180 voters go to the polls and 50 cast votes for the Republican candidate, give a point estimate of p, the population fraction casting votes for the Republican candidate.

10 Use the information in Exercise 9 to place a bound on the error of estimation.

11 Using the data in Example 6.3, assume that a second sample was drawn with $X = 20$ students who plan to attend a graduate school. Estimate p for the entire class, and place a bound on the error of estimation.

6.6 ESTIMATION OF THE
BINOMIAL PARAMETER p:
AN INTERVAL ESTIMATOR

large-sample 95 percent confidence interval for estimating p can be constructed more accurately by measuring $1.96\sigma_{\hat{p}}$ above and below the point estimator $\hat{p} = X/n$. Note that this is identical to the procedure employed in acquiring a large-sample confidence interval for μ.

Large-Sample 95 percent Confidence Interval for *p*

$$\left[\hat{p} - 1.96\sqrt{\frac{\hat{p}\hat{q}}{n}} \right] \quad \text{to} \quad \left[\hat{p} + 1.96\sqrt{\frac{\hat{p}\hat{q}}{n}} \right]$$

We illustrate with an example.

Example 6.4   *A sample of 1000 working-class people in Great Britain was interviewed to determine their political party affiliation. If 680 identified with the major left-of-center party, use a 95 percent confidence interval to estimate the true fraction, p, of Great Britain's working class that has identified with the left-of-center party.*

Solution   *The point estimator of p is*

$$\hat{p} = \frac{X}{n} = \frac{680}{1000} = .68$$

The corresponding 95 percent confidence interval has a lower confidence limit (LCL) of

$$LCL = \hat{p} - 1.96\sqrt{\frac{\hat{p}\hat{q}}{n}}$$

$$= .68 - 1.96\sqrt{\frac{(.68)(.32)}{1000}}$$

$$= .68 - 1.96\sqrt{.000218} = .68 - 1.96(.0148)$$

$$= .68 - .03 = .65$$

and upper confidence limit (UCL) of

$$UCL = \hat{p} + 1.96\sqrt{\frac{\hat{p}\hat{q}}{n}}$$

$$= .68 + 1.96\sqrt{\frac{(.68)(.32)}{1000}}$$

$$= .68 + .03 = .71$$

The appropriate 95 percent confidence for p is then .65 to .71. Recall that this interval may or may not enclose the parameter, p. However,

95 percent of the intervals constructed in this way would do so in repeated sampling. Hence we are 95 percent confident that the interval .65 to .71 does enclose the true fraction of Great Britain's working class that identified with the left-of-center political party.

EXERCISES

12   A survey was conducted to investigate the proportion of registered nurses in a particular state that are actively employed. A random sample of $n = 400$ nurses selected from the state registry showed 274 actively employed. Find a 95 percent confidence interval for the proportion of registered nurses actively employed.

13   A survey conducted to determine the proportion of college students favoring "more than equal" job rights opportunities for women (to offset past injustices) showed 258 of a random sample of 1000 favoring the proposal. Estimate the proportion of students in the entire population favoring the proposal, using a 99 percent confidence interval.

14   A random sample of 300 families in a city showed 23 earning salaries that placed them in a "poverty" category. Estimate the proportion of "poverty" families in the city, using a 90 percent confidence interval.

15   Refer to Exercise 14. Estimate the proportion, using a 99 percent confidence interval. Compare the properties of the two interval estimators and note the relation between interval width and confidence coefficient.

6.7   CHOOSING THE SAMPLE SIZE TO ESTIMATE A POPULATION MEAN μ OR PROPORTION p

*H*ow can we determine the number of observations to be included in the sample? The implications of such a question are clear. Observations cost money. If the sample is too large, time and talent are wasted. Conversely, it is wasteful if the sample is too small because inadequate information has been purchased for the time and effort expended. Also, it may be impossible to increase the sample size at a later point in time. Hence the number of observations the experimenter should buy will depend upon the amount of information he wants to buy.

Suppose that we wish to estimate the average amount for accident claims filed against an insurance company. In order to determine how many claims to examine (or to sample) we would have to determine how accurate the company wants us to be. Thus they might specify that the bound on the error of estimation must equal $5. We would then determine the required sample size.

The methods of choosing the sample size for estimating a population mean μ and a population proportion p are identical. The researcher must first specify a desired bound on the error of estimation. We shall call this bound B. Thus if he uses $\overline{X}$ to estimate a population mean μ, he must specify that μ and $\overline{X}$ differ by less than some value B. He then sets two standard deviations of the estimator equal to B and solves this expression for the desired sample size n. Using the calculated sample size, the error of estimation should be less than the specified value, B, approximately 95 percent of the time.

Recall that the standard deviation of $\overline{X}$ is $\sigma/\sqrt{n}$. Hence to determine the sample size required to estimate μ with a bound on the error of estimation equal to B we must solve the expression

$$\frac{2\sigma}{\sqrt{n}} = B$$

for n. The required sample size is as follows:

Sample Size Required to Estimate μ

$$n = \frac{4\sigma^2}{B^2}$$

where B is the bound on the error of estimation.

You will note that determining a sample size to estimate μ requires knowledge of the population variance σ^2 (or standard deviation σ). We can obtain an approximate sample size by estimating σ^2 using one of two methods:

1   Employ information from a prior experiment to calculate a sample variance s^2. This value is used to approximate σ^2.
2   Use information on the range of the observations (and the empirical rule) to obtain an estimate of σ. (This method was employed in Section 4.13 as a check on the calculation of s.)

We would then substitute the estimated value of σ^2 in the sample size equation to determine an approximate sample size n.

We illustrate the procedure for choosing a sample size with the following examples.

Example 6.5 *Union officials are concerned about reports of inferior wages paid to employees of a company under their jurisdiction. It is decided to take a random sample of n wage sheets from the company to estimate the average hourly wage. If it is known that wages within the company have a range of $10 per hour, determine the sample size required to estimate μ, the average hourly wage, with a bound on the error of estimation equal to $.60.*

Solution *The desired bound on the error of estimation is B = $.60. Before using the sample-size equation to determine n, we must estimate the population variance σ^2. We do so using a range estimate of σ. Since the range of hourly wages is $10, an estimate of σ (using the empirical rule) is*

$$\frac{range}{4} = \frac{10}{4} = 2.5$$

An approximate sample size can then be found by substituting $(2.5)^2$ for σ^2

$$n = \frac{4\sigma^2}{B^2} \approx \frac{4(2.5)^2}{(.6)^2} = 69.44$$

To be on the safe side, round this number to the next largest integer. Hence we would recommend that union officials sample $n = 70$ wage sheets to estimate the mean hourly wage to within $.60.

Example 6.6 *A federal agency has decided to investigate the advertised weight displayed on cartons of a certain brand of cereal. The company in question periodically samples cartons of cereal coming off the production line to check their weight. A summary of 1500 of the weights made available to the agency indicates a mean weight of 11.80 ounces per carton and a standard deviation of .75 ounce.*

Use this prior information to determine the number of cereal cartons the federal agency must examine to estimate the average weight of cartons produced now with a bound on the error of estimation equal to .25 ounce.

Solution   *The federal agency has specified that the bound on the error of estimation must be B = .25 ounce. In order to determine the sample size required to achieve this bound, we must obtain an estimate of the population variance, σ^2. Assuming that the weights made available to the federal agency by the company are accurate, we can use the standard deviation of these weights to form an estimate of σ^2. Thus*

$$\sigma^2 \approx (.75)^2 = .5625$$

An approximate sample size can now be found

$$n = \frac{4\sigma^2}{B^2} = \frac{4(.75)^2}{.0625} = 36$$

Hence the federal agency must obtain a random sample of n = 36 cereal cartons to estimate the mean weight to within .25 ounce.

Selecting the sample size to estimate a binomial parameter p is accomplished in a manner similar to the method employed when estimating μ. The estimator of p is $\hat{p} = X/n$ and its standard deviation (Section 6.5) is $\sqrt{pq/n}$. Then setting two standard deviations equal to the desired bound on the error of estimation B we have $2\sqrt{pq/n} = B$ or

---

Sample Size Required to Estimate p

$$n = \frac{4pq}{B^2}$$

where B is the bound on the error of estimation.

You will note that we must know p to solve for n and that this creates a circular problem since our final objective is to estimate p. Actually it is not as complicated as it appears.

The researcher often knows before the experiment commences that p will lie in a fairly narrow range. For example, the fraction of popular vote for a presidential candidate in a national election is often close to .50. Thus he will substitute the value dictated by experience for p. A second method for finding p is to use data collected from a prior study to estimate p. Finally, if you have no prior information, substitute $p = .50$. This will yield the largest possible sample size for the bound that you have specified and will thus give a conservative answer to the required sample size. The sample will likely be larger than required but you will be on the safe side.

We shall illustrate the selection of the sample size for estimating a binomial parameter p with an example.

Example 6.7   *In a national election poll, we wish to estimate the fraction p of voters in favor of candidate A. How many people should be polled in order to estimate p with a bound on the error of estimation equal to $B = .02$?*

Solution   *The researcher has specified that $B = .02$. The estimator of p is $\hat{p} = X/n$, which has a standard deviation*

$$\sigma_{\hat{p}} = \sqrt{\frac{pq}{n}}$$

The sample size n necessary to achieve the desired bound is found by solving

$$n = \frac{4pq}{B^2}$$

However, we must first estimate p. If a similar survey was run recently we could use the sample proportion to estimate p. Otherwise, we substitute $p = .5$ to obtain a conservative sample size (one that is likely to be larger than required). Assuming that no prior survey was run, the sample size is

$$n = \frac{4pq}{B^2}$$

$$= \frac{4(.50)(50)}{.0004} = 2500$$

That is, 2500 potential voters must be polled to estimate the proportion favoring candidate A with a bound on the error of estimation of .02.

We have discussed choosing the sample size required to estimate either μ or p with a bound on the error of estimation equal to a specified value B. The desired sample size can be determined by setting two standard deviations of the estimator equal to B and solving for n. When estimating μ we must supply some estimate of the population variance σ^2. This can usually be done by employing a value of s^2 obtained from a prior study or by using a range estimate to approximate σ. When estimating p we must supply some value for the binomial parameter appearing in the standard deviation of $\hat{p}$. If prior data are available from a previous study, we could use this information to estimate p in the formula; otherwise we substitute $p = .5$ into the formula for n to obtain a conservative sample size.

16   Using the empirical rule, determine the sample size (n) required to estimate the mean murder rate for all U.S. cities if we assume that the range is 26 (see Table 4.2, page 100). Use $B = .50$.

17   Assume that σ^2 is approximately 7.26 for the data of Table 4.2 (page 100). Determine the sample size (n) required to estimate the mean murder rate for all U.S. cities with $B = .50$.

18   If 40 percent of the voters usually cast ballots for the Democratic candidate in Atlanta, determine the sample size (n) required to estimate p, the fraction of all voters favoring a particular Democratic candidate. Set $B = .03$.

19   In estimating sample sizes to determine μ, describe the two methods for estimating the variance. Develop a problem of interest to the social sciences, first employing information from a prior experiment and then employing the range estimate.

6.8   SUMMARY

Social scientists frequently sample to assess public opinions concerning a key political issue or sample to determine a characteristic of a social system. Such assessments often are phrased as an estimate of a population proportion or a population mean. Thus we might wish to know the proportion of alcoholics who successfully complete a rehabilitation program or the mean number of days of employment per year for blacks with a high school education. This chapter discusses how these estimates can be acquired and what they mean.

Given the appropriate data, anyone can construct estimates of the desired population parameters by the use of statistics, clairvoyance, or any other subjective procedure. Regardless of how an estimate is obtained, the big question concerns how much reliance we can place upon it. Is the estimate reasonably close to the population parameter?

Subjective procedures based on experience or clairvoyance may or may not yield good estimates. They may be given by people who truly possess sufficient experience to yield satisfactory estimates (but how will you know when you have a person with this ability and how will you know how accurate the estimate is?) or they may be presented by people who are proud—those who are inadequately prepared to do the job—and know it. But how will you know?

One of the major contributions of statistics is that all inferences are accompanied by a measure of their goodness. We talk of point estimates with a "bound on the error of estimation" or confidence intervals with specified coefficients. This is the great advantage of statistics. By use of statistical estimation procedures, we make an estimate and we know how much we can rely upon it.

Chapter 6 emphasized the basic concepts of statistical estimation and illustrated these concepts by considering inferences based on single samples, estimation of population means, and proportions. Having given a brief introduction to estimation, we turn now to a second method of making inferences—tests of hypotheses.

QUESTIONS AND PROBLEMS

20 What are parameters and how do they differ from statistics?

21 Do we use estimators when the sample data collected constitute the population of interest? Why or why not?

22 Distinguish between point and interval estimators. What are the two basic procedures for making inferences? How do we measure the goodness of a point estimator? An interval estimator?

23 What is an unbiased point estimator? A biased point estimator?

24 What are the two properties we require of a good point estimator?

25 A random sample of $n = 500$ insurance records of physicians, selected from the files of an insurance company, shows that 10 percent have been involved in one or more lawsuits. Estimate the proportion of all physicians covered by the insurance company who have been involved in lawsuits. Place a bound on the error of estimation.

26 A social scientist reports that the average length of stay in 100 mental hospitals is 120 days. His sample involves hospitals rather than individuals. The standard

deviation for the hospitals is 5.50. Find a 95 percent confidence interval for the average length of stay. Interpret your result.

27 Use the data in Exercise 26 to construct a 99 percent confidence interval for the average length of stay.

28 A researcher reports that 45 percent of a random sample of 1000 college students (sample equals 1000) believe that the penalties for the use of marijuana should be reduced. Give a point estimate of the proportion of all college students who favor this proposition. Place a bound on your error of estimation.

29 Use the data contained in Exercise 28 to find a 99 percent confidence interval for the population proportion.

30 Explain in your own words the justification for the following statements: "If the sample is too large, time and talent are wasted; if the sample is too small, time and talent are wasted."

31 An industrial sociologist, researching work characteristics in a manufacturing plant, is interested in the average amount of time required by work groups to complete a complicated task. The desired bound on the error of estimation is .5 minute. A previous study suggests that the population variance is approximately 4.85 minutes. Use this information to determine the number of groups to be included in the study.

32 If a public opinion pollster is interested in estimating the proportion of registered voters favoring his candidate and if he sets $B = .05$ and $p = .5$, how large should his sample be?

33 Refer to Exercise 32. A fellow public opinion pollster argues that the population proportion is really in the neighborhood of .40. Does this produce a substantial change in the required sample size?

REFERENCES

Anderson, T. R., and M. Zelditch. *A Basic Course in Statistics*, 2nd ed. New York: Holt, Rinehart and Winston, Inc., 1968. Chapter 10.

Blalock, H. M. *Social Statistics*, 2nd ed. New York: McGraw-Hill Book Company, 1972. Chapters 11 and 12.

Champion, D. J. *Basic Statistics for Social Research*. Scranton, Pa.: Chandler Publishing Company, 1970. Chapter 6.

Mendenhall, W., and L. Ott. *Understanding Statistics*. North Scituate, Mass.: Duxbury Press, 1972. Chapter 7.

Mueller, J. H., K. F. Schuessler, and H. L. Costner. *Statistical Reasoning in Sociology*, 2nd ed. Boston: Houghton Mifflin Company, 1970. Chapter 13.

Palumbo, D. J. *Statistics in Political and Behavioral Science*. New York: Appleton-Century-Crofts, 1969. Chapter 6.

7

STATISTICAL TESTS OF HYPOTHESES: THE ONE-SAMPLE CASE

7.1 INTRODUCTION: A STATISTICAL TEST OF AN HYPOTHESIS

J n Chapter 5 we were concerned with the probability of an observed sample or, equivalently, reasoning from a known population to the outcome of a sample. In doing so, we took the opportunity to use the calculated probability to make an inference about the population from which the sample was drawn. Our method was intuitive and based on the following reasoning. If we thought a population parameter was a specific value and then observed a highly contradictory sample, we concluded that the hypothesized value of the parameter was incorrect. Contradictory sample results are those that are highly improbable. The more improbable, the more contradictory are the results.

To illustrate, you recall the sample poll of 20 precinct voters in Section 5.1. Not 1 of the 20 favored Jones, and hence we concluded that Jones would not win the precinct. Jones's claim to victory could be phrased as an hypothesis about the parameter p, the fraction of voters in the precinct favoring his election. The statement that he would win is equivalent to saying that p is equal to or greater than .5. Observing none out of 20 in a sample favoring Jones is highly improbable even if he has a bare margin of victory (i.e., p is very close to .5). Therefore, on the basis of this small probability and hence highly contradictory evidence, we rejected the hypothesis that $p \geqslant .5$ or, equivalently, we rejected his claim of victory.

The preceding example illustrates the reasoning behind a statistical test of an hypothesis. We commence by stating an hypothesis about one or more parameters of the population. This is commonly called the *null hypothesis.* The decision to accept or reject the null hypothesis is based on the value of some quantity observed or computed from the sample data. Certain values contradictory to the null hypothesis will imply rejection and the remainder will imply acceptance. Thus the observed or computed quantity functions as a statistical decision maker and is known as the *test statistic.*

The *research hypothesis* (what we wish to prove) for Jones's survey is that he will lose (i.e., that the fraction of voters favoring Jones, p, is less than .50). We shall attempt to prove this research hypothesis by rejecting a null hypothesis that is contrary to the research hypothesis. We attempt to prove the research hypothesis by contradiction of the null hypothesis.

The null hypothesis for Jones's survey is that 50 percent of the voters favor his election (the least that he could expect if he would win) or, equivalently, that $p = .50$. The test statistic is X, the number of prospective voters in the sample favoring his election. Values of X that favor the research hypothesis in contradiction to the null hypothesis are those that are highly improbable assuming the null hypothesis to be true. Recall from Chapter 5 that the mean value of X, assuming that $p = .50$, is

$$\mu = np = 20(.50) = 10$$

Then contradictory values are those lying too far away from μ.

How do we decide whether a particular X is too far away from μ? We simply resort to our knowledge of the empirical rule and the properties of the normal distribution. The probability that X will lie more than 1.96 standard deviations away from μ is .05 (see Figure 7.1). For the research hypothesis $p < .5$, contradictory values of X that lead to rejection of the null hypothesis are those lying in the left-hand tail of Figure 7.1 more than 1.96 standard deviations below μ. Contradictory values of X, known as the rejection region for a test, are shown in Figure 7.1 for Jones's survey. Note that we use a smooth normal curve to approximate the exact probability distribution for X.

Figure 7.1 *Rejection region for a test of the hypothesis $p = .5$, $n = 20$, $\mu = np = 10$, $\sigma = \sqrt{npq} = \sqrt{20(.5)\,(.5)} \approx 2.24$*

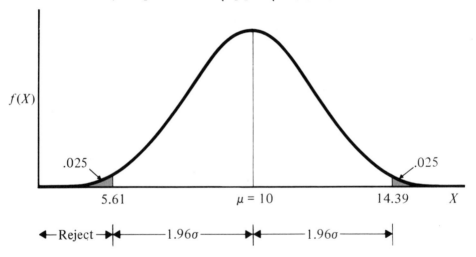

Since $X = 0$ (for our example) falls in the rejection region, we reject the null hypothesis in favor of the research hypothesis. That is, we conclude that Jones's claim of victory is false and he will lose the election.

What is the probability that X will accidentally fall in the rejection region when the null hypothesis is true? We can see from Figure 7.1 that this probability, lying in the left tail of the distribution of X, is equal to .025.

The decision makers for most statistical tests employ the same principle as the test of the hypothesis concerning the fraction p. They possess mound-shaped probability distributions with a mean that is known when the null hypothesis is true. Then contradictory values are those lying too many standard deviations away from the expected value. Most distributions of test statistics are mound-shaped, so we can use the empirical rule and the normal curve to approximate the probability of observing contradictory values of the test statistic.

You will note that a statistical test of an hypothesis is a decision-making process. We shall explain the factors that affect the goodness of this process in Section 7.2 and in particular show how varying the rejection region affects the properties of the test.

7.2   ELEMENTS OF A STATISTICAL TEST

*T*he procedure for testing and rejecting a null hypothesis was presented for Jones's survey in Section 7.1. Let us now formalize these ideas.

The objective of a statistical test is to test an hypothesis concerning the values of one or more population parameters. It is composed of four parts:

Elements of a Statistical Test

1 An hypothesis about a population parameter (called the *null hypothesis*).
2 A *research hypothesis* (sometimes called an *alternative hypothesis*) that we shall accept if the null hypothesis is rejected.
3 A decision maker (also called a *test statistic*) that is computed from sample data.
4 A *rejection region* that gives a set of values of the decision maker which are contradictory to the null hypothesis and hence imply its rejection.

Selecting the rejection region for a statistical test is more easily explained if we draw an analogy between a test of an hypothesis and a court trial. The trial hypothesis is that the defendant is innocent, a perfect analogy to the null (not guilty) hypothesis. The alternative or research hypothesis is that he is guilty. The jury functions as a decision maker (test statistic) after examining evidence (corresponding to sample data) presented by the defense and prosecuting attorneys. The judge's charge to the jury provides guidelines for reaching a decision, and this corresponds to the rejection region of the statistical test. The jury must then decide on the guilt or innocence of the defendant based on both factual and circumstantial evidence.

The goodness of a decision-making procedure, statistical or any other, can be measured by the probability of making an incorrect decision. The possible outcomes of the court trial are shown in Table 7.1. An error can be made by either convicting an innocent man or letting a guilty one go free. We want to keep the probabilities for both types of incorrect decisions small but, in particular, wish to protect the rights of an innocent defendant.

Table 7.1 *Possible Outcomes of a Court Trial*

Jury's Decision	*Unknown Truth*	
	Guilty	*Not Guilty*
Guilty	Correct decision	Incorrect
Not Guilty	Incorrect	Correct decision

The researcher tests an hypothesis concerning the value of one or more population parameters in much the same manner. He first specifies the research and null hypotheses. A random sample is drawn from the population of interest and a single number (value of a decision maker) is then computed from the sample values. The decision of whether to accept or reject the null hypothesis depends upon the computed value of the decision maker.

To assist in reaching a decision, we consider all values that the decision maker could possibly assume and divide the set into two regions, one corresponding to a rejection region and the other to an acceptance region. This situation is shown symbolically in Figure 7.2. A point on the horizontal line corresponds to a possible value of the decision maker. We symbolically divide these values with the vertical divider to obtain two sets, one corresponding to rejection and the other to acceptance of the null hypothesis. If the computed value of the decision maker falls in the rejection region, the null hypothesis is rejected. Otherwise it is accepted. Note that Figure 7.2 is just

Figure 7.2 *Possible values a decision maker may assume*

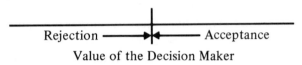

Rejection ———→|◄——— Acceptance
Value of the Decision Maker

an *example* of a rejection region. For some tests the rejection region might lie to the left (or right) of the acceptance region (as it did for Jones's survey); in others it might be divided on either side.

The decision-making procedure described here is subject to the same two types of errors that face the jury in a trial. We may reject the null hypothesis when it is, in fact, true, or we may accept the null hypothesis when it is false and some alternative is true. These errors are called type I and type II errors, respectively (see Table 7.2).

Definition 7.1

A type I error is committed if the null hypothesis is rejected when it is, in fact, true. The probability of a type I error is denoted by the Greek letter α.

Definition 7.2

A type II error is committed if the null hypothesis is accepted when it is, in fact, false. The probability of a type II error is denoted by the Greek letter β.

Table 7.2 *Decision Table*

	Null Hypothesis Is:	
Decision	*False*	*True*
Reject the null hypothesis	Correct	Type I error
Accept the null hypothesis	Type II error	Correct

The probability of making a type I error, α, is the probability of rejecting the null hypothesis when it is true. The probability of a type II error, β, is the probability of accepting the null hypothesis when it is false and the alternative hypothesis is true. For a given sample size, α and β are inversely related. As one goes up, the other goes down in a seesaw manner, as shown in Figure 7.3.

Figure 7.3 *Relation between α and β: (a) large α, small β; (b) small α, large β*

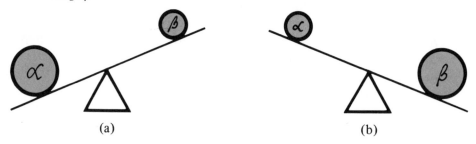

(a)　　　　　　　　　　　　　　　　　　(b)

Returning to the court trial, rejecting the null hypothesis when it is true (a type I error) is equivalent to convicting an innocent man. Thus we like α to be small. Accepting the null hypothesis when false (a type II error) implies freeing a guilty man, and this decision occurs with probability β. As the laws protecting individual rights are strengthened, the probability α of convicting an innocent man decreases. Simultaneously, the probability β of letting a guilty man go free increases. Relaxing the laws protecting individual rights would increase α and decrease β.

So much for the two types of error associated with a statistical test. We shall show you that finding the value of α is quite easy for a statistical test. Evaluating β is much more difficult and beyond the scope of this text. Hence we shall resort to our earlier and simpler procedure of seeking evidence to contradict the null hypothesis. If the decision maker assumes an improbable value (falls in the rejection region) we shall reject the null hypothesis. Having determined α, we shall know the risk of having made an incorrect decision. If the decision maker does not fall in the rejection region, we shall withhold judgment and hence eliminate the possibility of making a type II error. This decision is similar to the one that a jury must face. If there is any doubt that the defendant is guilty, the jury must free him. In such a case the defendant is not said to be innocent; instead, the verdict is not guilty, which implies that there was insufficient evidence to justify the rejection of the defendant's claim of innocence.

A discussion of several useful statistical tests should clarify what we have said about the decision-making process.

m

7.3   TESTING AN HYPOTHESIS
ABOUT A BINOMIAL
PARAMETER p

any surveys are conducted to make inferences about the fraction of people favoring a particular issue. A random sample of n people is selected from the total and each is interviewed to determine his position, pro or con, on the issue. The number in favor X, divided by the total n, represents the sample fraction and this quantity should come close to the true population fraction (which we shall call p). We shall denote the sample fraction as

$$\hat{p} = \frac{X}{n}$$

You will note that this sampling procedure is a binomial experiment when the number of people in the group surveyed is large. The true fraction in the population "in favor," p, then represents the probability that the first person interviewed favors the issue. If the population is large relative to the sample size, the probability of interviewing a person "in favor" remains constant, for all practical purposes, as additional people are selected for the sample. A random selection of the sample produces near independence for the binomial trials and ensures the validity of the statistical procedures that follow. Since many surveys satisfy these conditions, they can be viewed as binomial experiments.

You will note that the procedure employed to test the hypothesis that $p = .5$ for Jones's political survey, Section 7.1, was a test of an hypothesis concerning a binomial proportion p. We shall now formalize this procedure.

We can use either X or $\hat{p} = X/n$ as a decision maker to test the hypothesis that p equals some specified value, say p_0. Both lead to the same result. Since we used X for Jones's survey, we shall continue to do so in this section. The four elements of the test of an hypothesis concerning the binomial parameter p are:

Statistical Test for a Binomial Proportion, p, with $\alpha = .05$

Null Hypothesis: $p = p_0$. (*Note: p_0 is some specified value of p.*)

Alternative Hypothesis: $p \neq p_0$.

Test Statistic: X

*Rejection Region: **Reject if X lies more than** 1.96 **standard deviations** **away from its hypothesized mean. The hypothesized mean and** **standard deviation of X are***

$$\mu = np_0$$

$$\sigma = \sqrt{np_0q_0}$$

where $q_0 = 1 - p_0$.

(Note: Both np_0 and nq_0 must be 10 or more.)

Example 7.1   *The following situation actually occurred. At* 7:20 *in the evening of the* 1972 *presidential elections, we observed that the television returns for a particular state showed* 31,000 *in favor of Nixon,* 29,000 *for McGovern. Do these data, presented only* 20 *minutes after the closing of the polls, provide sufficient evidence to project a winner in the state?*

Solution   *Snatching a tablet, we made the following quick calculations. If either candidate possesses more than* 50 *percent of the final vote, we conclude that he is a winner. Letting p denote the fraction of voters that will vote for Nixon and X the number of votes he received in the sample, we wish to test the research hypothesis*

$$H_a: p \neq .5 \text{ (i.e., one of the candidates is a winner)}$$

The appropriate null hypothesis is

$$H_0: p = .5$$

We calculate

$$\mu = np_0 = 60,000\,(.5) = 30,000$$

$$\sigma = \sqrt{np_0q_0} = \sqrt{60,000\,(.5)\,(.5)} \approx 122.5$$

and

$$1.96(122.5) = 240.1$$

The rejection region would be as shown in Figure 7.4. You will note that the observed value of X falls well into the upper portion of the rejection region. In fact, it falls so far to the right of μ that it misses the page! Thus we have ample evidence to reject the hypothesis that $p = .5$ and we project a winner. In fact, we conclude that p is greater than .5 and that Nixon will win the state.

Figure 7.4   *Rejection region for the 1972 returns*

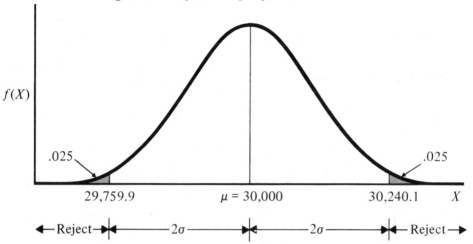

You are probably thinking that there is a flaw in our reasoning and you are correct. We are not certain that the n = 60,000 early returns constitute a random sample from the total of all votes cast in the state. They might well come from one or two urban areas where voter machines provide an early tabulation of voting results and the proportion favoring Nixon might not be representative of the total fraction in the state. Considering this occurrence to be somewhat improbable and noting the large distance (many standard deviations) between X and μ, we concluded that Nixon would win. History shows that Nixon carried the state and won the election.

Now that we have rejected the null hypothesis, we wish to know the value of α, the probability of accidentally rejecting the null hypothesis when, in fact, it is true. This probability that X will fall in the rejection region is the combined shaded areas ($\alpha = .05$) in Figure 7.4.

Note that in some situations our research hypothesis might have been one-sided in the sense that we only wish to show that p is greater than p_0. Then we would want to locate the rejection region only in the upper tails of the distribution of X. Similarly, if we wished to show that p is less than p_0, we would locate the rejection region in the lower tail of the distribution of X. These tests, called one-tailed statistical tests, will be discussed in detail in Section 7.5. Similarly, in Section 7.6 we will show how to vary the risk of making a type I error, α. You will see that we could have used $\alpha = .01$, $\alpha = .10$, or any other value. Of course, as we noted in Section 7.2, as α decreases, the probability of failing to reject the null hypothesis when it is false, β, will increase. For this reason α is usually chosen to equal .10, .05, or .01.

EXERCISES

1 Specify the four elements of a statistical test. Compare your answer with the text.

2 Early research suggests that the proportion of men age 30 or less who migrate from small communities in a particular state to urban areas is .38. A study of a particular small town showed a total of 92 migrators of a random sample of 200 males. Do the data suggest that the migratory proportion for this community differs from the value .38? (Use $\alpha = .05$.)

3 In a study to investigate the effect of the dollar exchange rate on travel to a particular country, researchers examined the proportion of people issued passports. Of those issued passports in prior years, records indicate that 10 percent visited the country in question. One year after substantial changes in the exchange rate, a random sample of 400 holders of newly issued passports indicated that 23 planned to visit the country. Do these data provide sufficient evidence to indicate a difference in the proportions desiring to visit the country, before and after the change in the exchange rate? (Use $\alpha = .05$.)

7.4 TESTING AN HYPOTHESIS
ABOUT A POPULATION
MEAN μ

he object of many studies is the population mean μ. Thus we may be interested in the mean reaction time to respond to a particular stimulus, the mean income for indigent families in a given community, the mean yield of oxygen in a chemical plant, or the mean level of impurities in a water supply. In this section we explain how to make inferences about a mean based on a random sample of n measurements selected from a population. As in preceding sections, we shall phrase the inference as a decision about μ (i.e., whether it is equal to some specific value). Therefore, we shall test a null hypothesis that μ equals a value denoted by the symbol μ_0.

Suppose that a union claims that the average annual wage for their craftsmen is $12,000 per year. If we doubt the veracity of the claim, we would select a random sample of n craftsmen from the total membership and, using this information, decide whether sufficient evidence exists to indicate that μ is different from $12,000.

A logical decision maker for the test is the sample mean $\bar{X}$. We know that $\bar{X}$ possesses a probability distribution that is approximately normal with mean μ and standard deviation $\sigma/\sqrt{n}$. Recall that σ is the standard deviation of the population from which the sample was drawn and n is the number of measurements in the sample. Like the test of the hypothesis for p, we shall reject the hypothesis $\mu = \mu_0$ if $\overline{X}$ lies more than two standard deviations (more specifically $1.96\sigma/\sqrt{n}$) away from the hypothesized mean. The four elements of the statistical test are:

Test of an Hypothesis Concerning μ with α = .05

Null Hypothesis: $\mu = \mu_0$. (*Note:* μ_0 *is a specified value of* μ.)
Alternative Hypothesis: $\mu \neq \mu_0$.
Test Statistic: $\overline{X}$
Rejection Region: Reject if $\overline{X}$ *lies more than* 1.96 *standard deviations* $(1.96\sigma/\sqrt{n})$ *away from* μ_0.

(*Note: The sample size n should be* 30 *or more if s is used to approximate* σ.)

Note that the test procedure requires that we know σ. Since this will rarely be true, you must either have a good estimate of σ based on prior studies or must use s, the standard deviation computed from the sample data. In order that s be a good approximation to σ, we require that the number of measurements in the sample be 30 or more. Inferences about μ based on small samples will be discussed in Chapter 8.

Example 7.2   *A random sample of* 30 *SMSA's was selected from the South, and the fertility ratio for each area was recorded. Use the data given in Table 7.3 to test the null hypothesis that* $\mu = 360$, *the same as the previous year's mean fertility ratio, against the alternative that* μ *differs from* 360.

Table 7.3   *Fertility Ratio Data*

346	326	319	352	313
357	334	370	284	310
347	395	360	386	373
337	365	301	313	331
324	401	319	356	315
344	315	314	336	358

Solution   *One can readily verify that the sample mean and standard deviation are, respectively,*

$$\bar{X} \approx 340.03$$

and

$$s \approx 28.41$$

The stated null hypothesis is

$$H_0 : \mu = 360$$

which is to be tested against the alternative

$$H_a : \mu \neq \mu_0.$$

The test statistic is $\bar{X}$ and we shall reject H_0 if the observed value of $\bar{X}$ lies more than 1.96 standard deviations away from μ_0. For our data, substituting s for σ we have

$$\sigma_{\bar{X}} = \frac{\sigma}{\sqrt{n}} \approx \frac{28.41}{\sqrt{30}} = 5.19$$

and

$$1.96\sigma_{\bar{X}} \approx 1.96(5.19) = 10.17$$

The rejection region for this test is shown in Figure 7.5. Since $\bar{X} = 340.03$ lies more than 1.96 standard deviations below the hypothesized mean, $\mu = 360$, we reject the null hypothesis and conclude that the mean fertility ratio throughout the South is different from 360. Practically we would conclude that the mean fertility ratio is less than 360.

Figure 7.5   *Rejection region for Example 7.2*

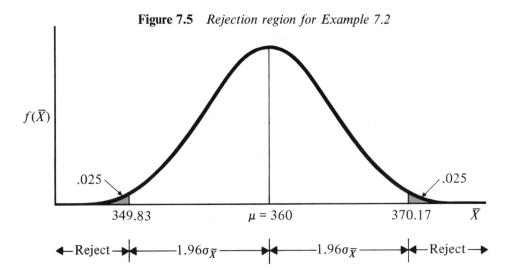

4 Use the data in Table 7.3 (page 239) to test the null hypothesis that $\sigma = 348$. (Use $\alpha = .05$.)

5 If $\bar{X} = 8.3$ and $s = 5.38$ for the murder rates of 90 cities, test the null hypothesis that the mean murder rate for all metropolitan areas is $\mu = 7.0$. (Use $\alpha = .05$.)

6 Repeat Exercise 5, but assume that the sample size is 60. What would this do to your test and conclusion? Explain.

7 A study was conducted to determine whether the average amount of money expended per household per week for food in a particular community differed from the national average (\$52.00 per week). A random sample of $n = 100$ households in the community gave a mean and standard deviation equal to \$56.00 and \$10.28, respectively. Do these data provide sufficient evidence to indicate that the mean household per-week expenditure for the community is different from the national average? (Use $\alpha = .05$.)

7.5 HOW TO SELECT THE RESEARCH AND NULL HYPOTHESES

As we have stated in Sections 7.1 through 7.4, the researcher formulates two hypotheses when conducting a statistical test concerning a population parameter. The research hypothesis is the motivating hypothesis behind our test procedure. It is the hypothesis that we think is true. Once the research hypothesis is formulated we specify an hypothesis (the null hypothesis) that is contradictory to the research hypothesis. The test procedure is then designed to verify the research hypothesis by showing that the null hypothesis is false. A statistical test of an hypothesis employs the technique of proof by contradiction; that is, we try to show that the research hypothesis is true by showing that the null hypothesis is false.

Example 7.3 *In order to evaluate the success of a 1-year experimental program designed to increase the mathematical achievement of underprivileged high-school seniors, the statewide mathematics test scores for a sample of $n = 100$ underprivileged seniors were tested against the previous* *year's statewide average of 525 for underprivileged students. If we wish to test an hypothesis about the mean mathematics test score for all underprivileged students who might participate in the program, state the research and null hypotheses.*

Solution *The research (motivating) hypothesis for this study is that the experimental program will indeed improve mean test scores above the comparable figure of 525 for the previous year. The research hypothesis or alternative hypothesis (denoted by H_a) is*

$$H_a : \mu > 525$$

An hypothesis contradictory to the research hypothesis would be that the experimental program had no effect on mathematics achievement; that is, the mean test score for students under the test program is identical to the previous year's average, 525. Thus the contradictory hypothesis or null hypothesis (denoted by H_0) is

$$H_0 : \mu = 525$$

We shall try to verify the research hypothesis ($\mu > 525$) by showing that the null hypothesis ($\mu = 525$) is false.

Example 7.4 *A professor wishes to determine whether a student's selection ability on a true–false test is better (or worse) than could be obtained by flipping a coin for each question. An examination is composed of 50 questions. State the research and null hypotheses for this statistical test.*

Solution *If a student flips a coin for each question and answers "true" if a head appears and "false" for a tail, we could conduct a statistical test concerning p, the fraction of questions students will answer correctly using a coin-tossing approach. In contrast, if the student actually answers these questions based on his knowledge of the subject, the research hypothesis would be that p, the fraction of questions correctly answered, will be either greater or less than .5 (the fraction that we would expect to be correct under a coin-tossing approach). Hence the research hypothesis is*

$$H_a : p \text{ is different from } .5 \text{ or simply } p \neq .5$$

The null hypothesis is then

$$H_0 : p = .5$$

which would be contradictory to the research hypothesis.

One final note deserves comment at this point. Researchers frequently distinguish between *"one- and two-tailed"* statistical tests. Perhaps these tests should be more appropriately named "one- and two-directional" tests.

Definition 7.3

If the research hypothesis states that the parameter under test is less than (or greater than) a specified value, the statistical test is called a one-tailed or one-directional test.

Definition 7.4

If the research hypothesis states that the parameter under test is not equal to a specified quantity, the statistical test is called a two-tailed or two-directional test.

In Example 7.3 the research hypothesis was $H_a: \mu > 525$; in other words, μ is greater than 525. Consequently, the test procedure is a one-tailed test. Similarly, in Example 7.4 we found the research hypothesis to be that p is different from .5. That is, the research hypothesis was verified if p was different from .5. Hence this test is called a two-tailed test.

At this point the reader should *not* become disturbed with the formulation of a null hypothesis and the designation of one- and two-tailed tests. It suffices to know that the research hypothesis dictates both the form of the null hypothesis and whether we have a one- or two-tailed test. More facility at hypothesis formulation will be obtained as we continue to discuss appropriate statistical tests for particular population parameters.

7.6   THE *z* STATISTIC AND THE CHOICE OF α

An easy way to measure the distance between the observed value of $\overline{X}$ and the hypothesized value of μ is to use the z statistic, which was first discussed in Section 5.11. Thus

$$z = \frac{distance}{standard\ deviation\ of\ \overline{X}} = \frac{\overline{X} - \mu}{\sigma/\sqrt{n}}$$

expresses the distance between $\overline{X}$ and μ in units of $\sigma/\sqrt{n}$, the standard deviation of the probability distribution of $\overline{X}$. If $z = 2$, $\overline{X}$ lies two standard deviations above the hypothesized mean. Similarly, $z = -1$ indicates that $\overline{X}$ is only one standard deviation below μ.

Consider the data of Example 7.2, where we took a random sample of 30 SMSA's from the South and computed the fertility ratio for each area. Recall that we were to test

$$H_0 : \mu = 360$$

against the alternative

$$H_a : \mu \neq 360$$

using $\overline{X}$ as the test statistic. For the probability of a type I error set at $\alpha = .05$, we were to reject the null hypothesis if the observed value of $\overline{X}$ was more than 1.96 standard deviations ($1.96\sigma_{\overline{X}}$) away from the hypothesized mean, 360. Similarly, we could use z as the test statistic and reject the null hypothesis if the computed value of z is greater than 1.96 or less than -1.96. That is, we can reject if the value of z irrespective of the sign, called the *absolute value* of z, exceeds 1.96. For the data of Example 7.2,

$$z = \frac{\overline{X} - \mu_0}{\sigma_{\overline{X}}}$$

$$= \frac{340.03 - 360}{5.19} = \frac{-19.97}{5.19} = -3.85$$

Since the absolute value of z, 3.85, exceeds 1.96, we reject the null hypothesis and conclude that the mean fertility rate is other than 360.

Suppose that the researchers of Example 7.2 were interested only in detecting whether the mean fertility rate was less than the hypothesized mean 360. For this situation we would reject the null hypothesis only for small values of $\overline{X}$, and hence we would locate all of the rejection region in the lower tail of the distribution of $\overline{X}$. For $\alpha = .05$ we would reject the null hypothesis

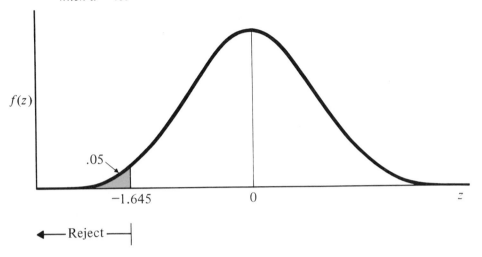

Figure 7.6   *Rejection region for a one-tailed test with* $H_a : \mu < \mu_0$ *when* $\alpha = .05$

if $\overline{X}$ lies more than $1.645\sigma_{\overline{x}}$ below the hypothesized mean. This would be equivalent to a rejection for values of z less than -1.645 (see Figure 7.6). In setting up a rejection region for any value of α (not necessarily .05), it becomes useful to define the quantity z_a to be the value in the z table (Table 1 in the Appendix) with an area "a" to its right (see Figure 7.7). Since entries in the table represent areas from $z = 0$ out to a specified value of z, we can determine z_a by finding the z value corresponding to a table entry of $(.5 - a)$. Typical values of a, $(.5 - a)$, and z_a are given in Table 7.4.

Figure 7.7   *Area under the curve and to the right of* z_a *is equal to a*

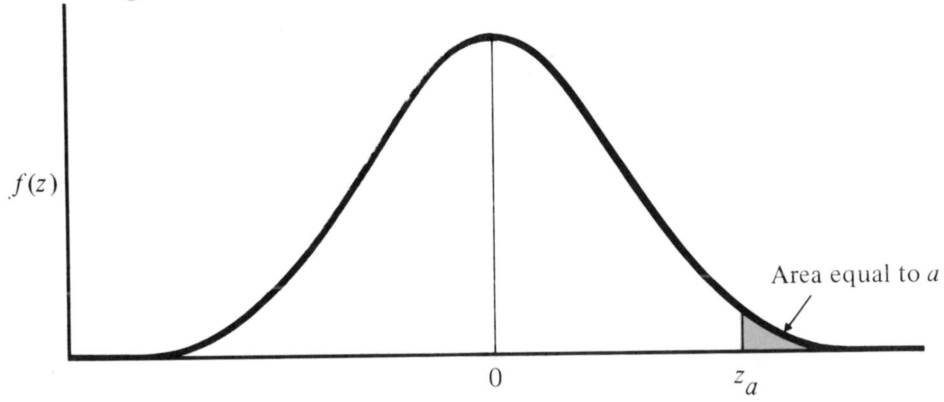

Table 7.4   *Common Values of z_a*

a	$(.5 - a)$	z_a
.005	.495	2.58
.01	.490	2.33
.025	.475	1.96
.05	.450	1.645
.10	.400	1.28

In general we may run a statistical test of the null hypothesis $\mu = \mu_0$ using either a one- or a two-tailed test with a specified value of α. The rejection regions are listed in Table 7.5.

Table 7.5   *Rejection Regions for the Null Hypothesis $\mu = \mu_0$ Using a Specified Value of α*

Alternative Hypothesis	Rejection Region
$H_a : \mu$ is less than μ_0 $(\mu < \mu_0)$	Reject H_0 if the computed value of z is less than $-z_\alpha$.
$H_a : \mu$ is greater than μ_0 $(\mu > \mu_0)$	Reject H_0 if the computed value of z is greater than z_α.
$H_a : \mu$ differs from μ_0 $(\mu \neq \mu_0)$	Reject H_0 if the absolute value of z is greater than $z_{\alpha/2}$.

Example 7.5   *Set up a rejection region for the mean fertility rate problem, Example 7.2, to test the null hypothesis*

$$H_0 : \mu = 360$$

against the alternative

$$H_a : \mu < 360$$

using $\alpha = .01$.

Solution   *Using Table 7.5 for the one-tailed alternative, $\mu < \mu_0$, we shall reject if the observed value of z is less than $-z_\alpha$. Substituting $\alpha = .01$, we shall reject the null hypothesis if z is less than $-z_{01}$. Referring to*

Table 7.4 (or Table 1 of the Appendix)

$$z_{.01} = 2.33$$

For H_a: $\mu < 360$, we shall reject H_0:$\mu = 360$ if the computed value of z is less than -2.33.

As noted previously, as the probability of a type I error, α, decreases, the probability of a type II error, β, increases. For this reason α is usually chosen to be .01, .05, or .10 for a statistical test.

EXERCISES

8   In testing the null hypothesis that $\mu = \mu_0$, the standard deviation of the distribution of $\overline{X}$ is $\sigma/\sqrt{n}$. How can we test the null hypothesis since we do not know the variance of the population?

9   Set up a rejection region for the mean of family income to test the null hypothesis H_0:$\mu = \$9860$ against the research hypothesis H_a:$\mu \neq \$9860$. (Use $\alpha = .02$.)

10   If the rejection region for a statistical test is located at $z > 1.645$ or $z < -1.645$, what is the probability of a type I error?

11   If the rejection region for a statistical test is located at $z > 1.645$, what is the probability of a type I error?

7.7   A ONE-SAMPLE GOODNESS-OF-FIT TEST: THE CHI-SQUARE TEST

*M*any studies particularly in the social sciences result in enumerative (or frequency) data for variables measured on normal or ordinal scales. For instance the classification of people into five income brackets would result in an enumeration or count corresponding to the number of people assigned to each of the five income classes. Or, we might be interested in studying the reactions of mental patients to a particular stimulus in a psychological experiment. If a patient will react in one of three ways when the stimulus is applied and if a large number of patients were subjected to the stimulus, the experiment would yield three counts indicating the number of patients falling in each of the reaction classes. Similarly, a traffic study might require a count and classification of the type of motor vehicles using a section of highway. An industrial process manufactures items that fall into one of three quality classes: acceptables, seconds, and rejects. A student of the arts might classify paintings in one of several categories according to style and period in order to study trends in style over time. We might wish to classify ideas in a philosophical study or style in the field of literature. The results of an advertising campaign would yield count data that indicate a classification of consumer reaction. Indeed, many observations in the social sciences are not amenable to measurement on a continuous scale and hence result in count data.

The illustrations in the preceding paragraph exhibit, to a reasonable degree of approximation, the following characteristics, which define a multinomial experiment:

The Multinomial Experiment

1   The experiment consists of n identical trials.
2   The outcome of each trial falls into one of k classes or cells.
3   The probability that the outcome of a single trial will fall in a particular cell, say cell i, is p_i $(i = 1, 2, \ldots, k)$, and remains the same from trial to trial. Note that

$$p_1 + p_2 + p_3 + \cdots + p_k = 1$$

4   The trials are independent.
5   We are interested in $n_1, n_2, n_3, \ldots, n_k$, where n_i $(i = 1, 2, \ldots, k)$ is equal to the number of trials in which the outcome falls in cell i. Note that

$$n_1 + n_2 + n_3 + \cdots + n_k = n$$

The above experiment is analogous to tossing n balls at k boxes where each ball must fall in one of the boxes. The boxes are arranged such that the probability that a ball will fall in a box varies from box to box but remains the same for a particular box in repeated tosses. Finally, the balls are tossed in such a way that the trials are independent. At the conclusion of the experiment, we observe n_1 balls in the first box, n_2 in the second, and n_k in the kth. The total number of balls is equal to

$$\Sigma n_i = n$$

You will note the similarity between the binomial and multinomial experiments and, in particular, that the binomial experiment represents the special case for the multinomial experiment when $k = 2$. The single parameter of the binomial experiment p is replaced by the k parameters $p_1, p_2, \ldots, p_k$ of the multinomial. In this chapter inferences concerning $p_1, p_2, \ldots, p_k$ will be expressed in terms of a statistical test of an hypothesis concerning their specific numerical values or their relationship one to another.

Suppose that we knew the probability that a ball lands in cell 1, p_1, is .1. Then if we toss $n = 100$ balls at the cells we would expect $100(.1) = 10$ balls to land in cell 1. Indeed the expected number in cell i after n trials is np_i.

Definition 7.5

In a multinomial experiment where each trial can result in one of k outcomes, the expected number of outcomes of type i in n trials is np_i, *where p_i is the probability that a given trial results in outcome i.*

In 1900 Karl Pearson proposed the following test statistic, which is a function of the squares of the deviations of the observed cell counts from their expected value. Letting O denote the observed cell count and E the expected count, Pearson's statistic was

$$\chi^2 = \sum \frac{(O - E)^2}{E}$$

Suppose that we hypothesize values for the cell probabilities $p_1, p_2, \ldots, p_k$ and calculate the expected cell counts using Definition 7.5 to examine how well the data fit or agree with the hypothesized cell probabilities. Certainly if our hypothesized values of the p's are true, the cell counts should not deviate greatly from the expected cell counts. Large values of χ^2 would imply rejection of an hypothesis. How large is large? The answer to this question can be found by examining the probability distribution of χ^2.

The quantity χ^2 will possess approximately a chi-square (χ^2) probability distribution in repeated sampling when n is large. For this approximation to be good, it is desirable that the expected number falling into each cell be equal to or greater than 5. We shall not give the formula for the chi-square probability distribution but rather will characterize it with the following properties:

1   The chi square is not a symmetrical distribution (see Figure 7.8).
2   There are many chi-square distributions. We obtain a particular one by specifying a number called the degrees of freedom (d.f.) associated with the chi-square distribution (this number will change depending upon the application).

Thus large values of χ^2 are those that fall in the upper (right-hand) tail of the chi-square probability distribution.

Figure 7.8   *Chi-square probability distribution; d.f.* $= 4$

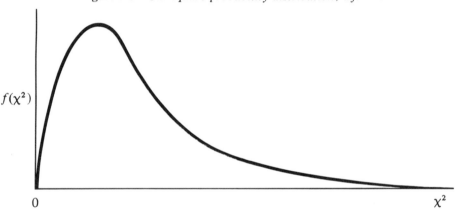

The chi-square test concerning k specified cell probabilities, often called a chi-square goodness-of-fit test, will be based on $(k - 1)$ degrees of freedom. Upper-tail values of

$$\chi^2 = \sum \frac{(O - E)^2}{E}$$

are shown in Table 3 of the Appendix. Entries in the table are chi-square values such that an area of size a lies to the right of χ_a^2 under the curve. We specify the degrees of freedom in the left-hand column of the table and the specified value of a in the top row of Table 3. For $a = .10$ and d.f. $= 14$, the critical value of the chi-square distribution is 21.0642 (see Figure 7.9).

     The rejection region for the one-tailed test concerning k cell probabilities can be determined for a specified value of a type I error α using Table 3 of the Appendix. If the observed value of χ^2 falls beyond the critical value, we

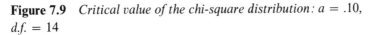

Figure 7.9   *Critical value of the chi-square distribution: a = .10, d.f. = 14*

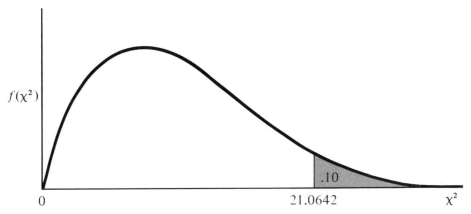

reject the null hypothesis which specifies the *k* cell probabilities. We summarize the test procedure.

Chi-Square Goodness-of-Fit Test

*Null Hypothesis: **Each of k cell probabilities is specified.***
*Alternative Hypothesis: **At least one of the cell probabilities differs from the hypothesized value.***
Test Statistic

$$\chi^2 = \sum \frac{(O - E)^2}{E}$$

*Rejection Region: **Reject the null hypothesis if χ^2 exceeds the tabulated value of χ^2 for a = α, d.f. = (k − 1).***

(Note: This test requires all expected cell frequencies to be 5 or more.)

Example 7.6   *Investigators were interested in evaluating the social–psychological aspects of a social experiment. Integrated interracial projects were sponsored in several towns in Minnesota. A sample of 99 housewives were interviewed in Koaltown, Minnesota, to examine changes in the attitudes toward Negroes. These data are listed in Table 7.6. Previous experience with segregated social projects suggests that the* *fraction of housewives in the project reporting favorable change, no change, and unfavorable change are, respectively, .22, .67, and .11. Test the hypothesis that the cell probabilities for the interracial projects do not differ from .22, .67, and .11. (Use $\alpha = .05$.)*

Table 7.6   *Number of Housewives in the Project Reporting Favorable Change, No Change, or Unfavorable Change in Their Attitudes Toward Negroes*

	Koaltown
Favorable change	58
No change	38
Unfavorable change	3
Total	99

SOURCE: Morton Deutch and Mary E. Collins, *Interracial Housing: A Psychological Evaluation of a Social Experiment* (Minneapolis, Minn.: University of Minnesota Press, 1951), p. 97, by permission.

Solution   *This experiment possesses the characteristics of a multinomial experiment with $n = 99$ trials and $k = 3$ outcomes:*

Outcome 1:   Favorable change in attitude with probability p_1.
Outcome 2:   No change in attitude with probability p_2.
Outcome 3:   Unfavorable change in attitude with probability p_3.

The null hypothesis is

$$H_0 : p_1 = .22, \quad p_2 = .67, \quad p_3 = .11$$

The observed and expected cell counts are given in Table 7.7.

Table 7.7   *Observed and Expected Cell Counts for Example 7.6*

	Observed	*Expected*
Favorable change	58	$99(.22) = 21.78$
No change	38	$99(.67) = 66.33$
Unfavorable change	3	$99(.11) = 10.89$

Note that the expected cell counts all exceed 5 and that the observed cell counts differ markedly from the expected cell counts.

The test statistic is then

$$\chi^2 = \sum \frac{(O - E)^2}{E}$$

$$= \frac{(58 - 21.78)^2}{21.78} + \frac{(38 - 66.33)^2}{66.33} + \frac{(3 - 10.89)^2}{10.89}$$

$$= \frac{(36.22)^2}{21.78} + \frac{(28.33)^2}{66.33} + \frac{(7.89)^2}{10.89} = 78.05$$

The tabulated value of χ^2 for $a = .05$ with $(k - 1) = (3 - 1) =$ 2*d.f.* is 5.99147. Since 78.05 exceeds 5.99147 we reject the null hypothesis concerning the specified cell probabilities. Certainly it appears that the interracial project has had more success in obtaining favorable attitudinal changes toward Negroes.

Example 7.7 *The observed and expected number of 356 mental patients classified by social class, based on the social class distribution of the city of New Haven, Connecticut, are given in Table 7.8. Perform a chi-square goodness-of-fit test on these data. (Use α = .10.)*

Table 7.8

Social Class	Observed	Expected
Upper	18	31
Middle	46	59
Lower-middle	127	173
Lower	165	93

SOURCE: A. B. Hollingshead and F. C. Hedlich, *Social Class and Mental Illness* (New York: John Wiley & Sons, Inc., 1958), p. 203, by permission.

Solution *Using the observed and expected cell counts in Table 7.8 we can compute χ^2.*

$$\chi^2 = \sum \frac{(O - E)^2}{E}$$

$$= \frac{(18 - 31)^2}{31} + \frac{(46 - 59)^2}{59} + \frac{(127 - 173)^2}{173} + \frac{(165 - 93)^2}{93}$$

$$= \frac{(13)^2}{31} + \frac{(13)^2}{59} + \frac{(46)^2}{173} + \frac{(72)^2}{93}$$

$$= 76.28$$

The tabulated value of χ^2 for $a = .10$ and d.f. $= 3$ is 6.25139. Since $\chi^2 = 76.28$ exceeds 6.25139, we reject the null hypothesis concerning the cell probabilities. In this example the specified probabilities were based on the social class distribution of New Haven, so we have shown that the mental health patient distribution by social class is different than the social class distribution for the entire city.

EXERCISES

12 Using the national percentage distribution for males in Exercise 6, Chapter 3 (page 56), to determine the expected frequencies, perform a chi-square goodness of fit on the actual frequency count of the size of four racial–ethnic groupings in a suburb of San Francisco listed below. (Use $\alpha = .05$.)

18 Indians   60 Japanese   80 Chinese   42 Filipinos

13 Repeat Exercise 12 using the national percentage distribution for females in Exercise 6, Chapter 3 (page 56). Actual frequencies in the suburb are the same. (Use $\alpha = .01$.)

14 Must the total expected frequency always equal the total actual frequencies?

7.8   THE KOLMOGOROV–SMIRNOV ONE-SAMPLE TEST: ANOTHER GOODNESS-OF-FIT TEST

The Kolmogorov–Smirnov (K–S) one-sample test, like the chi-square test of Section 7.7, is a statistical method for testing the degree of agreement between a sample of observed values and an hypothesized probability distribution. We wish to determine whether the observations in the sample came from the hypothesized population. A discussion of the K–S statistic will be helpful because some authors employ this test rather than the chi-square test. Instead of working with the observed and expected cell counts for a multinomial distribution, the K–S test utilizes the cumulative relative frequency distribution of the sample data measured on an ordinal scale and compares this to an hypothesized cumulative relative frequency distribution.

Before defining the K–S test procedure, we need the following notation. Let

$F_0(X) =$ theoretical cumulative relative frequency distribution specified under H_0

$S_n(X) =$ observed cumulative relative frequency distribution for the n sample measurements [if X denotes a particular measurement, then $S_n(X) = k/n$, where k is the number of observations less than or equal to X]

If we hypothesize that the sample measurements have been drawn from the population identified under H_0, we would expect $F_0(X)$ and $S_n(X)$ to be in close agreement for all values of X. Hence the differences

$$F_0(X) - S_n(X)$$

should be small.

The Kolmogorov–Smirnov one-sample test is based on the largest absolute difference between $F_0(X)$ and $S_n(X)$. That is, we compute $S_n(X)$ for all values of X and compare these with the corresponding values of the hypothesized cumulative distribution, $F_0(X)$. For some value of X, this difference (ignoring the sign) will be larger than all others. This is the test statistic used for the Kolmogorov–Smirnov test.

Definition 7.6

The maximum deviation between $F_0(X)$ and $S_n(X)$ is the largest absolute difference and is designated by the symbol D. Thus

$$D = maximum|F_0(X) - S_n(X)|$$

The distribution of D in repeated sampling is known and critical values have been tabulated in Table 7 of the Appendix for all possible sample sizes. The table gives the value of D for a specified sample size n which has area a to its right under the distribution of D. Values of a from .01 to .20 appear across the top of the table.

We can summarize the test procedure as follows:

Kolmogorov–Smirnov One-Sample Test

H_0: *The theoretical relative frequency distribution for the sample observations measured on an ordinal scale is $F_0(X)$.*

H_a: *The sample measurements were drawn from a population whose*

frequency distribution differed from the one specified under H_0.
(*Note that this is a two-tailed test.*)
Test Statistic: $D = maximum \left| F_0(X - S_n(X) \right|$.
Rejection Region: *For specified* α *and sample size n, reject* H_0 *if the observed value of D exceeds the value tabulated in Table 7 with*
$a = \alpha$.

Example 7.8   *Twenty college professors affiliated with small liberal arts colleges were interviewed concerning administrative demands upon them for scholarly publications. Their responses are grouped into four categories and these categories are rank-ordered from "none" to "quite a lot." The frequencies, relative frequencies, and cumulative relative frequencies are shown in Table 7.9. Test the hypothesis that the relative frequencies for the four categories are identical, namely 5/20. This would be equivalent to hypothesizing that the cumulative relative frequency distribution* $F_0(X)$ *would be 5/20, 10/20, 15/20, and 20/20. (Use* $\alpha = .05$.)

Table 7.9   *Observed Data for College Survey*

Categories	Frequency f	Relative Frequency f/n	Cumulative Relative Frequency $S_n(X)$
None	5	5/20	5/20
Very little	10	10/20	15/20
Some	3	3/20	18/20
Quite a lot	2	2/20	20/20

Solution   *The Kolmogorov–Smirnov one-sample test is appropriate because we want to determine if the observed data measured on an ordinal scale came from a population with the hypothesized cumulative relative frequency distribution specified above. We now compute the absolute differences* $F_0(X) - S_n(X)$:

	$F_0(X)$	$S_n(X)$	$\left| F_0(X) - S_n(X) \right|$
None	5/20	5/20	0/20
Very little	10/20	15/20	5/20
Some	15/20	18/20	3/20
Quite a lot	20/20	20/20	0/20

Note that D, the maximum absolute difference, is 5/20. For α = .05 and n = 20 the critical value of D is .294. Since the observed value of D, 5/20 = .25, is less than the critical value, .294, we have insufficient evidence to reject the null hypothesis that there are an equal number of small liberal arts colleges in each of the four categories.

EXERCISES

15 A survey was conducted to study the attitudes of corporate executives to the initiation of a mandatory retirement policy at age 60. The attitudes of 20 randomly selected executives are

> disfavor   5
> neutral   4
> favor   11

Test the hypothesis that executive attitudes are equally distributed in the three categories. (Use $\alpha = .05$.)

16 Ten newsmen were questioned about their opinions concerning the level of violence at a recent prison riot. Their responses were

> little violence   7
> moderate violence   1
> extreme violence   2

Do the data provide sufficient evidence to indicate a preference for one of the three categories? (Use $\alpha = .05$.)

7.9   SUMMARY

Recall that the objective of statistics is to make an inference about a population based on information contained in a sample. Chapters 2, 3, and 4 showed how we describe a set of measurements, thus providing us with a way to phrase an inference about a population. Chapter 5 introduced the concept of probability and probability distributions, providing us with the mechanism for making inferences. Chapters 6 and 7 utilized this information in presenting the reasoning and methodology employed in statistical inference.

Since populations are described by parameters, we can make inferences about them in two ways—we can test hypotheses about their values or we can estimate them. Chapter 6 illustrated the reasoning employed in estimating population parameters based on single samples selected from a population. Methods for testing hypotheses about population parameters, again based on single samples, were presented in Chapter 7.

The two methods for making inferences about population parameters are related to two of the steps in the scientific method. The scientific method involves observation of nature to formulate a theory, followed by reobservation to test the theory against reality. The testing of hypotheses is a vital tool for determining whether or not observation agrees with theory. Estimation, the second method for making inferences about population parameters, attempts to determine the value of one or more parameters based on information contained in a sample. Unlike hypothesis testing, estimation does not ask "Do the data provide sufficient evidence to indicate that μ differs from 80?" Rather, it asks: "What is μ?" Hence estimation is concerned with the first step in the scientific method, the formulation of a theory.

A statistical test is composed of four parts: a null hypothesis, an alternative hypothesis, a decision maker, and a rejection region. The rejection region is a set of values of the decision maker that are contradictory to the null hypothesis. Contradictory values are those that lie too many standard deviations away from the hypothesized mean.

The goodness of a statistical test is measured by the sizes of α and β, the probability of rejecting the null hypothesis when it is true, and the probability of accepting when the null hypothesis is false.

The rejection region for most tests is easy to find because we employ our knowledge of the empirical rule. If we know that the probability distribution for a decision maker is mound-shaped, improbable values will lie roughly two standard deviations away from the mean. The rejection region is then located in the tails of the probability distribution. The selection of the rejection region for the z test is based on the same line of reasoning.

Note that the sample size n plays an important role in both estimation and testing hypotheses because it measures the amount of data (and hence information) contained in the sample. As n increases, the bound on the error of estimation (and the width of a confidence interval) decreases. When testing hypotheses where the data are quite variable and n is small, it is unlikely that we shall reject the null hypothesis even when the null hypothesis is false. That is, the probability of making a type II error, β, will be large. This is important because one frequently hears that "experts," panels, and high-level government commissions have found little effect of some drug on humans, that there is no difference in the effect of some sociological factor on the behavior or condition of various groups of people in our society, or that certain conditions have no effect on sociological behavior. These reports, which may vitally affect our society, are often based on pitifully small quantities of very variable data which do not support the experts' conclusions. In fact, sample sizes employed are rarely revealed in the reports.

QUESTIONS AND PROBLEMS

17   Define what is meant by a type I error and a type II error.

18   What is the relationship between α and β for a fixed sample size?

19   How does increasing the sample size affect α and β?

20   During the 1972 Democratic primary campaign in California, a poll of 1000 registered Democrats showed 55 percent favoring McGovern and 45 percent favoring Humphrey. Do these data provide sufficient evidence to indicate that McGovern would win the primary? Would you use a one-tailed or two-tailed test? (Use at $\alpha = .05$.)

21   A sample of 30 SMA's located in the West shows a mean fertility ratio of 350.4 and a standard deviation of 25.7. Test the null hypothesis $H_0: \mu = 360$ using a two-tailed test. (Use $\alpha = .01$.)

22   A study shows that 80 percent of college freshmen scoring above a certain cutoff point on a social science orientation test also passed principles of anthropology. This year 600 freshmen were tested, and of those scoring at or above the cutoff point, 450 passed and 150 failed principles of anthropology. Use the chi-square test of goodness of fit. Do the observed frequencies provide sufficient evidence to indicate that the failure rate in anthropology differs from that for the past year? (Use $\alpha = .05$.)

23   In a study of premarital pregnancy, a research team reports 30 such pregnancies in the lower social class, 20 in the middle social class, and 10 in the higher social class. If the expected frequencies are 20, 20, and 20, test the goodness of fit of the observed frequencies with the expected. What do these data suggest? (Use $\alpha = .05$.)

24   Test the null hypothesis that in a particular sorority there are no differences among the members as far as social class background is concerned. Use the .05 level of significance and the Kolmogorov–Smirnov one-sample test. Comment on your findings.

Social Class	Frequency
Lower-lower	1
Upper-lower	1
Lower-middle	2
Upper-middle	10
Lower-upper	8
Upper-upper	2

25   Test the null hypothesis that $p = .5$ for the reform party candidate. A sample of 1000 voters indicates that the candidate received 550 votes. Do the data provide sufficient evidence that the reform party candidate will poll *more than* 50 percent of the votes?

26   To determine consistency in evaluating student behavior, two evaluators were presented with a group of 200 students for examination. Each student was examined by both of the evaluators. The evaluators agreed on 133 of the evaluations. Does this indicate that their agreement is due to reasons other than pure chance?

27   A hospital claims that the average length of patient confinement is 5 days. A study of the length of patient confinements on $n = 36$ people showed that $\bar{X} = 6.2$ and $s = 5.2$. Do these data present sufficient evidence to contradict the hospital's claim?

REFERENCES

Anderson, T. R., and M. Zelditch. *A Basic Course in Statistics*, 2nd ed. New York: Holt, Rinehart and Winston, Inc., 1968. Chapters 10, 11, and 12.

Blalock, H. M. *Social Statistics*, 2nd ed. New York: McGraw-Hill Book Company, 1972. Chapters 10 and 11.

Champion, D. J. *Basic Statistics for Social Research*. Scranton, Pa.: Chandler Publishing Company, 1970. Chapters 6 and 7.

Mendenhall, W., and L. Ott. *Understanding Statistics*. North Scituate, Mass.: Duxbury Press, 1972. Chapter 6.

Mueller, J. H., K. F. Schuessler, and H. L. Costner. *Statistical Reasoning in Sociology*. 2nd ed. Boston: Houghton Mifflin Company, 1970. Chapter 14.

Palumbo, D. J. *Statistics in Political and Behavioral Science*. New York: Appleton-Century-Crofts, 1969. Chapter 7.

8

ESTIMATION AND STATISTICAL TESTS OF HYPOTHESES: TWO SAMPLES

R

arely do you read one of the popular news magazines or Sunday editions of a leading newspaper that you do not find one or more articles dealing with the comparison of two populations. We read that factory orders in July rose 1.7 percent (in comparison with June), car production for September is scheduled to drop 4.5 percent (in comparison with August), the public school teachers of a certain state receive salaries less than the national average, and the percent of people suffering from arteriosclerosis is higher for individuals with cadmium in their water supply than for those without. All these examples imply the comparison of two populations based on information contained in samples selected from each.

How can you tell whether the observed difference in the above comparisons is real or whether it is due to random variation? People unfamiliar with statistics frequently answer this question by saying: "But you can see the difference, can't you? There is no question about it!" They forget that the difference that they observe is based on samples that yield means and other estimates that vary in a random manner about the true population parameters. They confuse the observed difference between the *sample* means or percentages with those between the *populations*.

To further emphasize this point, many researchers have been particularly interested in studying the impact of "labeling" upon behavior. More specifically, it has been suggested that the performance of a deviant act, in and of itself, is not really important to the individual's self-concept but rather the labeling of the person as deviant by others is crucial. For purposes of illustration, let us compare two groups of individuals in which each individual has performed a deviant act. Those individuals in group *A* have been publicly labeled as deviant, whereas those in group *B* have escaped such labeling. As a hypothetical case, the results might show that 19 persons of 22 in group *A* have committed additional deviant acts and 14 of 21 in group *B* have committed additional deviant acts. What do we conclude about the impact of labeling upon deviant acts?

The proportions of deviants who commit additional deviant acts in the two groups appear to be substantially different, .86 versus .67; but keep in mind that these proportions are estimates of binomial parameters based on relatively small samples. As a consequence, the sample estimates vary substantially about the true binomial parameters. Indeed, if p, the likelihood of performing additional deviant acts, were identically the same for individuals in groups A and B, the difference between the two sample proportions, $(.86 - .67) = .19$, could occur just due to chance.

It can be shown that the probability that the two estimates could differ by as much as .19 is rather large. Hence there *is not* sufficient evidence to indicate a difference in the proportion of individuals who have performed additional deviants acts for the two groups. The true proportion for group *A* may be larger than *B*, *B* larger than *A*, or they may be equal. The relevant fact is that there *is not* sufficient data in this study to indicate that the proportions of repeated deviant acts differ between groups *A* and *B*. The technique employed in reaching this decision will be presented in Section 8.3.

The comparison of two populations is another practical example of statistical inference.

8.2 A COMPARISON OF TWO POPULATION MEANS

opulations are most frequently compared by examining the difference in their means. We may wish to compare the mean suicide rates between two regions of a country or compare mean divorce rates between two racial classifications. Or we might want to compare the mean strengths of two economic systems before proceeding to adopt a constitution.

For each of these problems we assume that we are sampling from two populations, the first with mean μ_1 and variance σ_1^2, the second with mean μ_2 and variance σ_2^2. Independent random samples of n_1 and n_2 measurements are then drawn from populations I and II, respectively. Finally, the estimates $\overline{X}_1$, s_1^2, $\overline{X}_2$, and s_2^2 of the corresponding population parameters are computed from the sample data.

It can be shown that the difference in sample means $(\overline{X}_1 - \overline{X}_2)$ possesses a probability distribution that is approximately normal with mean and standard deviation

$$\mu = (\mu_1 - \mu_2)$$

$$\sigma_{(\overline{X}_1 - \overline{X}_2)} = \sqrt{\frac{\sigma_1^2}{n_1} + \frac{\sigma_2^2}{n_2}}$$

That is, if we draw samples of n_1 and n_2 observations, respectively, from the two populations, calculate the sample means and their difference $(\overline{X}_1 - \overline{X}_2)$, and repeat this process over and over again a very large number of times, then the histogram of the differences in sample means would appear as shown in Figure 8.1. The distribution will be approximately normal with mean and standard deviation as noted above. This standard deviation is often called the standard error of the difference between means.

Figure 8.1 *Distribution of* $(\bar{X}_1 - \bar{X}_2)$

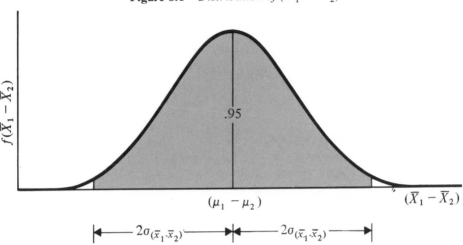

The behavior of $(\bar{X}_1 - \bar{X}_2)$ in repeated sampling, as shown in Figure 8.1, can be used to evaluate the error of estimation when $(\bar{X}_1 - \bar{X}_2)$ is used to estimate the difference in population means, $(\mu_1 - \mu_2)$. Particularly we do not expect $(\bar{X}_1 - \bar{X}_2)$ to lie more than two standard deviations, $2\sqrt{\sigma_1^2/n_1 + \sigma_2^2/n_2}$, away from $(\mu_1 - \mu_2)$. Thus $2\sqrt{\sigma_1^2/n_1 + \sigma_2^2/n_2}$ provides a bound on the error of estimation when the difference in sample means, $(\bar{X}_1 - \bar{X}_2)$, is used to estimate the difference in population means, $(\mu_1 - \mu_2)$. We can summarize the point estimation of $(\mu_1 - \mu_2)$ as follows:

Point Estimation of $(\mu_1 - \mu_2)$

Point Estimator of $(\mu_1 - \mu_2)$: $(\bar{X}_1 - \bar{X}_2)$

Standard Deviation of $(\bar{X}_1 - \bar{X}_2)$: $\sigma_{(\bar{X}_1 - \bar{X}_2)} = \sqrt{\dfrac{\sigma_1^2}{n_1} + \dfrac{\sigma_2^2}{n_2}}$

Bound on the Error of Estimation: $2\sigma_{(\bar{X}_1 - \bar{X}_2)}$

In the same way we can use the point estimator, $(\bar{X}_1 - \bar{X}_2)$, to form a confidence interval for $(\mu_1 - \mu_2)$. Since $(\bar{X}_1 - \bar{X}_2)$ will be within $2\sigma_{(\bar{X}_1 - \bar{X}_2)}$ or really $1.96\sigma_{(\bar{X}_1 - \bar{X}_2)}$ of $(\mu_1 - \mu_2)$ 95 percent of the time, the interval

$$(\bar{X}_1 - \bar{X}_2) \pm 1.96\sigma_{(\bar{X}_1 - \bar{X}_2)}$$

forms a 95 percent confidence interval for $(\mu_1 - \mu_2)$.

95 Percent Confidence Interval for $(\mu_1 - \mu_2)$

$$(\overline{X}_1 - \overline{X}_2) \pm 1.96\sigma_{(\overline{X}_1 - \overline{X}_2)}$$

where

$$\sigma_{(\overline{X}_1 - \overline{X}_2)} = \sqrt{\frac{\sigma_1^2}{n_1} + \frac{\sigma_2^2}{n_2}}$$

(Note: If σ_1^2 and σ_2^2 are unknown, we can substitute s_1^2 and s_2^2, respectively, provided the sample sizes, n_1 and n_2, are equal to 30 or more. Also, for a 90 or 99 percent confidence interval, substitute 1.645 or 2.58, respectively, for the 1.96 that appears in the 95 percent confidence interval shown above.)

Example 8.1   *A study was conducted to determine if persons in suburban district I have a different mean income than those in district II. A random sample of 50 homeowners was taken in district I. Although we also wanted to interview 50 homeowners in district II, one man refused to provide the information requested even though we promised to keep it confidential, so only 49 observations were obtained from district II. The data, recorded in units of thousands of dollars, produced sample means and variances as follows:*

District I	District II
$n_1 = 50$	$n_2 = 49$
$\overline{X}_1 = 14.27$	$\overline{X}_2 = 12.78$
$s_1^2 = 8.74$	$s_2^2 = 6.58$

Use these data to place a 95 percent confidence interval on $(\mu_1 - \mu_2)$.

Solution   *The difference in the sample mean is*

$$\overline{X}_1 - \overline{X}_2 = 14.27 - 12.78 = 1.49$$

Since the sample variances are good estimates of σ_1^2 and σ_2^2 (which are unknown), we can substitute them into the formula for $\sigma_{(\overline{X}_1 - \overline{X}_2)}$ to obtain

$$\sqrt{\frac{\sigma_1^2}{n_1} + \frac{\sigma_2^2}{n_2}} \approx \sqrt{\frac{8.74}{50} + \frac{6.58}{49}} = .56$$

Hence $1.96\sigma_{(\bar{X}_1 - \bar{X}_2)} = 1.10$. *A 95 percent confidence interval for the difference in mean incomes for the two districts has a lower confidence limit*

$$LCL = (\bar{X}_1 - \bar{X}_2) - 1.96\sigma_{(\bar{X}_1 - \bar{X}_2)}$$
$$= 1.49 - 1.10 = .39$$

and an upper confidence limit

$$UCL = (\bar{X}_1 - \bar{X}_2) + 1.96\sigma_{(\bar{X}_1 - \bar{X}_2)}$$
$$= 1.49 + 1.10 = 2.59$$

We are 95 percent confident that the difference in the two means $(\mu_1 - \mu_2)$ *lies in the interval .39 to 2.59.*

The corresponding test of an hypothesis concerning the difference between two population means, μ_1 and μ_2, follows the logic developed for a test concerning a single population mean. The test procedure requires adequate sample sizes to estimate the population variances σ_1^2 and σ_2^2, because they will rarely be known. We suggest the requirement that n_1 and n_2 both be 30 or more. Then you can use s_1^2 and s_2^2 to approximate σ_1^2 and σ_2^2. What test can you use if one or more of the sample sizes is less than 30? You can employ a small-sample test as described in Section 8.8. This test is more sensitive but also more complicated to use, so we delay its discussion.

A summary of the elements in the large-sample test for the difference in means follows:

The Test for Comparing Two Population Means (n_1 and n_2 equal to 30 or more)

Null Hypothesis: $(\mu_1 - \mu_2) = 0$ (*i.e.,* $\mu_1 = \mu_2$).
Alternative Hypothesis: $\mu_1 \neq \mu_2$ (*for a two-tailed test*).

Test Statistic: $z = \dfrac{\bar{X}_1 - \bar{X}_2}{\sqrt{\dfrac{\sigma_1^2}{n_1} + \dfrac{\sigma_2^2}{n_2}}}.$

Rejection Region: Reject the null hypothesis if z is larger than $z_{\alpha/2}$ *or less than* $-z_{\alpha/2}$. *Use* s_1^2 *and* s_2^2 *to approximate* σ_1^2 *and* σ_2^2, *respectively.*

(*Note: We can select any value of* α *we desire, but generally we adopt* $\alpha = .10, .05,$ *or* $.01$. *These* $z_{\alpha/2}$ *values are 1.645, 1.96, and 2.58, respectively.*)

Example 8.2 *A sample of 30 SMSA's was collected from both the North and South. Use the sample data of Table 8.1 to test the hypothesis of equality of means for the fertility ratios for the North and South. (Use* $\alpha = .05$.)

Table 8.1 *Fertility Ratios for 30 SMSA's from the North and South*

North	Fertility Ratio	South	Fertility Ratio
Albany, N.Y.	353	Atlanta, Ga.	346
Allentown, Pa.	319	Augusta, Ga.	357
Atlantic City, N.J.	352	Baton Rouge, La.	347
Canton, Ohio	352	Beaumont, Tex.	337
Chicago, Ill.	356	Birmingham, Ala.	324
Cincinnati, Ohio	372	Charlotte, N.C.	344
Cleveland, Ohio	345	Chattanooga, Tenn.	326
Detroit, Mich.	369	Columbia, S.C.	334
Evansville, Ind.	330	Corpus Christi, Tex.	395
Grand Rapids, Mich.	382	Dallas, Tex.	365
Johnstown, Pa.	333	El Paso, Tex.	401
Kalamazoo, Mich.	318	Fort Lauderdale, Fla.	315
Kenosha, Wis.	384	Greensboro, N.C.	319
Lancaster, Pa.	356	Houston, Tex.	370
Lansing, Mich.	355	Jackson, Miss.	360
Lima, Ohio	400	Knoxville, Tenn.	301
Madison, Wis.	321	Lexington, Ky.	319
Mansfield, Ohio	366	Lynchburg, Va.	314
Milwaukee, Wis.	364	Macon, Ga.	352
Newark, N.J.	337	Miami, Fla.	284
Paterson, N.J.	311	Monroe, La.	386
Philadelphia, Pa.	344	Nashville, Tenn.	313
Pittsfield, Mass.	358	Newport News, Va.	356
Racine, Wis.	402	Orlando, Fla.	336
Rockford, Ill.	394	Richmond, Va.	313
South Bend, Ind.	351	Roanoke, Va.	310
Springfield, Ill.	344	Shreveport, La.	373
Syracuse, N.Y.	377	Washington, D.C.	331
Vineland, N.J.	377	Wichita Falls, Tex.	315
Youngstown, Ohio	339	Wilmington, Del.	358

Solution *Using the sample data from Table 8.1, we find the following means and variances:*

North	South
$\overline{X}_1 = 355.4$	$\overline{X}_2 = 340.0$
$s_1^2 = 579.70$	$s_2^2 = 807.10$

To investigate the research hypothesis that the mean fertility ratios differ between the North and South, we use the null hypothesis that the mean fertility ratios for the North and South, μ_1 and μ_2, respectively, are equal. Our test statistic is

$$z = \frac{\overline{X}_1 - \overline{X}_2}{\sqrt{\dfrac{\sigma_1^2}{n_1} + \dfrac{\sigma_2^2}{n_2}}}$$

$$= \frac{355.4 - 340.0}{\sqrt{\dfrac{579.70}{30} + \dfrac{807.10}{30}}} = 2.26$$

For $\alpha = .05$ we shall reject the null hypothesis if z exceeds 1.96 or is less than -1.96. Since the observed value, $z = 2.26$, falls in the rejection region, we reject the null hypothesis that the mean fertility ratios are identical for the North and South and accept the research hypothesis that they differ. In fact, clearly it appears that the mean fertility ratio for the North is higher than for the South.

EXERCISES

1   Two hospitals wished to study the average number of days required for treatment of patients between the ages of 25 and 34. Random samples of 500 hospital patients were selected from each of the two hospitals. The sample means were 5.4 and 6.8 and the sample standard deviations were 3.1 and 3.7 days, respectively. Estimate the difference in mean stay, $(\mu_1 - \mu_2)$, and place a bound on the error of estimation.

2   Two random samples of 50 each were selected to compare the mean hourly income of nonunionized migrant workers in two different states. Those selected from state I possessed a sample mean and variance of $2.80 and .09, respectively. Those for state II were $2.45 and .07, respectively. Do these data present sufficient evidence to indicate a difference in mean hourly wages for migrant workers employed in two states? (Use $\alpha = .05$.)

3   Random samples of 100 adults, selected from two ethnic groups in a large city, were questioned concerning the number of years they attended public schools.

Samples for the two ethnic groups gave $\overline{X}_1 = 7.4$, $s_1 = 2.1$, $\overline{X}_2 = 8.2$, and $s_2 = 2.4$. Test to see if there is evidence of a real difference between the two population means. (Use $\alpha = .05$.)

4   Refer to Exercise 13, Chapter 3 (page 66). A researcher argues that in Colombia males experience their first heterosexual activity at a younger age, on the average, than females. Compute the means and standard deviations. Do the data support the researcher's position? Explain.

8.3   A COMPARISON OF TWO BINOMIAL PARAMETERS

*m*any practical problems require the comparison of two binomial parameters. We might wish to compare the fractions of housewives who utilized prenatal health services before and after a campaign to publicize the services, the fractions of households in two states that are entirely supported by welfare, or the fractions of voters favoring candidate A in a suburban versus a rural area.

We assume that independent random samples are drawn from two binomial populations with parameters p_1 and p_2. Further assume that the samples contain n_1 and n_2 observations, respectively. If X_1 represents the number of successes in n_1 trials and X_2 the number of successes in n_2 trials, then

$$\hat{p}_1 = \frac{X_1}{n_1} \quad \text{and} \quad \hat{p}_2 = \frac{X_2}{n_2}$$

the sample proportions of successes, provide point estimators of p_1 and p_2, respectively.

The probability distribution of $(\hat{p}_1 - \hat{p}_2)$ will be approximately normal for large values of n_1 and n_2 (see **Figure** 8.2) with mean and standard deviation (standard error of the difference between proportions)

$$\mu = (p_1 - p_2)$$

$$\sigma_{(\hat{p}_1 - \hat{p}_2)} = \sqrt{\frac{p_1 q_1}{n_1} + \frac{p_2 q_2}{n_2}}$$

When large samples are involved, the sample size and value of p for each population should satisfy the requirement that *both* np and nq equal 10 or more. Then the difference between the estimate, $(\hat{p}_1 - \hat{p}_2)$, and the true difference in population proportions, $(p_1 - p_2)$, called the *error of estimation,* will be less than $2\sigma_{(\hat{p}_1 - \hat{p}_2)}$ approximately 95 percent of the time.

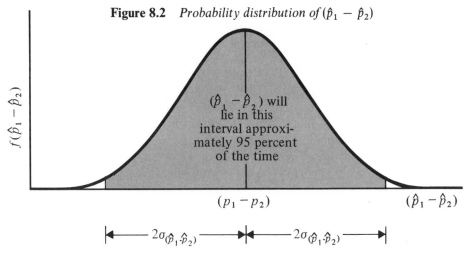

Figure 8.2 *Probability distribution of* $(\hat{p}_1 - \hat{p}_2)$

We can summarize point and interval estimation of $(p_1 - p_2)$ based on large samples.

Point Estimator of $(p_1 - p_2)$

Point Estimator of $(p_1 - p_2)$: $(\hat{p}_1 - \hat{p}_2)$
Standard Deviation of $(\hat{p}_1 - \hat{p}_2)$:

$$\sigma_{(\hat{p}_1 - \hat{p}_2)} = \sqrt{\frac{p_1 q_1}{n_1} + \frac{p_2 q_2}{n_2}}$$

Bound on the Error of Estimation: $2\sigma_{(\hat{p}_1 - \hat{p}_2)}$

[Note: Since we do not know p_1 and p_2 in the standard deviation of $(\hat{p}_1 - \hat{p}_2)$, we replace them with $\hat{p}_1$ and $\hat{p}_2$.]

---

95 Percent Confidence Interval for $(p_1 - p_2)$

$$(\hat{p}_1 - \hat{p}_2) \pm 1.96\sigma_{(\hat{p}_1 - \hat{p}_2)}$$

where we approximate $\sigma_{(\hat{p}_1 - \hat{p}_2)}$ *using*

$$\sigma_{(\hat{p}_1 - \hat{p}_2)} = \sqrt{\frac{\hat{p}_1 \hat{q}_1}{n_1} + \frac{\hat{p}_2 \hat{q}_2}{n_2}} \quad \text{with } \hat{q}_1 = (1 - \hat{p}_1) \text{ and } \hat{q}_2 = (1 - \hat{p}_2)$$

Also, for a 90 percent or 99 percent confidence interval, substitute 1.645 or 2.58, respectively, for the 1.96 that appears in the 95 percent confidence interval above.

Example 8.3   *In a survey to analyze the cost of funeral expenditures for various social classes, a random sample of 162 families from the lower and working classes were interviewed to determine the funeral expenses for a recent family death. Of the 162 families contacted, 61 spent over $800 on the funeral. In a sample of 189 middle- and upper-class families experiencing a recent family death, 106 spent over $800 on the funeral. Estimate $(p_1 - p_2)$, the difference in the proportion of families paying over $800 for funeral expenses, for the two social classifications. Place a bound on the error of estimation.*

Solution   *The point estimator of $(p_1 - p_2)$ is the difference in the sample fraction, $(\hat{p}_1 - \hat{p}_2)$.*

$$(\hat{p}_1 - \hat{p}_2) = \left(\frac{X_1}{n_1} - \frac{X_2}{n_2}\right) = \left(\frac{61}{162} - \frac{106}{189}\right)$$

$$= .376 - .561 = -.185$$

The standard deviation of $(\hat{p}_1 - \hat{p}_2)$ is

$$\sigma_{(\hat{p}_1 - \hat{p}_2)} = \sqrt{\frac{\hat{p}_1\hat{q}_1}{n_1} + \frac{\hat{p}_2\hat{q}_2}{n_2}}$$

$$= \sqrt{\frac{(.376)(.624)}{162} + \frac{(.561)(.439)}{189}}$$

$$= .052$$

A bound on the error of estimation for $(p_1 - p_2)$ is, therefore,

$$2\sigma_{(\hat{p}_1 - \hat{p}_2)} = 2(.052) = .104$$

We can readily formulate a statistical test to test the equality of two binomial parameters. A good decision maker for this test is one that would immediately occur to you. That is, we use the difference in the point estimators, $(\hat{p}_1 - \hat{p}_2)$, where $\hat{p}_1 = X_1/n_1$, $\hat{p}_2 = X_2/n_2$, and X_1 and X_2 are the number of successes observed in samples I and II, respectively. The greater the difference between $\hat{p}_1$ and $\hat{p}_2$, the greater is the evidence to indicate that p_1 does not equal p_2. How large is large? That is, how far away from zero must $(\hat{p}_1 - \hat{p}_2)$ depart before we have evidence to indicate a difference between p_1 and p_2? The answer is "too many standard deviations." When $p_1 = p_2$, the mean and standard deviation of the distribution of $(\hat{p}_1 - \hat{p}_2)$ will be

$$(p_1 - p_2) = 0 \qquad (\text{i.e., } p_1 = p_2)$$

$$\sigma_{(\hat{p}_1 - \hat{p}_2)} = \sqrt{pq\left(\frac{1}{n_1} + \frac{1}{n_2}\right)}$$

so the z score is

$$z = \frac{\hat{p}_1 - \hat{p}_2}{\sqrt{pq(1/n_1 + 1/n_2)}}$$

You will need to use the data to approximate p in the formula for $\sigma_{(\hat{p}_1 - \hat{p}_2)}$. The best estimate of p, the parameter common to both populations, is

$$\hat{p} = \frac{\text{total number of successes}}{\text{total number of trials}} = \frac{X_1 + X_2}{n_1 + n_2}$$

We summarize the test procedure as follows:

Large-Sample Test Comparing Two Binomial Proportions
[np and $n(1 - p)$ must both be 10 or larger]

Null Hypothesis $:(p_1 - p_2) = 0.$
Alternative Hypothesis $: p_1 \neq p_2$ *(for a two-tailed test).*
Test Statistic:

$$z = \frac{\hat{p}_1 - \hat{p}_2}{\sqrt{pq(1/n_1 + 1/n_2)}}$$

Rejection Region: Reject H_0 if z is larger than $z_{\alpha/2}$ or less than $- z_{\alpha/2}$.
Note: p is approximated by

$$\hat{p} = \frac{X_1 + X_2}{n_1 + n_2}$$

The comparison of two binomial fractions is illustrated by the following example.

Example 8.4   *Two sets of $n_1 = n_2 = 60$ ninth graders were taught high school algebra by two different methods. The experimental group used a programmed learning text with no formal lectures; the control group was given formal lectures by a teacher. At the conclusion of a 4-month period, a comprehensive test was given to both groups to determine the proportion of students in each group that obtained a score of 85 (out of 100) or better.*

The results were as follows:

Experimental Group	Control Group
$n_1 = 60$	$n_2 = 60$
$X_1 = 41$	$X_2 = 24$

Test the hypothesis that the two population fractions, p_1 and p_2, are equal. (Use a two-tailed test with $\alpha = .01$.)

Solution *For all practical purposes, sampling from both populations satisfies the requirements of a binomial experiment. We wish to test the hypothesis that the proportions of students scoring 85 or better are the same for both teaching techniques. Thus the null hypothesis is that $p_1 = p_2$, or $(p_1 - p_2) = 0$. We shall reject the hypothesis if the z score lies more than 2.58 away from zero. Then the rejection region for the test will be as shown in Figure 8.3.*

Figure 8.3 *Rejection region for Example 8.4*

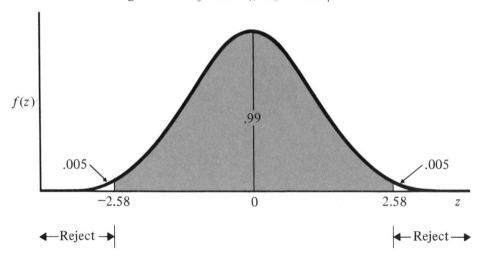

We must first compute

$$\sigma_{(\hat{p}_1 - \hat{p}_2)} = \sqrt{pq\left(\frac{1}{n_1} + \frac{1}{n_2}\right)}$$

Substituting the values of n_1 and n_2 and the approximation for p,

$$\hat{p} = \frac{X_1 + X_2}{n_1 + n_2} = \frac{41 + 24}{60 + 60} = .54$$

into the formula, we obtain

$$\sigma_{(\hat{p}_1 - \hat{p}_2)}\sqrt{(.54)(.46)\left(\frac{1}{60} + \frac{1}{60}\right)} = .091$$

Similarly, we must compute the observed sample proportion difference to be

$$(\hat{p}_1 - \hat{p}_2) = \left(\frac{X_1}{n_1} - \frac{X_2}{n_2}\right) = \frac{41}{60} - \frac{24}{60} = .28$$

Then

$$z = \frac{\hat{p}_1 - \hat{p}_2}{\sqrt{pq\left(\frac{1}{n_1} + \frac{1}{n_2}\right)}} = \frac{.28}{.091} = 3.08$$

Since this value falls in the rejection region (i.e., exceeds 2.58) (Figure 8.3) we reject the hypothesis that $p_1 = p_2$ and conclude that the two population proportions are different. Practically we would conclude that the programmed teaching technique produces a higher fraction of students scoring 85 or more than for the standard lecture-teaching method. What is the probability that our test procedure would lead us to an incorrect rejection of the null hypothesis? The answer is $\alpha = .01$.

EXERCISES

5 A law student believes that the proportion of Republicans in favor of the unrestricted right of executive privilege is greater than the proportion of Democrats in favor of the unrestricted right. He acquired independent random samples of 200 Republicans and 200 Democrats, respectively, and found 46 Republicans and 37 Democrats in favor of the unrestricted right of executive privilege. Does this evidence provide statistical support for the law student's belief? (Use $\alpha = .05$.)

6 A survey is conducted to determine whether a difference exists between the proportion of married and the proportion of single persons in the 20–29 age group who smoke. A sample of 200 persons from each group is polled and 64 married and 80 single persons are found to smoke. Do the data provide sufficient evidence to indicate a difference in the proportion of smokers for the two populations? (Use $\alpha = .05$.)

7 Sixty of 87 Protestant house wives prefer contraceptive technique A; 40 of 100 Catholic housewives prefer the same technique. Estimate the difference in proportions of preference using a 95 percent confidence interval.

8 Refer to Exercise 7. Test the null hypothesis that Protestant housewives and Catholic housewives do not differ in their preference for the contraceptive technique. (Use $\alpha = .01$.)

9 Independent random samples of 800 Republicans and 800 Democrats showed 40 and 32 percent, respectively, in favor of the death sentence for major crimes. Do these data provide sufficient evidence to indicate a difference in the population proportions favoring the death sentence? (Use $\alpha = .01$.)

8.4 ANOTHER TWO-SAMPLE TEST: THE MANN–WHITNEY U TEST

Some studies often yield only ordinal-level data because of the crudeness of the measuring instruments employed by the investigator. Although measurements of this type occur in almost all fields of study, they are particularly evident in the social sciences. For example, when the variables of interest are anomie, social differentiation, prestige, power, social control, and alienation, they are often measured by crude scales and the researcher does not feel he can assume ratio or even interval-level data. When this occurs, the estimation and comparative test procedures presented thus far in this chapter are of little use. Nonparametric statistical techniques are appropriate, however, and they are very useful for analyzing these types of data. Nonparametric statistical methods are also appropriate for ratio data when the assumptions underlying the parametric test procedures (for example, the assumption of normal populations) fail to hold.

The word "nonparametric" evolves from the type of hypothesis usually tested when dealing with nominal or ordinal data. Thus instead of hypothesizing that two populations have the same mean (as in Section 8.2) we could hypothesize that the two populations of measurements are identical. Note that the practical implications of these two hypotheses are not equivalent because the latter hypothesis is less clearly defined. Two distributions could be different and still have the same mean. We shall now consider a procedure that utilizes the ranks of measurements to test the hypothesis that two populations are identical.

In 1947 Mann and Whitney proposed a nonparametric test statistic for the comparison of two populations. Since the two populations are assumed to be identical, samples taken from the respective populations should be "similar." One way to measure the similarity of the two samples is to count the number of observations in sample 1 (drawn from the first population) that precede each observation in sample 2 (drawn from the second population). If this number, U, is extremely large or extremely small, intuitively we would reject the hypothesis that the populations are identical.

A shortcut procedure for calculating U can be obtained by ranking the observations from the two samples as if they were one, letting the smallest observation from the two samples have a rank of 1, the next smallest a rank of 2, and so on. Then for samples of size n_1 and n_2 the number of observations from sample 1 preceding each observation from sample 2 is

$$U = n_1 n_2 + \frac{n_1(n_1 + 1)}{2} - T_1$$

Similarly, the number of observations from sample 2 preceding each observation in sample 1 is

$$U = n_1 n_2 + \frac{n_2(n_2 + 1)}{2} - T_2$$

where T_1 and T_2 are the sum of the ranks for observations in samples 1 and 2, respectively.

It can be shown that when the two populations are identical and when n_1 and n_2 are both larger than 10, U is approximately normally distributed with mean and standard deviation

$$\mu = \frac{n_1 n_2}{2}$$

$$\sigma_U = \sqrt{\frac{n_1 n_2(n_1 + n_2 + 1)}{12}}$$

We can summarize the test procedure as follows:

Large-Sample Test for Identical Populations

(n_1 and n_2 are both 10 or more)

Null Hypothesis : The two populations are identical.
Alternative Hypothesis : The two populations are different (two-tailed).
Decision Maker :

$$z = \frac{U - \mu}{\sigma_U}$$

Region Region : Reject if z is larger than $z_{\alpha/2}$ or less than $-z_{\alpha/2}$.

Example 8.5 *A sample* $n_1 = 25$ *Catholic priests and a sample of* $n_2 = 25$ *Methodist ministers were randomly selected to determine if there is a difference between the two groups with respect to knowledge about the causes of mental illness. Each of the ministers and priests was administered a standard questionnaire that could be scored from 0 to 40. The sample test score data arranged in ascending order appear in Table 8.2.*

Table 8.2 *Mental Health Test Scores*

Methodist Ministers		Catholic Priests	
5	17	5	19
6	19	7	19
8	19	8	20
8	20	8	20
10	21	8	21
10	22	9	22
11	22	12	24
11	22	13	24
13	23	13	26
13	24	14	26
14	28	14	27
16	28	18	27
17		19	

Jointly rank the 50 test scores and use the Mann–Whitney test to check the hypothesis that there is no difference between Methodist ministers and Catholic priests with respect to their knowledge about the causes of mental illness. (Use $\alpha = .05$.)*

Solution *Recall that any test procedure developed for any one scale of measurement can be applied to data collected on a scale of higher quantitative sophistication. Thus, although the test score data are measured on an interval scale, we can apply the Mann–Whitney U test which was constructed for ordinal data.*

Before computing U, we must first jointly rank all 50 observations, by assigning the lowest observation the rank of 1, the second lowest observation the rank of 2, and so on. When two or more measurements are the same, we assign all of them a rank equal to the average of the ranks they occupy. The smallest observation from either sample is 5. But two ministers received a score of 5. These two measurements occupy ranks 1 and 2, so they both receive a rank equal to the average of the occupied ranks, namely 1.5. The next smallest score is a 6 and since there are no other scores of 6, we assign it the rank of 3. Similarly, only one priest scored 7; he received a rank of 4. There were five ministers or priests that received a score of 8, the next lowest measurement. The average of the occupied ranks is

$$\frac{5 + 6 + 7 + 8 + 9}{5} = 7$$

So all five observations receive a rank of 7. The remaining scores are ranked in the same way. These results appear in Table 8.3.

Table 8.3   *Mental Health Test Scores and Ranks*

Methodist Ministers		Catholic Priests	
Score	Rank	Score	Rank
5	1.5	5	1.5
6	3	7	4
8	7	8	7
8	7	8	7
10	11.5	8	7
10	11.5	9	10
11	13.5	12	15
11	13.5	13	17.5
13	17.5	13	17.5
13	17.5	14	21
14	21	14	21
16	23	18	26
17	24.5	19	29
17	24.5	19	29
19	29	19	29
19	29	20	33
20	33	20	33
21	35.5	21	35.5
22	38.5	22	38.5
22	38.5	24	43
22	38.5	24	43
23	41	26	45.5
24	43	26	45.5
28	49.5	27	47.5
28	49.5	27	47.5
	$T_1 = 621.5$		$T_2 = 653.5$

The quantities μ and σ_U are easily computed for $n_1 = n_2 = 25$:

$$\mu = \frac{n_1 n_2}{2} = \frac{25(25)}{2} = 312.5$$

$$\sigma_U = \sqrt{\frac{n_1 n_2 (n_1 + n_2 + 1)}{12}}$$

$$= \sqrt{\frac{25(25)(51)}{12}} = \sqrt{2656.25}$$

$$= 51.54$$

and for $\alpha = .05$ we shall reject the null hypothesis if the computed value of z is greater than $z_{\alpha/2} = 1.96$ or less than $-z_{\alpha/2} = -1.96$.
It makes no difference whether we compute

$$U = n_1 n_2 + \frac{n_1(n_1 + 1)}{2} - T_1$$

or

$$U = n_1 n_2 + \frac{n_2(n_2 + 1)}{2} - T_2$$

from our data; the conclusions will be identical. From Table 8.3 we see that the sum of the ranks for observations in sample 1 is $T_1 = 621.5$. Therefore,

$$U = n_1 n_2 + \frac{n_1(n_1 + 1)}{2} - T_1$$

$$= 25(25) + \frac{25(26)}{2} - 621.5$$

$$= 328.5$$

Since the computed value of z,

$$z = \frac{U - \mu}{\sigma_U} = \frac{328.5 - 312.5}{51.54} = .31$$

does not exceed 1.96, we conclude there is insufficient evidence to reject the hypothesis of no difference between Catholic priests and Methodist ministers with respect to their knowledge about causes of mental illness.

The three test procedures that we have discussed so far in Chapter 8 have all assumed that an independent sample is drawn from each of the two populations of interest in an experiment prior to running a statistical test. We shall now consider an experimental situation where *pairing* of the observations from the two samples will result in more information to the experimenter. This procedure illustrates the usefulness of designing an experiment.

EXERCISES

10 What are the assumptions that should be met before performing a Mann–Whitney U test?

11 A sample of $n_1 = 11$ groups of student activists and a sample of $n_2 = 11$ groups of nonactivists were randomly selected to determine if there is a difference between the two kinds of groups in the efficient use of time to complete the drafting of a student-government constitution. Each group was rated on a 10-point scale, where a low score denotes inefficient use of time. Test the null hypothesis of no difference between the two populations using the Mann–Whitney U test. (Use $\alpha = .05$.)

Groups of Student Activists	*Groups of Student Nonactivists*
1	5
2	7
4	8
6	9
5	2
7	6
2	7
3	8
2	6
4	6
8	5

12 A conference was held for all governmental personnel managers to make them aware of the hiring inequities of the government toward several minority groups (including women). Random samples of $n_1 = 20$ women and $n_2 = 20$ men were selected from the conference participants and each person sampled was asked to rate the success of the conference on a 1 to 10 point scale (high scores denote success). These data are recorded in the table. Use the Mann–Whitney U test to

Females		Males	
1	4	7	5
3	4 .	9	10
4	6	8	6
3	3	5	1
6	0	10	2
8	1	9	3
9	6	10	9
7	4	6	7
8	5	5	4
1	6	2	2

test the null hypothesis of no difference in perception of success for the populations of male and female personnel managers. (Use $\alpha = .05$.)

8.5   THE PAIRED-DIFFERENCE EXPERIMENT: AN EXAMPLE OF A DESIGNED EXPERIMENT

*T*he design of an experiment is essentially a plan for purchasing a specified quantity of information and, as you might suspect, we hope to do this at the lowest possible cost. The amount of information available in a sample to make an inference about a population parameter can be measured by the bound on the error of estimation that could be constructed from the sample data. The smaller the bound on the error, the more we know about the parameter of interest. Since the bounds on the errors of estimation for most commonly used estimators are dependent on the population standard deviation σ and the sample size n, the experimenter can control the quantity of information contained in the sample by determining the number of observations to include in his sample and by controlling the variation in the data. This is an oversimplification of the problems encountered in designing an experiment, but it summarizes the essential points. Recall from Section 6.3 that for n greater than or equal to 30 a bound on the error of estimation for a population mean is $2\sigma/\sqrt{n}$. Thus we obtain more information (a reduction in the bound on error) by either increasing the sample size n, or decreasing the population standard deviation σ.

A strong similarity exists between the audio theory of communication and the theory of statistics. Both are concerned with the transmission of a message (signal) from one point to another, and consequently both are theories of information. For example, the telephone engineer is responsible for transmitting a verbal message that might originate in New York City and be received in New Orleans. Equivalently, a speaker may wish to communicate with a large and noisy audience. If static or background noise is sizable for either example, the receiver may acquire only a sample of the complete signal and from this partial information must infer the nature of the complete message. Similarly, scientific experimentation is conducted to verify certain theories about natural phenomena. Sometimes, we want simply to explore some aspect of nature and hopefully to deduce, with a good approximation, the relationships of certain natural variables. One might think of experimentation as the communication between nature and the scientist. The message about the natural phenomenon is contained, in garbled form, in the experimenter's sample data. Imperfections in his measuring instruments, nonhomogeneity of experimental material, and many other factors contribute background noise (or static) that tends to obscure nature's signal and cause the observed response to vary in a random manner. For both the communications engineer and the statistician, two factors affect the quantity of information in an experiment, namely the magnitude of the background noise (variation) and the volume of the signal (size of the sample). The greater the noise (variation), the less information will be contained in the sample. On the other hand, the louder the signal (large-sample size), the greater the amplification will be and hence the more likely the message will penetrate the noise and be received.

The design of experiments is a very broad subject concerned with methods of sampling that will reduce the variation in an experiment, amplify nature's signal and thereby acquire a *specified quantity* of information at minimum cost. Despite the complexity of the subject, some of the important considerations in the design of good experiments can be easily understood and should be presented to the beginner. We shall illustrate the concept of noise reduction by use of the paired-difference experiment.

We observed that two factors affect the quantity of information in an experiment pertinent to making inferences about a population parameter: the variation in the population (as measured by σ) and the size of the sample. Experimental designs may be classified as either *noise reducers* or *volume increasers*, depending on whether the primary effect on the quantity of information in an experiment is to reduce the variation (σ) in the data or increase the sample.

The paired-difference experiment is an example of a noise-reducing (variation-reducing) design. Suppose that we wish to test the difference between the mean monthly salaries for men and women sales representatives. One way to select the sample would be to randomly select 30 male and 30 female sales representatives at random from a large list of companies. Then the salary of each person would be recorded.

You might detect a disadvantage to this sampling design because each company has a different mean salary level for their sales representatives. The salaries for representatives would vary greatly from company to company, making the respective sample variances s_1^2 and s_2^2 large. Recall that for both samples of size 30 or more the decision maker for the equality of two means μ_1 and μ_2 is

$$ z = \frac{\overline{X}_1 - \overline{X}_2}{\sqrt{\sigma_1^2/n_1 + \sigma_2^2/n_2}} $$

and we reject if the observed value of z exceeds $z_{\alpha/2}$ or is less than $-z_{\alpha/2}$. For almost all examples, σ_1^2 and σ_2^2 will be unknown and we shall estimate them using s_1^2 and s_2^2.

If the sample variances s_1^2 and s_2^2 are large, it takes large differences in the sample means, $\overline{X}_1$ and $\overline{X}_2$, to reject the null hypothesis and declare a difference; hence the company-to-company variability, which inflates the sample variances, makes it difficult to detect differences in the mean monthly salaries $(\mu_1 - \mu_2)$, and thereby reduces the quantity of information in the experiment.

We can improve on the design just mentioned by reducing the variability in the sample data. To do this we *filter out* the variability due to companies by making male–female comparisons within each company. We could select one female and one male sales representative from each company and measure the difference in monthly salaries (see Table 8.4). This process would be repeated for each of the companies and we would estimate the difference in mean income, $(\mu_1 - \mu_2)$, by averaging the salary differences.

The results of such a salary survey are given in Table 8.4. The monthly salaries for one male and one female (X_1 and X_2) selected from each company are shown in the second and third columns of the table. The difference in salary for each pair, denoted by $d = X_2 - X_1$, is recorded in the fourth column. The sample averages for X_1, X_2, and d are shown at the bottom of the table. Note that $\overline{d} = \overline{X}_2 - \overline{X}_1$.

We would like to use these data to determine if there is sufficient evidence to indicate a difference in the mean monthly salaries for males and females, $(\mu_1 - \mu_2)$.

At first glance we might be tempted to employ the methods of comparing two means in Section 8.2; however, this would not be the appropriate analysis because one of the assumptions of the previous test procedure has been

Table 8.4   *Monthly Salaries (in dollars)*

Company	Female X_1	Male X_2	Difference $d = X_2 - X_1$
1	730	820	90
2	690	770	80
3	750	790	40
4	620	680	60
5	710	840	130
6	780	790	10
7	610	650	40
8	580	590	10
9	540	540	00
10	630	610	−20
11	750	780	30
12	730	790	60
13	700	710	10
14	690	750	60
15	590	620	30
16	750	770	20
17	800	760	−40
18	710	790	80
19	640	680	40
20	530	550	20
21	550	650	100
22	770	820	50
23	750	760	10
24	790	810	20
25	700	770	70
26	620	730	110
27	680	790	110
28	570	560	−10
29	690	720	30
30	670	750	80
	$\bar{X}_1 = 677.33$	$\bar{X}_2 = 721.33$	$\bar{d} = 44.00$

violated. The two samples are not independent because the pairs of observations are linked. They read high or low, depending upon the company from which the male–female pair was drawn. Hence the analysis for an experiment is dictated by the design used. If the design employed was the one obtained by random selection of 30 male and 30 female sales representatives, disregarding their company affiliation, we would analyze the data using the methods of Section 8.2. But the experimenter realized that salaries varied greatly from company to company, so he could filter or block out this variability if he made a male–female comparison within each company. The restricted random assignment (Table 8.4) dictates that we must perform another analysis to determine whether the mean monthly salaries differ for males and females. The appropriate method is to compare the 30 difference measurements given in Table 8.4.

The proper analysis utilizes the 30 difference measurements to test the hypothesis that the mean difference between males and females equals zero. This is equivalent to testing the hypothesis that the difference in the means for males and females, $(\mu_2 - \mu_1)$, is equal to zero. The testing procedure will be identical to that presented in Section 7.4 for a test of an hypothesis concerning a population mean μ. We wish to test the null hypothesis $\mu_d = 0$, where μ_d denotes the mean difference, $(\mu_1 - \mu_2)$. Elements of the test are as follows:

Paired-Difference Test

Null Hypothesis: $\mu_d = (\mu_1 - \mu_2) = 0$.
Alternative Hypothesis: $\mu_d \neq 0$ *(for a two-tailed test).*
Test Statistic:

$$z = \frac{\bar{d}}{\sigma_d / \sqrt{n}}$$

Rejection Region: **Reject the null hypothesis if z is larger than** $z_{\alpha/2}$ **or less than** $-z_{\alpha/2}$.

(Note: We can replace σ_d by s_d, the sample standard deviation of the differences d for n equal to 30 or more.)

The notation is slightly different from that used in Section 7.4, but the meaning is the same. We use the symbol $\bar{d}$ to denote the average of sample differences. We observe differences $d_1, d_2, \ldots, d_n$ and compute

$$\bar{d} = \frac{\Sigma d}{n}$$

This is equivalent to observing n sample measurements $X_1, X_2, \ldots$, and computing

$$\bar{X} = \frac{\Sigma X}{n}$$

We state one word of warning: n refers to the number of sample differences (or number of pairs) rather than the total number of measurements and s_d to the standard deviation of the differences.

$$s_d = \frac{\Sigma d^2 - (\Sigma d)^2/n}{n-1}$$

Now consider the data in Table 8.4 and test the null hypothesis that $\mu_d = 0$ against the alternative that μ_d is not equal to zero. (Use $\alpha = .05$.) We must first compute $\bar{d}$ and s_d. For these data

$$\Sigma d = 1320$$

and

$$\Sigma d^2 = 107{,}800$$

Hence for $n = 30$ differences

$$\bar{d} = \frac{\Sigma d}{n} = \frac{1320}{30} = 44.00$$

$$s_d^2 = \frac{\Sigma d^2 - (\Sigma d)^2/n}{n-1} = \frac{107{,}800 - (1320)^2/30}{29}$$

$$= \frac{107{,}800 - 58{,}080}{29} = 1714.48$$

and

$$s_d = \sqrt{1714.48} = 41.41$$

Then

$$z = \frac{\bar{d}}{s_d/\sqrt{n}} = \frac{44.00}{41.41/\sqrt{30}} = 5.82$$

For $\alpha = .05$, $z = 5.82$ exceeds $z_{\alpha/2} = 1.96$ and we reject the null hypothesis of equality of mean salaries for male and female sales representatives. In fact, it appears that males possess a higher mean monthly salary than females.

The statistical design employed in this experiment represents a simple example of a randomized block design. The resulting statistical test is often called a paired-difference test. Several points should be emphasized. First, pairing of the measurements occurred when the study was planned. Comparisons of monthly salaries between males and females were then made within each corporation to eliminate the variability between corporations. Second, pairing will not always provide more information for testing the difference between two population means. If there were actually no differences between the salary structures of corporations, we lose information because instead of having 60 sample measurements (one on each of 30 males and females), we have only 30 differences. This makes it more difficult to detect a difference in means when it exists if no corporation-to-corporation variability is present. We therefore recommend pairing only if we can filter out undesirable background noise.

Planned use of paired observations to control background variability is not the only way we may make use of the paired-difference experiment. Sometimes experiments are conducted where pairs occur naturally such as where we make an observation on an object at two different points in time.

Example 8.6   *The murder rates from a sample of 30 Southern SMSA's are recorded for 1960 and 1970. Use the data in Table 8.5 to test the null hypothesis that the mean murder rates for 1960 and 1970 are identical. (Use $\alpha = .01$.)*

Solution   *Using the summary data in Table 8.5 we have $\bar{d} = 4.07$ and $s_d = 3.86$. The null hypothesis is*

$$H_0 : \mu_d = 0$$

which we shall test against the alternative

$$H_a : \mu_d \neq 0$$

The decision maker is

$$z = \frac{\bar{d}}{s_d/\sqrt{n}} = \frac{4.07}{3.86/\sqrt{30}}$$

$$= 5.78$$

For $\alpha = .01$, $z = 5.77$ exceeds $z_{\alpha/2} = 2.58$ and we reject the hypothesis of no difference in the mean murder rates. We conclude that a difference exists between the mean murder rates from 1960 to 1970 in the South; in fact, it appears that the rate has increased over this period of time.

In each of the two paired-difference examples there were 30 sample differences. Hence we could employ the methods of Section 7.4 to test the hypothesis $\mu_d = 0$. What procedure can we use if the number of differences is

Table 8.5 *Southern SMSA's Murder Rates, 1960 and 1970*

SMSA	1960 x_1	1970 x_2	Diff. $d = x_2 - x_1$
Atlanta, Ga.	10.1	20.4	10.3
Augusta, Ga.	10.6	22.1	11.5
Baton Rouge, La.	8.2	10.2	2.0
Beaumont, Tex.	4.9	9.8	4.9
Birmingham, Ala.	11.5	13.7	2.2
Charlotte, N.C.	17.3	24.7	7.4
Chattanooga, Tenn.	12.4	15.4	3.0
Columbia, S.C.	11.1	12.7	1.6
Corpus Christi, Tex.	8.6	13.3	4.7
Dallas, Tex.	10.0	18.4	8.4
El Paso, Tex.	4.4	3.9	−.5
Fort Lauderdale, Fla.	13.0	14.0	1.0
Greensboro, N.C.	9.3	11.1	1.8
Houston, Tex.	11.7	16.9	5.2
Jackson, Miss.	9.1	16.2	7.1
Knoxville, Tenn.	7.9	8.2	.3
Lexington, Ky.	4.5	12.6	8.1
Lynchburg, Va.	8.1	17.8	9.7
Macon, Ga.	17.7	13.1	−4.6
Miami, Fla.	11.0	15.6	4.6
Monroe, La.	10.8	14.7	3.9
Nashville, Tenn.	12.5	12.6	.1
Newport News, Va.	8.9	7.9	−1.0
Orlando, Fla.	4.4	11.2	6.8
Richmond, Va.	6.4	14.9	8.5
Roanoke, Va.	3.8	10.5	6.7
Shreveport, La.	14.2	15.3	1.1
Washington, D.C.	6.6	11.4	4.8
Wichita Falls, Tex.	6.2	5.5	−.7
Wilmington, Del.	3.3	6.6	3.3

$$\overline{X}_1 = 9.28 \qquad \overline{X}_2 = 13.36 \qquad \overline{d} = 4.07$$
$$s_d = 3.86$$

less than 30? Two possibilities exist: we can use a small-sample test that will be formulated in Section 8.7, or a nonparametric test, the sign test. We shall consider the latter alternative first.

13 In doing a paired-difference test, can the pairing take place after the data are obtained? Explain.

14 Why would a researcher want to use the paired-difference test?

15 Use the murder rates for Western cities for 1970 (see Table 3.4, page 60). From your library, locate the *Uniform Crime Reports* that contain the 1960 information. Test the null hypothesis that $\mu_d = 0$ for Western cities between 1960 and 1970. (Use $\alpha = .05$.)

16 Compare your answer of Exercise 15 with the solution contained in the text for Southern cities. Can you provide any possible explanations or interpretations of these findings?

8.6 A SIMPLE COMPARATIVE
TEST: THE SIGN TEST

hy use the sign test to make a comparison between two populations? First, some studies yield responses that are hard to quantify at the interval level. Examples are ranking and evaluating power, prestige, social differentiation, social control, or alienation: for example, ranking the moods of individuals who for the first time have been placed on welfare and evaluating the performance of the faculty within a large history department. Both would be difficult to quantify. The sign test works particularly well for these types of data because we do not need to know the exact value of each measurement, only whether one is larger or smaller than the other. Second, the sign test is easy to conduct. Third, we need make no assumptions about the form of the population probability distributions. For example, we need not assume that the populations are normal or mound-shaped.

The sign test is a procedure for testing whether two populations have identical probability distributions. Since we shall make no assumptions concerning the parameters of the distributions, we refer to the sign test as a *nonparametric* statistical test.

Suppose that independent random samples of seven measurements each are selected from two different populations, 1 and 2. They are shown in Table 8.6. The sign test is based on the difference in pairs of observations, one from each sample. If the pairs are identified as those in the same row of Table 8.6, they would match as shown in the first column. Next note the difference

Table 8.6   *Two Independent Random Samples of $n_1 = n_2$ Measurements*

Pair	Population 1 X_1	Population 2 X_2	Sign of $X_2 - X_1$
1	10.2	10.3	+
2	10.1	10.0	−
3	10.3	10.2	−
4	10.4	10.2	−
5	10.3	10.0	−
6	10.2	10.1	−
7	10.2	10.0	−

between X_1 and X_2 for each pair, showing the sign of the difference in the fourth column. If X_2 is greater than X_1, we show a plus sign; otherwise, a minus sign.

The sign test of the null hypothesis, "there is no difference in the probability distribution for the two populations," utilizes the number of plus signs, X, as a decision maker. If the null hypothesis is true, the probability that X_1 is greater than X_2 is $p = .5$ for any pair. That is, there is a 50:50 chance that X_1 will be greater than X_2 or vice versa. Testing the null hypothesis that the distributions are identical is equivalent to testing the hypothesis that a binomial parameter p is equal to .5. This test was discussed in detail for large samples in Section 7.3.

In previous chapters we have used the normal approximation to the binomial if both np and nq are 10 or more. But because the normal distribution provides a good approximation to the binomial probability distribution for small samples when $p = .5$, we can employ the empirical rule to determine the rejection region for the sign test. This technique will be satisfactory for samples of size 10 or more and provides a crude test procedure for n as small as $n = 5$. For samples between $n = 5$ and $n = 9$, α will not equal .05 but will vary from approximately .01 to .07. For $n = 10$ or larger, α will be fairly close to .05. We shall illustrate the test procedure with an example.

Example 8.7   *Each of 20 young mothers was asked to compare two different approaches, A and B, to socializing their young children. After two weeks of employing each approach the mothers were asked to grade their satisfaction with each approach on an interval scale going from 0 to 5. The results are given in Table 8.7. Determine if there is evidence to indicate that the two populations of ratings differ and hence whether approach A is preferred to approach B or vice versa. (Use $\alpha = .05$.)*

Table 8.7

Housewife	Approach A X_1	Approach B X_2	Sign of $X_2 - X_1$
1	3	2	−
2	4	2	−
3	3	5	+
4	5	4	−
5	4	3	−
6	3	2	−
7	3	4	+
8	4	3	−
9	3	2	−
10	4	2	−
11	5	4	−
12	3	4	+
13	2	1	−
14	3	2	−
15	5	3	−
16	5	4	−
17	5	3	−
18	2	3	+
19	4	2	−
20	4	3	−

Solution   *Consider the pair of observations for each mother and let X be the number of times approach B has a higher score (satisfaction) than A. The null hypothesis is that the two approaches are equally preferred (and hence the distributions of scores are identical) or, equivalently, the probability that X_2 exceeds X_1 for any pair is p = .5. The alternative is that the two distributions are different; that is, p is greater than or less than .5.*

The test statistic is $z = (X - \mu)/\sigma$, where

$$\mu = np_0 = 20(.5) = 10$$
$$\sigma = \sqrt{np_0q_0} = \sqrt{20(.5)(.5)} \approx 2.24$$

For $\alpha = .05$ we shall reject the null hypothesis if z is greater than 1.96 or less than −1.96. Note that we are using a two-tailed test (as shown in Figure 8.4) because we wish to detect values of p that are either greater or smaller than p = .5.

The observed value of z is

$$z = \frac{X - \mu}{\sigma} = \frac{4 - 10}{2.24} = -2.7$$

Figure 8.4 *Rejection region for the sign test; n = 20, p = .5*

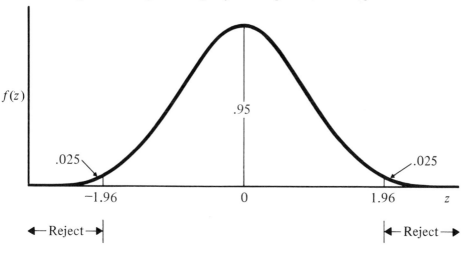

Since this value is less than − 1.96, z falls in the rejection region and we have evidence to indicate that the degree of satisfaction expressed by mothers differs between approaches A and B. Further examination of the data suggests that mothers prefer approach A.

Example 8.8 *A group of businessmen were invited by the local mayor to attend a three-day working conference on the need to improve job opportunities for Spanish-speaking Americans. Prior to the conference, each of a random sample of 9 businessmen was asked how he felt about the need to improve job opportunities. Possible responses were: No need, some need, or great need. After the conference, each of the sampled businessmen again was asked the same question. These results are listed in Table 8.8.*

Table 8.8 *Responses by Businessmen in the Job Opportunity Survey*

Businessman	Before	After	Sign of Change
1	No need	Some need	+
2	Some need	Great need	+
3	No need	Great need	+
4	Some need	Great need	+
5	Some need	No need	−
6	No need	Some need	+
7	Some need	Great need	+
8	No need	Great need	+
9	No need	Some need	+

What the researcher wishes to show, that is, his research hypothesis, is that the conference increases the proportion of businessmen who show greater concern for a need to improve job opportunities. Test the null hypothesis that, after the conference, there is no change in businessmen's judgments about the need to improve job opportunities for Spanish-speaking Americans against the alternative that the proportion has increased. Note that we only wish to detect an increase (not decrease) in the proportion showing greater concern to determine whether the conference achieved its aims. We shall run a one-tailed test of the null hypothesis of "no change" against the alternative that there was an improvement. (Use α = .05.)

Solution *If we let p be the probability that a businessman will change his response to a more sympathetic stand toward the need to improve job opportunities, our test procedure is as follows:*

$$H_0: p = .5$$
$$H_a: p \text{ is greater than } .5 \qquad (i.e., p > .5)$$

Test Statistic: $z = (X - \mu)/\sigma$ *where* $\mu = np_0$ *and* $\sigma = \sqrt{np_0 q_0}$.

Rejection Region: For α = .05 *we shall reject the null hypothesis if* $z > 1.645$.

For this example,

$$\mu = np_0 = 9\,(.5) = 4.5$$

$$\sigma = \sqrt{np_0 q_0} = \sqrt{9\,(.5)(.5)} = 1.5$$

and

$$z = \frac{X - \mu}{\sigma} = \frac{8 - 4.5}{1.5} = 2.33$$

Since the observed value of z, z = 2.33, exceeds the tabulated value, z = 1.645, we reject the null hypothesis that businessmen's judgments would not change and conclude that they become more sympathetic to the Spanish-speaking-American's problem.

One problem occasionally encountered when we use the sign test is that ties occur. That is, for one or more pairs, X_1 may equal X_2. When this happens, delete the pair(s). For example, if you have 20 pairs and one of these results in a tie, you delete the tied pair and work with the remaining $n = 19$.

A summary of the elements of the sign test is presented next:

Sign Test for Comparing Two Populations

Null Hypothesis: *The probability distributions for populations A and B are identical (i.e., p = .5).*

Alternative Hypothesis: *For a two-tailed test, the probability distributions are not identical (i.e., p ≠ .5).*

Test Statistic: $z = (X - \mu)/\sigma$, *where X equals the number of pairs for which* X_2 *is greater than* X_1 *(omit ties),* $\mu = np_0$, *and* $\sigma = \sqrt{np_0q_0}$.

Rejection Region: *Reject the null hypothesis if z is larger than* $z_{\alpha/2}$ *or less than* $- z_{\alpha/2}$. *We can select any value of α we desire, but generally we adopt α = .05 or .01. These* $z_{\alpha/2}$ *values are 1.96 and 2.58, respectively.*

To summarize, the sign test is very useful because it can be applied to ordinal data if comparisons between observations from the two samples are made in pairs.

It can also be used in comparing interval or ratio data based on independent random samples containing the same number of observations. (In this case, observations from one sample are randomly paired with those of the other. The test procedure is unchanged from that described above.) The advantages of the sign test are that it makes no assumptions concerning the nature of the underlying populations, it can be applied to small samples, and, most important, it is very easy to apply.

The disadvantage of the sign test is instinctively clear. Since we do not use the exact value of the measurements X_1 and X_2, we lose information. Consequently, the sign test based on small samples may not detect a difference between two populations when other tests do. Although this is a serious disadvantage, it does not detract from the value of the sign test as a rapid test to detect a difference between two populations.

The second method for working with small samples in the paired-difference experiment utilizes the general approach to small-sample parametric test procedures. These small-sample results are presented in Section 8.7.

EXERCISES

17   Use the sign test on the data in Table 8.5 (page 289). (Use α = .05.)

18   Corporate executives were invited to participate in a conference on "Racism and Sexism, a Two-Edged Sword." The purpose of the conference was to familiarize

the participants with employment problems in each of these areas. Ten participants were interviewed before and after the conference and rated on a basis of 1 to 10 concerning their acceptance of the hiring of blacks and women in their corporations (1 implies least acceptance, 10 most). The data are given in the table. Use the sign test to test the hypothesis that the conference produced no change in the attitudes of executives. (Use $\alpha = .05$.)

Executive	Before	After
1	6	10
2	5	7
3	2	1
4	6	8
5	4	5
6	4	8
7	8	8
8	9	10
9	6	7
10	2	5

19 Two psychiatrists were asked to rate each of 12 prison inmates concerning their rehabilitative potential. Do the data shown suggest a difference in the rating scales employed by the two psychiatrists? (Use $\alpha = .05$.)

Inmate	Psychiatrist 1	2
1	6	5
2	12	11
3	3	4
4	9	10
5	5	2
6	8	6
7	1	2
8	9	12
9	6	5
10	7	4
11	6	6
12	9	8

20 A comparison of rates of alcoholism for white collar versus blue collar workers shows the white collar rate exceeding the blue collar rate in 14 of 20 randomly

selected cities. Does this result provide sufficient evidence to indicate that city rates of alcoholism are greater for white than for blue collar workers (for the populations associated with the set of all cities in the United States). (Use $\alpha = .05$.)

8.7 A SMALL-SAMPLE TEST CONCERNING A POPULATION MEAN

You will recall that the decision to reject an hypothesis concerning μ (Section 7.4) was based on the distance that $\overline{X}$ departed from the hypothesized value of μ. Too large a distance meant that $\overline{X}$ was too many standard deviations away from μ, based on the probability distribution of $\overline{X}$.

As noted in previous sections, an easy way to measure the distance between the observed value of $\overline{X}$ and the hypothesized value of μ is to use the z statistic.

The quantity

$$z = \frac{\text{distance}}{\text{standard deviation of } \overline{X}} = \frac{\overline{X} - \mu}{\sigma/\sqrt{n}}$$

expresses the distance in units of $\sigma/\sqrt{n}$, the standard deviation of the probability distribution of $\overline{X}$. If $z = 2$, it means that $\overline{X}$ lies two standard deviations above the hypothesized mean. Similarly, $z = -1$ indicates that $\overline{X}$ is only one standard deviation below μ. Using $\overline{X}$ or z as the decision maker leads us to the same conclusion; z merely counts the number of standard deviations that $\overline{X}$ lies away from μ.

This large-sample method for testing an hypothesis about μ is inappropriate for samples containing fewer than 30 measurements unless prior information is available concerning the value of σ. Since we must sometimes use small samples because of cost, time, or other restrictions, we need a small-sample procedure to test an hypothesis about μ. To illustrate, consider the following example.

A pharmaceutical firm has been conducting restricted studies on small groups of people to determine the effectiveness of a measles vaccine. The following measurements are readings on the antibody strength for five individuals injected with the vaccine.

1.2   2.5   1.9

3.0   2.4

Use the sample data to test the hypothesis that the mean antibody strength for individuals vaccinated with the new drug is 1.6.

Problems of a similar nature were encountered by experimenters early in the twentieth century when the only available test statistic was

$$z = \frac{\overline{X} - \mu}{\sigma/\sqrt{n}}$$

Faced with a small sample randomly selected from a normal population, they computed s to approximate σ and substituted it into the formula for z. Although aware that their procedure might be invalid, they took the only course open to them. That is, they used the z test to test μ.

Many researchers of that day were quite concerned that their statistical tests were leading to incorrect decisions. One of these, W. S. Gosset, translated his curiosity into action. Gosset was a chemist for Guiness Breweries and, as you might suspect, he was only provided small samples for use in tests of quality. He was faced with the problem of finding the probability distribution for

$$\frac{\overline{X} - \mu}{s/\sqrt{n}}$$

a small-sample z statistic with s substituted for σ. Gosset found that this decision maker, which he called t, possessed a probability distribution similar in appearance to the normal but with a much wider spread. In fact, the smaller the sample size n, the greater the spread in the probability distribution for t.

Figure 8.5   *Standard normal z and the t distribution for n = 6 measurements*

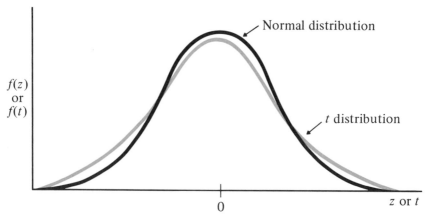

The standard normal distribution and a t distribution based on a sample of six measurements are shown in Figure 8.5.

Gosset published his work on the t distribution in 1908 under the pen name Student, because of company policy. His publication included the exact form of the distribution as well as the tail-end values of t which are helpful in locating rejection regions. Gosset's statistic has many other applications in statistical decision making and has achieved a position of major importance in the field of statistics. His unique choice of a pen name has caused the statistic to be called Student's t.

The t distribution possesses the following characteristics:

1   It, like z, is mound-shaped and symmetrical about $t = 0$ (see Figure 8.6).
2   It is more variable than z since both $\overline{X}$ and s change from sample to sample drawn from a population.
3   There are many t distributions. We determine a particular one by specifying a parameter, called the degrees of freedom (d.f.), that is directly related to the sample size, n.
4   As the sample size, n (or d.f.), gets large, the t distribution becomes a standard normal (z) distribution. Intuitively this is reasonable since s provides a better estimate of σ as n increases.

The values of t used to locate the rejection region for a statistical test are presented in Table 2 of the Appendix. Since the t distribution is symmetrical about $t = 0$, we give only right-tail values. A value in the left tail is simply the negative of the corresponding right-tail value. An entry in the table that corresponds to a certain number of degrees of freedom specifies a value of t,

Figure 8.6   *Use of the t tables*

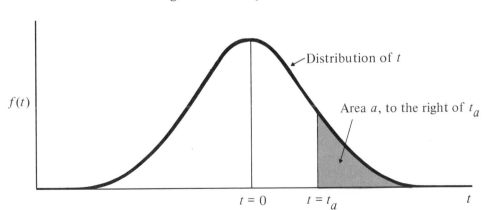

$f(t)$

Distribution of t

Area a, to the right of t_a

$t = 0$ $t = t_a$ t

say t_a, such that an area a lies to its right (see the shaded portion of Figure 8.6). We shall indicate how to determine the degrees of freedom for a specific problem, but first we shall consider an example.

Example 8.9   *Return to the pharmaceutical data on the measle vaccine and suppose that the regulations require that the mean antibody strength for vaccines exceeds 1.6 before it can be placed on the market. Test the research hypothesis that the firm's vaccine is marketable, using the following five measurements recorded for five people injected with the vaccine: 1.2, 3.0, 2.5, 2.4, and 1.9. (Use α = .05.)*

Solution   *The null hypothesis for this example should be*

$$H_0 : \mu = 1.6$$

since we wish to establish the hypothesis

$$H_a : \mu > 1.6$$

The decision maker is

$$t = \frac{\bar{X} - \mu}{s/\sqrt{n}}$$

Before computing t, we must first obtain $\bar{X}$ and s for the sample data. We find

$$\Sigma X = (1.2) + (3.0) + (2.5) + (2.4) + (1.9) = 11.00$$

and

$$\Sigma X^2 = (1.2)^2 + (3.0)^2 + (2.5)^2 + (2.4)^2 + (1.9)^2 = 26.06$$

Hence

$$\bar{X} = \frac{\Sigma X}{n} = \frac{11.0}{5} = 2.2$$

$$s^2 = \frac{\Sigma X^2 - (\Sigma X)^2/n}{n-1} = \frac{26.06 - (11)^2/5}{4}$$

$$= \frac{26.06 - 24.20}{4} = .465$$

and

$$s = \sqrt{.465} = .682$$

Substituting into the decision maker, we calculate

$$t = \frac{2.2 - 1.6}{.682/\sqrt{5}} = \frac{.6(2.236)}{.682}$$

$$= 1.97$$

To locate the rejection region for our test statistic we must specify the degrees of freedom for our test. We simply state that d.f. = n − 1 for a small-sample test concerning μ. For a one-tailed test with probability of type I error, α = .05, we can locate the rejection region from Table 2 of the Appendix for a = .05 and d.f. = 4. The critical value of t is 2.132 (see Figure 8.7). Since the observed value of t does not fall in the rejection region, we have insufficient evidence to reject the null hypothesis that the mean antibody strength is 1.6.

Figure 8.7   *Rejection region for α = .05 and d.f. = 4*

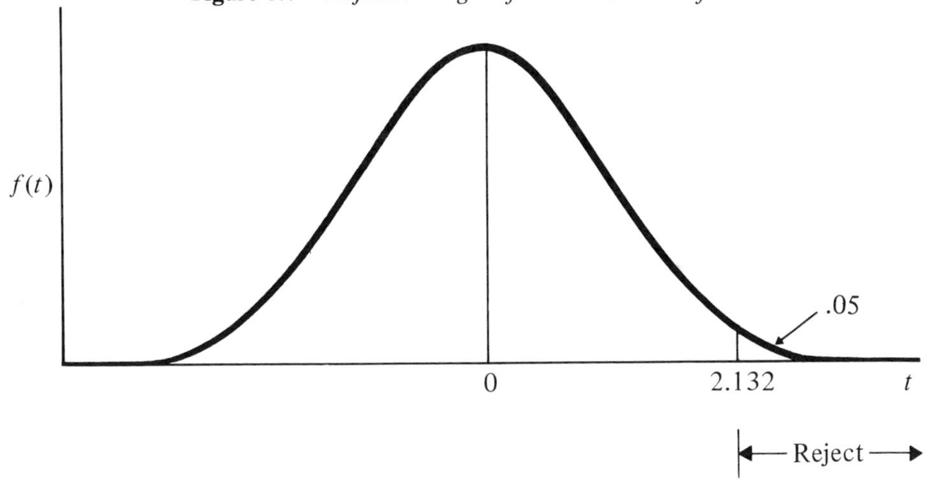

A summary of the elements of the Student's *t* test for *μ* is as follows:

Student's *t* Test for μ

Null Hypothesis: $\mu = \mu_0$ *(where μ_0 is specified).*

Alternative Hypothesis: $\mu \neq \mu_0$ *(for a two-tailed test).*

Test Statistic:

$$t = \frac{\overline{X} - \mu_0}{s/\sqrt{n}}$$

> *Rejection Region: Reject if the absolute value of the computed t is greater than the tabulated value. The tabulated value will be based on $(n - 1)$ degrees of freedom with $a = \alpha/2$.*

A similar test can be constructed for comparing the means of two normal populations based on paired observations. This test accomplishes the same inferential objective of the sign test, Section 8.6, but utilizes more information. If the underlying populations are normal (or near normal!), the use of ratio data and the following test procedure is more likely than the sign test to detect a difference in population means if it really exists.

You will recall that in conducting a paired difference test of $H_0 : \mu_d = 0$ for large samples (n_1 and n_2 both 30 or more) we made use of the corresponding large-sample test concerning a population mean, μ. Similarly the paired-difference test for small samples is identical to the small-sample test concerning μ. We can summarize our results for a small-sample (the number of differences, n, less than 30) paired-difference test as follows:

Small-Sample Paired-Difference Test

Null Hypothesis: $\mu_d = 0$ *(i.e.,* $\mu_1 - \mu_2 = 0$*).*

Alternative Hypothesis: $\mu_d \neq 0$ *(two-tailed test).*

Decision Maker:

$$t = \frac{\bar{d} - 0}{s_d / \sqrt{n}}$$

Rejection Region: Reject if the absolute value of t is greater than the tabulated value of t for $a = \alpha/2$ and $(n - 1)$ degrees of freedom.

(Note: n is the number of differences.)

Example 8.10 *Psychologists wish to examine the amount of learning exhibited by schizophrenics after taking a specified dose of a tranquilizer. Ten schizophrenics are randomly selected from a patient ward and before the drug is administered each is given a standard examination. The amount of time required to complete the exam is recorded for each individual. One hour after the specified drug dose is administered, each patient is given the same standard exam as previously. Again completion times are recorded (see Table 8.9). If these patients exhibit any learning we*

Table 8.9 *Before and After Dosage Completion Times (in minutes)*

Patient	Before X_1	After X_2	Difference $d = X_2 - X_1$
1	10	8	-2
2	15	13	-2
3	30	29	-1
4	29	25	-4
5	26	21	-5
6	28	28	0
7	19	15	-4
8	13	10	-3
9	14	12	-2
10	21	17	-4

would expect the mean completion time prior to receiving the drug to be more than that for 1 hour after receiving the drug. This is our research hypothesis. Use the data in Table 8.9 to test the null hypothesis that the mean difference for before and after dosing test scores is zero (i.e., $H_0: \mu_d = 0$) against the observation $H_a: \mu_d < 0$.

Solution   *Utilizing Table 8.9 we have*

$$\Sigma d = -27$$

and

$$\Sigma d^2 = 95$$

Hence

$$\bar{d} = \frac{\Sigma d}{n} = \frac{-27}{10} = -2.7$$

$$s_d^2 = \frac{\Sigma d^2 - (\Sigma d)^2/n}{n-1} = \frac{95 - (27)^2/10}{9}$$

$$= \frac{22.1}{9} = 2.46$$

and

$$s_d = 1.57$$

The decision maker is

$$t = \frac{\bar{d} - \mu_d}{s_d/\sqrt{n}}$$

and we shall reject the null hypothesis for large negative values of t (or, equivalently, d). The critical value of t for $\alpha = .05$ *is 1.833, which is found in Table 2 of the Appendix, with a* $= .05$ *and d.f.* $= 9$. *Substituting in t we have*

$$t = \frac{-2.7}{1.57/\sqrt{10}} = \frac{-2.7(3.16)}{1.57} = -5.43$$

Since the absolute value of t exceeds the critical value 1.833, we reject the null hypothesis and conclude that learning has occurred.

EXERCISES

21   Why is z usually an inappropriate test statistic when the sample size is small?

22   The voter turnout in upper-middle-class areas of a city has averaged 650 voters for every 1000 registered voters. This year a random sample of five precincts shows turnouts of 635, 655, 640, 643, and 620 per 1000 registered voters. Do these data indicate a final average of less than 650? (Use a one-tailed test with $\alpha = .05$.)

23   In a before-and-after study, a social psychologist reports the following data on eight small groups. Test $H_0 : \mu_d = 0$, $H_a : \mu_d \neq 0$, using $\alpha = .05$.

Groups	Before Instructions	After Instructions
A	8	5
B	10	10
C	7	9
D	9	5
E	5	5
F	9	7
G	13	8
H	10	8

8.8   A SMALL-SAMPLE TEST
OF A DIFFERENCE IN
MEANS

he small-sample test of an hypothesis concerning the difference
in two means is similar to the large-sample test except that we make a few
additional assumptions concerning the nature of the sampled populations.
We shall comment on the restrictive nature of these assumptions at the end of
this section.

As for large samples, the most logical decision maker to test the hypo-
thesis $(\mu_1 - \mu_2) = 0$ is based on the difference in the sample means, $(\overline{X}_1 - \overline{X}_2)$,
based on independent random samples of n_1 and n_2 observations, respectively,
from the two populations. If the sample difference is too far away from
$(\mu_1 - \mu_2) = 0$, we reject the null hypothesis. How far is too far? The answer
is: roughly more than two standard deviations of the probability distribution
for $\overline{X}_1 - \overline{X}_2$).

If we assume that both populations are normal and $\sigma_1^2 = \sigma_2^2 = \sigma^2$,
that is, both populations possess roughly the same amount of variation, we
can construct a Student's t statistic for a decision maker:

$$t = \frac{\overline{X}_1 - \overline{X}_2}{s\sqrt{1/n_1 + 1/n_2}}$$

measures the distance $(\overline{X}_1 - \overline{X}_2)$ differs from $(\mu_1 - \mu_2) = 0$ and expresses
it in units of the estimated standard deviation of the difference, $(\overline{X}_1 - \overline{X}_2)$.
This estimated standard deviation of $(\overline{X}_1 - \overline{X}_2)$, often called the estimated
standard error of the difference between two means, is

$$s\sqrt{\frac{1}{n_1} + \frac{1}{n_2}}$$

The common population standard deviation, σ, can be estimated by
using the individual sample variances to generate a pooled sample estimate s,

$$s = \sqrt{\frac{(n_1 - 1)s_1^2 + (n_2 - 1)s_2^2}{n_1 + n_2 - 2}}$$

The rejection region for the t test is selected in the same manner as for
the test of an hypothesis about a single mean, Section 8.7. For a two-tailed
test with $\alpha = .05$, we look up t_a in Table 2 corresponding to $a = .025$ and

$(n_1 + n_2 - 2)$ degrees of freedom (d.f.). For a one-tailed test, $\alpha = .05$, read the tabulated t value corresponding to $a = .05$ with $(n_1 + n_2 - 2)$ d.f. We shall illustrate the use of the t test with an example.

Example 8.11 *A study was conducted to investigate the effect of two diets on the weight gain of 14-year-old children suffering from malnutrition. Ten children were subjected to diet I and nine to diet II. The gains in weight over a 9-month period are shown in Table 8.10. Use the data in Table 8.10 to determine if there is evidence to indicate a difference between the mean gain in weight for children fed on the two diets.*

Table 8.10   *Gains in Weight for Malnutritioned Children*

Diet I	Diet II
14.0	14.4
12.5	18.2
10.2	19.5
9.8	21.2
10.5	15.3
11.2	11.6
15.0	12.8
22.0	13.1
13.0	11.3
9.6	

Solution *First we assume that weight gains for the two diets are normally distributed with means μ_1 and μ_2 and common unknown variance σ^2. We wish to test the hypothesis*

$$\text{null hypothesis: } (\mu_1 - \mu_2) = 0$$

against the alternative that μ_1 and μ_2 are different. Before calculating the value of t, we need to obtain $\overline{X}_1, \overline{X}_2, s_1^2$, and s_2^2. The necessary calculations are shown in Table 8.11. The estimate of σ is then

$$s = \sqrt{\frac{(n_1 - 1)s_1^2 + (n_2 - 1)s_2^2}{n_1 + n_2 - 2}}$$

$$= \sqrt{\frac{9(13.877) + 8(12.805)}{10 + 9 - 2}} = \sqrt{\frac{124.893 + 102.440}{17}}$$

$$= \sqrt{13.37}$$

Table 8.11 *Computations for Example 8.11*

Diet A	Diet B
$\Sigma X_1 = 127.80$	$\Sigma X_2 = 137.40$
$\Sigma X_1^2 = 1758.18$	$\Sigma X_2^2 = 2200.08$
$n_1 = 10$	$n_2 = 9$
$\bar{X}_1 = \dfrac{\Sigma X_1}{n_1}$	$\bar{X}_2 = \dfrac{\Sigma X_2}{n_2}$
$= \dfrac{127.80}{10} = 12.78$	$= \dfrac{137.40}{9} = 15.27$
$s_1^2 = \dfrac{\Sigma X_1^2 - (\Sigma X_1)^2/n_1}{n_1 - 1}$	$s_2^2 = \dfrac{\Sigma X_2^2 - (\Sigma X_2)^2/n_2}{n_2 - 1}$
$= \dfrac{1758.18 - (127.8)^2/10}{9}$	$= \dfrac{2200.08 - (137.40)^2/9}{8}$
$= 13.877$	$= 12.805$

and

$$s = \sqrt{13.37} = 3.66$$

The value of the test statistic, t, for this test is

$$t = \frac{\bar{X}_1 - \bar{X}_2}{s\sqrt{1/n_1 + 1/n_2}} = \frac{12.78 - 15.27}{3.66\sqrt{\frac{1}{10} + \frac{1}{9}}}$$

$$= \frac{-2.49}{3.66(.459)} = -1.48$$

The rejection region for $\alpha = .05$ *utilizes a t value corresponding to* $a = .025$ *and* $(n_1 + n_2 - 2) = (10 + 9 - 2) = 17$ *d.f. We reject if the computed value of t is greater than 2.110 or less than −2.110 (see Table 2). This rejection region is shown in Figure 8.8. Noting that the computed value of t, t = −1.48, does not fall in the rejection region, we conclude that there is insufficient evidence to indicate a difference in the mean weight gains for the two diets.*

Before concluding our discussion of the t test, we would wonder whether the assumptions of normal populations and equal variances ($\sigma_1^2 = \sigma_2^2$)

Figure 8.8   *Rejection region for t; α = .05 and d.f. = 17*

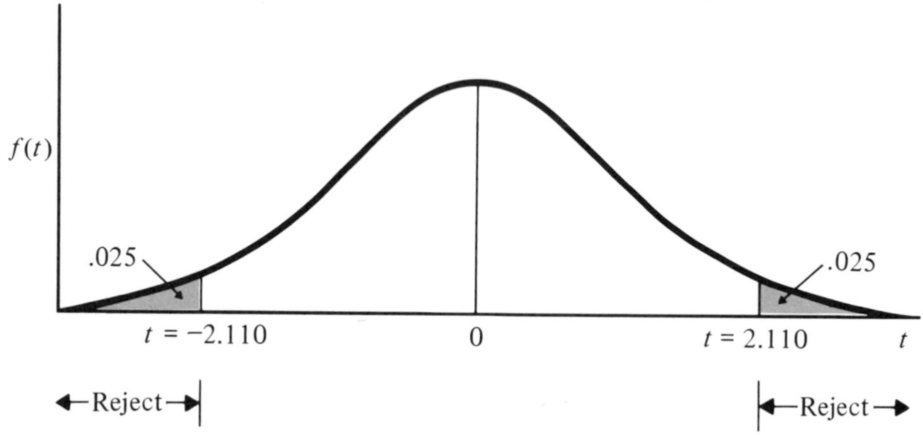

must hold in order for the *t* test to be valid. The test functions satisfactorily for populations possessing mound-shaped probability distributions. The assumption that $\sigma_1^2 = \sigma_2^2$ is more critical but again it does not seriously affect the properties of the test if n_1 and n_2 are approximately equal. Consequently, the assumptions are not too restrictive so that the test has wide applicability. A summary of the elements of the *t* test for comparing two means is shown below:

The *t* Test for Comparing Two Means

Null Hypothesis: $(\mu_1 - \mu_2) = 0$.

Alternative Hypothesis: $\mu_1 \neq \mu_2$ *(for a two-tailed test).*

Test Statistic:

$$t = \frac{\overline{X}_1 - \overline{X}_2}{s\sqrt{1/n_1 + 1/n_2}}$$

where

$$s = \sqrt{\frac{(n_1 - 1)s_1^2 + (n_2 - 1)s_2^2}{n_1 + n_2 - 2}}$$

Rejection Region: **Reject if t is greater than** t_a **or less than** $-t_a$, **where** $a = \alpha/2$ **and t is based on** $(n_1 + n_2 - 2)$ **d.f.**

---

24   When utilizing the *t* distribution in making small-sample inferences concerning the difference in population means, what assumptions are made about the populations from which the independent random samples are selected?

25   Four sets of identical twins (pairs *A*, *B*, *C*, and *D*) were selected at random from a population of identical twins. One child was taken at random from each pair to form an experimental group. These four children were sent to school. The other four children were kept at home as a control group. At the end of the school year the IQ scores listed were obtained. Does this evidence justify the conclusion that lack of school experience has a depressing effect on IQ scores? (Use a one-tailed test with $\alpha = .05$.)

Pair	*Experimental Group*	*Control Group*
A	110	111
B	125	120
C	139	128
D	142	135

26   Sixteen white, racially prejudiced, married couples were selected with eight randomly assigned to live in a highly integrated neighborhood and the other eight in a highly segregated (black) neighborhood. After the first 12 months had passed, the racial attitudes of the couples were compared, treating husband and wife as a unit. Assume interval-level data and that high scores reflect favorable attitudes toward minority-group members. Do the data present sufficient evidence to indicate a difference in the mean racial attitude for the two types of neighborhoods? (Use $\alpha = .05$.)

Racial Attitudes	
Segregated Neighborhood	*Integrated Neighborhood*
1	4
3	2
2	3
1	3
2	1
1	2
3	3
2	3

C 8.9 SUMMARY

hapter 8 presents statistical tests for comparing two population means (or two proportions) based on sample means (or proportions).
This chapter is concerned with the following questions: If two sample means
differ, say $\overline{X}_1 = 8.2$ and $\overline{X}_2 = 7.1$, does this imply that the corresponding
population means μ_1 and μ_2 differ? Similarly, if two sample proportions
$\hat{p}_1$ and $\hat{p}_2$ differ, do we have sufficient evidence to indicate a difference in the
corresponding population proportions? Or, rather than seek evidence of a
difference in two population parameters, we might wish to estimate the difference using a confidence interval or simply a point estimate with its associated
bound on the error of estimation. To summarize, we have learned how to
make inferences about the differences in population means or proportions
and, in doing so, have provided, along with the inference, a measure of its
goodness.

Statistical tests for comparing two population means or two proportions, based on independent (and relatively large) random samples from the
two populations, were presented, along with associated methods of estimation,
in Sections 8.2 and 8.3. These techniques were applicable only for interval or
ratio data. Similar tests and estimation procedures appropriate for small
samples, based on the Student's t statistic, were presented in Section 8.8.
A test for comparing two populations, suitable for ordinal (as well as interval
and ratio) data, utilized the Mann–Whitney U statistic of Section 8.4.

The paired-difference experiment, designed to increase information in
our data, makes comparisons between pairs of measurements, one each
selected from the two populations. If the variability of the difference of
observations *within* a pair is less than the variability of observations between
pairs, then the quantity of information in the sample will be greater than if
you had selected two independent random samples from the population.
For example, if you wished to determine the effect of two sociological conditions on humans, you could obtain more information per person by observing
the effect of these conditions on pairs of human twins. This type of comparison
could reduce variation by using pairs of genetically similar humans. The
observed differences in the X values, computed for each pair, could be analyzed using the Student's t statistic, which is appropriate for interval or ratio
data; in contrast, the sign test can be used for ordinal as well as interval and
ratio data.

Finally, note the utility of the sign test for comparing data contained
in any two samples, whether the samples have been collected in an independent random manner or whether they occur as a result of a paired-difference

experiment. The sign test may not be the best test to use (that is, it may not utilize all the available information in the data), but it is applicable to ordinal data and is easy and rapid to conduct.

QUESTIONS AND PROBLEMS

27 Seventy-two college students enrolled in French 101 were randomly divided into two equal groups and subjected to one of two teaching techniques: class 1: instructor spoke in both English and French to students; class 2: instructor spoke only in French to students. The mean and variances for the sample achievement test scores for the two groups were as given. State an appropriate research hypothesis that implies a two-tailed test. State the appropriate null hypothesis. Test the null hypothesis using $\alpha = .05$.

Group 1	Group 2
$n_1 = 36$	$n_2 = 36$
$\overline{X}_1 = 260$	$\overline{X}_2 = 294$
$s_1^2 = 3600$	$s_2^2 = 4300$

28 Based on two samples, each containing 1000 persons, a researcher reports that the average life expectancy (mean) is 65.8 for lower-class persons (sample A) and 69.4 for upper-class persons (sample B). If the standard deviations are 12.61 for sample A and 9.33 for sample B would you accept the argument that upper-class people, in general, can expect to live longer? What evidence would you provide to buttress your argument?

29 The lengths of time (in hours) that it took to complete a work assignment were recorded for two samples of 30 persons, one sample drawn from those working in plants with high morale, the other from those working in plants with low morale. Do the data present sufficient evidence to indicate a difference in the mean time to complete the work assignment of groups in two different environments? (Use $\alpha = .05$.)

High-Morale Atmosphere	*Low-Morale Atmosphere*
Mean = 115	Mean = 137
Variance = 595	Variance = 420

30 Two nonadjacent residential areas had homes randomly selected for tax purposes based on their marketability. The mean prices were the same. One year later, area *A* had changed its composition so that it was now highly integrated, whereas area *B* was basically all white. Two new random samples of homes, each containing 50 homes, were selected for evaluation. The sample means and variances for prices are shown in the table. Do the data provide sufficient evidence to indicate a difference in the mean market values of the homes in these two areas? Explain. (Use $\alpha = .05$.)

Area A	*Area B*
Mean = $29,840	Mean = $27,520
Variance = $25,000,000	Variance = $19,400,000

31 The percentage of D's and F's awarded by two college professors was noted by the student paper. Professor Smith achieved a rate of 43 percent as opposed to 30 percent for Professor Jones, based upon 200 and 180 students, respectively. Do the data indicate a difference in the rate of awarding D's and F's? (Use $\alpha = .01$.) How confident would you be in your conclusions? What are some of the explanations for your results?

32 The proportion of voters favoring Nixon in the rural area of a county is 48 percent versus 35 percent around the university district, based on randomly selected samples of 300 voters from each area. Do the data indicate a difference in the proportion of voters favoring Nixon in the two areas? (Use $\alpha = .01$.) How might you explain your findings?

33 When would you prefer to use the Mann–Whitney *U* test as opposed to the *z* or *t* test?

34 Use the data in Table 8.10 (page 306) and test the null hypothesis by using the Mann–Whitney *U*. Assume, for the sake of this example, that the distributions of weights are not normally distributed.

35 Ten sets of identical twins were separated at the age of 15. One of each set was sent abroad to study for 3 years while his or her identical twin remained in the United States. At the end of the 3-year-period, identical twin pairs were measured for the degree to which they were ethnocentric in their outlook. Low scores represent low ethnocentrism. Test the hypothesis of "no difference in mean scores," using the sign test. The research hypothesis states that individuals educated abroad are less ethnocentric.

Identical Twin Pair	Lived Abroad	Remained Home
1	12	19
2	14	13
3	8	6
4	11	24
5	14	12
6	12	15
7	13	10
8	15	18
9	18	21
10	17	22

36 A social scientist is interested in comparing social adjustment scores for prisoners of war who have participated in two rehabilitative programs. Fifteen returnees were randomly selected from each of the two programs and scored 6 months after the end of the program. The data are as given. Do the data present sufficient evidence to indicate a difference in mean scores for the two rehabilitative programs?

(a) Use a sign test with $\alpha = .05$.

(b) Use a Student's t test with $\alpha = .05$.

\multicolumn{2}{c}{Programs}	
I	*II*
65	67
87	88
94	76
74	94
52	31
42	48
55	99
91	98
50	80
65	40
37	80
46	91
30	74
63	94
82	42

37 A study was conducted to compare the average number of years of service at the age of retirement for military personnel, 1960 versus 1973. A random sample of 100 career records, for each of these two years, showed average lengths of service of 30.3 and 25.2, respectively. If the population standard deviations are both approximately equal to 4, do the data present sufficient evidence to indicate a difference in the mean length of service for 1960 versus 1973? (Use $\alpha = .01$.)

38 Discuss the differences in the assumptions required for the use of z and t.

39 Discuss the characteristics of the t distribution.

40 What assumptions are made when using a Student's t test for testing an hypothesis concerning μ?

REFERENCES

Blalock, H. M. *Social Statistics*, 2nd ed. New York: McGraw-Hill Book Company, 1972. Chapter 13.

Champion, D. J. *Basic Statistics for Social Research*. Scranton, Pa.: Chandler Publishing Company, 1970. Chapter 7.

Mendenhall, W., and L. Ott. *Understanding Statistics*. North Scituate, Mass.: Duxbury Press, 1972. Chapters 8 and 9.

Mueller, J. H., K. F. Schuessler, and H. L. Costner. *Statistical Reasoning in Sociology*, 2nd ed. Boston: Houghton Mifflin Company, 1970. Chapter 14.

Palumbo, D. J. *Statistics in Political and Behavioral Science*. New York: Appleton-Century-Crofts, 1969. Chapter 7.

Siegel, S. *Nonparametric Statistics for the Behavioral Sciences*. New York: McGraw-Hill Book Company, 1956. Chapter 5.

9

CROSS CLASSIFICATION OF BIVARIATE DATA

9.1 INTRODUCTION

We learned in Chapter 2 that data generated by a single variable —nominal, ordinal, interval, or ratio data—can be grouped into classes for descriptive or inferential purposes. We might classify murder rates for a collection of U.S. cities, enabling us to visualize the distribution of rates and permitting us to calculate median and mean murder rates and various other numerical descriptive measures. Very often this is not enough.

Research in the social sciences often requires the study and classification of more than one variable at a time. We are not only interested in the classification of a single variable, say murder rates, but we are interested also in the relationship between this variable and a second variable, say geographical location.

Recall that the object upon which a measurement is taken is called a sampling unit. If measurements are taken on two (or more) variables for each sampling unit, we say that we have bivariate (or multivariate) data. Then, just as univariate data are classified in a one-way table, bivariate data are classified using a two-way table. What is the objective of such a classification? Sometimes we might just wish to know whether the two variables are related, as a matter of intellectual curiosity, or, as frequently happens, we might wish to predict one variable based on knowledge of the other. This chapter deals primarily with tests of independence based on data arranged in a two-way contingency table. The tables are so called because the research hypothesis will be that a dependence (contingency) exists between the two variables. The techniques described are appropriate for all types of data.

9.2 DESCRIBING CROSS-CLASSIFICATION DATA: PERCENTAGE COMPARISONS

Udry, Bauman, and Chase studied the relationship between two qualitative variables, male skin color and job mobility, and reported their results in the *American Journal of Sociology*. A sample of 349 males was classified by mobility orientation (low, medium, or high) and skin color (light, medium, or dark). These results are presented in Table 9.1. A cell entry is the number of males classified in that particular category.

Table 9.1 *Relationship Between Male Skin Color and Job Mobility Orientation*

Job Mobility Orientation	Male Skin Color			Total
	Light	Medium	Dark	
High	35	84	51	170
Medium	49	78	23	150
Low	10	13	6	29
Total	94	175	80	349

SOURCE: J. Richard Udry, Karl E. Bauman, and Charles Chase, "Skin Color, Status, and Mate Selection," *American Journal of Sociology*, 76 (Jan. 1971), p. 728; by permission.

In studying the relationship between two variables it is sometimes useful to present the raw data (Table 9.1) as percentages. At this point we have a choice. We can either present the cell frequencies as a percentage of the row totals or the column totals. The accepted procedure is to base the percentages on the marginal totals of the independent (or causal) variable. Each category of the independent variable then provides a base sample size (n) from which we can compute percentages. In situations where either variable can be viewed as the independent variable, we can make two separate percentage comparisons, one with each variable as the independent variable.

A percentage comparison of the data of Table 9.1 can be made using either of the qualitative variables as the independent variable. In Table 9.2, with skin color treated as the independent variable, we base all percentages on the column totals. For example, there were 94 of the original sample of

Table 9.2 *Use of Skin Color as the Independent Variable by Percentages*

Job Mobility Orientation	Male Skin Color		
	Light	Medium	Dark
High	37.2	48.0	63.8
Medium	52.1	44.6	28.8
Low	10.6	7.4	7.5
Total	99.9	100.0	100.1
	$n = 94$	$n = 175$	$n = 80$

349 with light skin color. Of these 94, 35 (or 35(100)/94 = 37.2 percent) were classified as having high mobility orientation, 49 (or 49(100)/94 = 52.1 percent) as having medium mobility orientation, and 10 (or 10(100)/94 = 10.6 percent) with low mobility orientation. Percentages for the other columns of Table 9.2 were computed in a similar way.

Observation of the results in Table 9.2 leads to one obvious trend: the darker the male skin color, the higher the mobility orientation. For example, while only 37.2 percent of the light-skin-color group were classified as having high mobility orientation, 63.8 percent of the dark-skin-color group were labeled with high mobility orientation.

A similar percentage comparison can be made treating mobility orientation as the independent variable.

Table 9.3 *Mobility Orientation as the Independent Variable, by Percentage*

Job Mobility Orientation	*Male Skin Color*			*Total*
	Light	*Medium*	*Dark*	
High	20.6	49.4	30.0	100.0 ($n = 170$)
Medium	32.7	52.0	15.3	100.0 ($n = 150$)
Low	34.5	44.8	20.7	100.0 ($n = 29$)

Entries in Table 9.3 were computed using the row totals of Table 9.1 as the base. For example, from a total of 170 of the original 349 males classified as having high mobility orientation, 35 (or 35(100)/170 = 20.6 percent) were light-skinned, 84 (or 84(100)/170 = 49.4 percent) were medium-skinned, and 51 (or 51(100)/170 = 30.0 percent) were dark-skinned.

The data of Table 9.3 reveal less in the way of trends than Table 9.2, and it appears that the distribution of male skin color is approximately the same for all three mobility-orientation groups.

Percentage comparisons need not be limited to qualitative variables. We could compare two qualitative variables, one qualitative and one quantitative variable, or two quantitative variables. In Table 9.4, we have categorized a sample of 80 SMSA's according to region (North, South, and West) and murder rates. The variable, geographic regions, is a qualitative variable; the variable murder rate is quantitative.

Any study of murder rates would likely compare the effect of different geographic regions on the distribution of percentages in the murder-rate categories. A percentage comparison of the data, using geographic areas as

Table 9.4   *Murder Rates for 80 SMSA's Classified by Region*

Murder Rate	Geographic Region		
	North	South	West
0–6	1	20	16
7–12	7	5	6
13–18	16	4	3
19–24	2	0	0
Total	26	29	25

the independent variable, will provide a means for comparing the rates across geographic areas. We can readily detect a trend in the percentages of Table 9.5. SMSA's from the North have a different percentage distribution of murder rates than either the South or the West. In particular the percentage of SMSA'S in the North with a murder rate between 13 and 24 is 61.5 + 7.7 = 69.2 percent, whereas the South and West have only 13.8 percent and 12.0 percent, respectively.

Table 9.5   *Percentage Comparison of Murder Rates Using Geographic Regions as the Independent Variable*

Murder Rate	North	South	West
0–6	3.8	69.0	64.0
7–12	26.9	17.2	24.0
13–18	61.5	13.8	12.0
19–24	7.7	0.0	0.0
Total	99.9	100.0	100.0
	n = 26	n = 29	n = 25

Having illustrated percentage comparisons for describing bivariate (cross-classification) data, we offer the following guidelines for future use.

Guidelines for Percentage Comparison of Cross-Classified Data

1   When dealing with qualitative variables, the categories should satisfy the principles of inclusiveness and exclusiveness (Section 3.2).

2   Always include the marginal totals that serve as the sample sizes for the "independent" variable. This practice protects the reader from identifying possible trends when in fact the sample sizes are so small as to render percentage differences meaningless.
3   The number and size of the class interval (especially for qualitative variables) are usually predetermined by the comparisons one wishes to make. In situations where the experimenter has a choice, keep the comparison as simple as possible without destroying the relationship between the variables.
4   For quantitative data use the rules presented for frequency distributions, Section 3.6.

   In this section we have presented a way for describing cross-classification data. Although we tried to identify trends with the percentage comparisons we made, we did not draw specific statistical inferences. Identification of possible trends is useful for illustrative purposes but cannot substitute for a statistical test of significance when a decision with regard to the trend must be made. In Section 9.3 we shall present several tests of significance which are useful for making inferences from cross-classification data.

EXERCISES

1   Employ the data in Table 9.4 (page 320) to make a percentage comparison using murder rates as the independent variable. Compare the results with those in Table 9.5.

2   What are the guidelines for percentage comparisons of cross-classified data?

9.3   A TEST OF SIGNIFICANCE FOR
CROSS-CLASSIFICATION DATA:
THE CHI-SQUARE TEST OF INDEPENDENCE

A problem frequently encountered in the analysis of cross-classification data concerns the independence of two methods of classification of observed events. For example, in trying to determine an advertising strategy for an upcoming campaign, it would be useful to compare the proportion of voters favoring the incumbent councilman in the three different voting districts.

One method of comparison would be to test the independence of the qualitative variable, favor or do not favor the incumbent councilman, and the qualitative variable, voting districts. If the two classifications are dependent, this would imply that the proportion of voters favoring the councilman varies from district to district. Detecting this dependence would aid campaigners in allocating their election resources to obtain the best results.

We shall illustrate a test of the independence of two classifications using the following example: A county sociological survey was conducted to compare the proportion of families with more than two children for different income categories. A random sample of 150 families was questioned and each was classified into one of three income categories (under $7000, $7000 to $14,000, or over $14,000). The number of children was also recorded for each family. Using the data below, determine if there is sufficient evidence to indicate that the proportion of families having more than two children differs among the three income categories. In other words, is the classification "number of children per family" dependent upon the classification "annual income"? The results of the survey are presented in Table 9.6.

Table 9.6 *Summary of the Results for the Survey Relating Annual Income to the Number of Children*

Number of Children per Family	Annual Income			Total
	Less Than $7000	$7000–$14,000	More Than $14,000	
Two children or fewer	11	13	21	45
More than two children	68	28	9	105
Total	79	41	30	150

To answer the question of independence, we begin by defining two distributions of measurements. The first distribution corresponds to the classification "number of children per family." If we consider all families in the county of interest, a certain fraction or proportion, p_A, have two children or fewer, and the remaining proportion of the families, p_B, have more than two children. Since all families are classified into one of these two categories, $p_A + p_B = 1$. The second distribution corresponds to the annual income classification. If we again consider all families in the sampled county, a certain proportion, p_1, would have annual incomes of less than $7000; a second proportion, p_2, would have incomes in the range $7000 to $14,000; and the remaining proportion, p_3, would have annual incomes greater than $14,000. Again we have $p_1 + p_2 + p_3 = 1$.

It is interesting to note that the proportions for these distributions also represent the probabilities associated with each category in a classification. Thus the probability that a family chosen at random from the county will have an annual income of more than $14,000 is p_3. These probabilities (proportions) are given in Table 9.7. Now consider a combination of the

Table 9.7 *Probabilities Associated with the Classifications "Annual Income" and "Number of Children per Family"*

First classification: number of children per family	
Two children or fewer	p_A
More than two children	p_B
Second classification: annual income	
Less than $7000	p_1
$7000 to $14,000	p_2
More than $14,000	p_3

categories of the table, one category from each classification. The two classifications are independent if, for all combinations of categories, the probability that an item falls into a particular category combination is equal to the product of the respective category probabilities. For our example, the classification "number of children per family" is independent of the classification "annual income" if the cell probabilities for the cross-classified data are

Table 9.8 *Cell Probabilities When the Classifications "Number of Children" and "Annual Income" Are Independent*

Number of Children per Family	*Annual Income*		
	Less Than $7000 p_1	*$7000–$14,000* p_2	*More Than $14,000* p_3
Two children or fewer, p_A	$p_A p_1$	$p_A p_2$	$p_A p_3$
More than two children, p_B	$p_B p_1$	$p_B p_2$	$p_B p_3$

as given in Table 9.8. Note that the probability associated with each cell is found by multiplying the probabilities for the categories which are combined in that cell. The probability that a family chosen at random has more than two children and an annual income of more than $14,000 is the product $p_B p_3$, where p_B is the probability of having more than two children and p_3 is the probability of a family earning more than $14,000.

Table 9.9 *Observed and Expected Cell Counts*

Number of Children per Family	Annual Income			Total
	Less Than $7000	$7000–$14,000	More Than $14,000	
Two children or fewer	11 (23.7)	13 (12.3)	21 (9.0)	45
More than two children	68 (55.3)	28 (28.7)	9 (21.0)	105
Total	79	41	30	150

A test of the null hypothesis of independence of the number of children and annual income classifications makes use of Table 9.9. The observed cell counts are shown in the table. In addition we have given the *expected cell counts* (in parentheses), that is, the cell counts that we expect to obtain if the null hypothesis of independence of the two classifications is true. Without proof we state that the expected cell counts can be computed very simply as follows:

(2 children or fewer, less than $7000)
$$\text{expected cell count} = \frac{45(79)}{150} = 23.7$$

(2 children or fewer, $7000 to $14,000)
$$\text{expected cell count} = \frac{45(41)}{150} = 12.3$$

(2 children or fewer, more than $14,000)
$$\text{expected cell count} = \frac{45(30)}{150} = 9.0$$

(more than 2 children, less than $7000)
$$\text{expected cell count} = \frac{105(79)}{150} = 55.3$$

(more than 2 children, $7000 to $14,000)
$$\text{expected cell count} = \frac{105(41)}{150} = 28.7$$

(more than 2 children, more than $14,000)
$$\text{expected cell count} = \frac{105(30)}{150} = 21.0$$

Note that all we do to compute the expected cell count for a particular cell of Table 9.9 is to multiply the row total by the column total and divide by the total number of measurements in the sample.

We would suspect that the null hypothesis is false if the observed cell counts do not agree with the expected cell counts. To measure this agreement or disagreement we compute the quantity chi square,

$$\chi^2 = \sum \frac{(O - E)^2}{E}$$

That is, we subtract the expected cell count E from each observed cell count O. The square of this difference is then divided by E. We do this for all cells and add our results. Obviously, if the observed cell counts differ from the expected cell counts, the quantities $(O - E)^2$ will be large, and hence χ^2 will be large. Thus we shall reject our hypothesis of independence of the two classifications for large values of χ^2. How large is large? The answer to this question can be found by obtaining the distribution of χ^2 in many repetitions of an experiment or by examining the probability distribution of χ^2.

It can be shown that when n is large, χ^2 has a probability distribution in repeated sampling which is approximately a chi-square probability

Figure 9.1 *Chi-square probability distribution with 4 degrees of freedom*

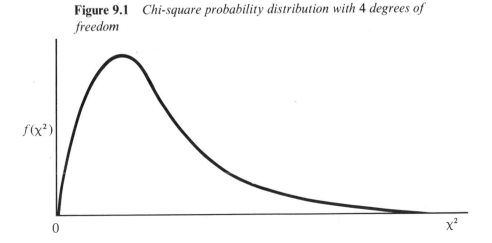

distribution. The requirement of a large n is satisfied if for each cell in the contingency table, the expected cell count is equal to or greater than 5. We shall not give the formula for the chi-square probability density but will characterize it with the following properties.

1 Chi square is not a symmetrical distribution (see Figure 9.1).
2 There are many chi-square probability distributions. We obtain a particular one by specifying the degrees of freedom associated with the chi-square distribution.

The chi-square test statistic for testing the hypothesis of independence of two classifications, one with r rows, the other with c columns, will be based on $(r - 1)(c - 1)$ degrees of freedom. Upper-tail values of

$$\chi^2 = \sum \frac{(O - E)^2}{E}$$

are shown in Table 3 of the Appendix. Entries in the table are chi-square values such that an area of size a lies to the right under the curve. We specify the degrees of freedom (d.f.) in the left-hand column of the table, and the specified value of a is given in the top row of the table. Thus for 14 d.f. and $a = .10$, the critical value of the chi-square distribution is 21.0642 (see Figure 9.2).

The rejection region for the one-tailed test of independence can then be determined for a specified probability of a type I error α by use of Table 3.

Figure 9.2 *Critical value of the chi-square distribution with d.f. = 14 and a = .10*

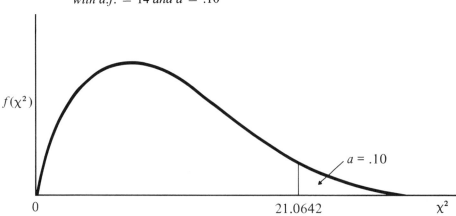

If the observed value of χ^2 falls beyond the critical value, we reject the null hypothesis of independence of the two classifications.

Chi-Square Test of Independence

Null Hypothesis: **The two classifications are independent.**

Alternative Hypothesis: **The two classifications are dependent.**

Test Statistic:

$$\chi^2 = \sum \frac{(O - E)^2}{E}$$

Rejection Region: **Reject the null hypothesis if χ^2 exceeds the tabulated value of χ^2 for $a = \alpha$ and $d.f. = (r - 1)(c - 1)$.**

$r = $ *number of rows in the cross-classification table.*

$c = $ *number of columns in the cross-classification table.*

The computation of

$$\chi^2 = \sum \frac{(O - E)^2}{E}$$

is sometimes simplified using an equivalent expression.

Computing Formula for χ^2

$$\chi^2 = \sum \left(\frac{O^2}{E} \right) - n$$

We shall illustrate the chi-square test and the use of the above computing formula with an example.

Example 9.1   *Use the sample count data in Table 9.9 to test the hypothesis of independence of the two classifications "number of children" and "annual income." Use $\alpha = .05$.*

Solution *The observed and expected cell counts are presented in Table 9.9. The value of the test statistic χ^2 will be computed and compared with the critical value $(a = .05)$ of the chi-square distribution possessing d.f. $= (r - 1)(c - 1) = 1(2) = 2$ degrees of freedom. To find the value of the test statistic, we use the formula*

$$\chi^2 = \sum \left(\frac{O^2}{E}\right) - n$$

where $n = 210$. Substituting, we obtain

$$\chi^2 = \frac{(11)^2}{23.7} + \frac{(13)^2}{12.3} + \frac{(21)^2}{9} + \frac{(68)^2}{55.3} + \frac{(28)^2}{28.7} + \frac{(9)^2}{21} - 150$$

$$= 5.10 + 13.74 + 49.00 + 83.62 + 27.32 + 3.86 - 150$$

$$= 182.64 - 150 = 32.64$$

The rejection region for this test can be obtained from Table 3 of the Appendix for d.f. $= 2$ and $a = .05$. This value is 5.99. Since the observed value of χ^2 exceeds the critical value 5.99, we conclude that the classifications "number of children" and "annual income" are dependent. That is, the distribution of families in the categories "two children or fewer" and "more than two children" depends upon the income category of interest.
It should be noted that the original formula

$$\chi^2 = \sum \frac{(O - E)^2}{E}$$

will give the same value for χ^2. For our data using the above formula

$$\chi^2 = \frac{(11 - 23.7)^2}{23.7} + \frac{(13 - 12.3)^2}{12.3} + \frac{(21 - 9.0)^2}{9.0}$$

$$+ \frac{(68 - 55.3)^2}{55.3} + \frac{(28 - 28.7)^2}{28.7} + \frac{(9 - 21.0)^2}{21.0}$$

$$= 6.80 + .040 + 16.0$$

$$+ 2.92 + .017 + 6.86$$

$$= 32.64$$

This agrees with our computational formula.

Example 9.2 *A poll of* 100 *Congressmen was taken to determine their opinion concerning a bill to raise the ceiling on our national debt. Each Congressman was then classified according to political party affiliation and opinion on the policy. The survey results are listed in Table 9.10.*

Table 9.10 *Congressional Opinion Survey Results*

Party	*Opinion*			*Total*
	Approve of Bill	*Do Not Approve of Bill*	*No Opinion Yet*	
Republican	28	14	5	47
Democrat	19	28	6	53
Total	47	42	11	100

At this point neither the opponents nor proponents can claim victory. Test the null hypothesis that these two classifications are independent of one another against the alternative that a Congressman's opinion in the national debt bill is related to his particular affiliation. (*Use* $\alpha = .05$.)

Solution *Before calculating* χ^2 *we must first obtain the expected cell frequencies. Using the row and column totals of Table 9.10 the expected cell counts are:*

$$\textit{Republican, approve} = \frac{47(47)}{100} = 22.09$$

$$\textit{Republican, disapprove} = \frac{47(42)}{100} = 19.74$$

$$\textit{Republican, no opinion} = \frac{47(11)}{100} = 5.17$$

$$\textit{Democrat, approve} = \frac{53(47)}{100} = 24.91$$

$$\textit{Democrat, disapprove} = \frac{53(42)}{100} = 22.26$$

$$\textit{Democrat, no opinion} = \frac{53(11)}{100} = 5.83$$

Note that all of these expected cell frequencies equal or exceed 5. Hence these data satisfy the assumptions required for the chi-square test. We now compute

$$\chi^2 = \sum \left(\frac{0^2}{E}\right) - n$$

where n = 100.

$$\chi^2 = \frac{(28)^2}{22.09} + \frac{(14)^2}{19.74} + \frac{(5)^2}{5.17} + \frac{(19)^2}{24.91} + \frac{(28)^2}{22.26} + \frac{(6)^2}{5.83} - 100$$

$$= 35.49 + 9.93 + 4.84 + 14.49 + 35.22 + 6.17 - 100$$

$$= 6.14$$

The rejection region for this test can be located using Table 3 with a = .05 and d.f. = (2 − 1)(3 − 1) = 1(2) = 2. This critical value is 5.99. Since the observed value of χ^2 exceeds the critical chi-square value, we reject the null hypothesis of independence of the classification and conclude that Congressmen seem to hold opinions along party lines. The probability of our making a wrong decision using the chi-square test is α = .05. Consequently, we feel fairly confident that congressional opinion is dependent on party affiliation.

The use of the chi-square probability distribution in analyzing cross-classification data can present a problem. Researchers have long recognized that the chi-square distribution may provide a poor approximation to the actual distribution of χ^2 where some or all of the expected cell frequencies are small. One attempt at improving the approximation was offered by F. Yates in 1934 for the 2 × 2 table and has become known as Yates' correction for continuity, or simply Yates' correction. The statistic suggested by Yates is identical to the chi-square test of independence just discussed with the exception that we use the statistic

$$\chi_c^2 = \sum \frac{(|0 - E| - .5)^2}{E}$$

Note that in using χ_c^2 we take the absolute value of all differences $(0 - E)$ and subtract .5. Each of these quantities is then squared and divided by the appropriate expected value, E.

We illustrate its use with an example.

Example 9.3 *John W. M. Whiting and associates compared the duration of the postpartum taboo restricting coitus with the presence or absence of initiation ceremonies at puberty in 56 cultures. Use these data to test the null hypothesis of independence of the two classifications. (Use* $\alpha = .05$.)

Table 9.11 *Data from Whiting's Study*

| Initiation Ceremonies | Duration of the Postpartum Taboo | | Total |
	Up to a Year	One Year or More	
Absent	26	10	36
Present	3	17	20
Total	29	27	56

SOURCE: J. W. M. Whiting, Richard Kluckhohn, and Albert Anthony, "The Function of Male Initiation Ceremonies at Puberty," in Eleanor E. Maccoby, Theodore M. Newcomb, and Eugene L. Hartley (eds.), *Readings in Social Psychology* (New York: Holt Rinehart and Winston, Inc., 1958), p. 365; by permission.

Solution *The expected cell counts for the data of Table 9.11 are given in Table 9.12. In addition we have included the differences $(O - E)$ in parentheses. For the absence of initiation ceremonies and duration of the postpartum taboo up to one year the expected cell count is $29(36)/56 = 18.6$ and the difference $(O - E)$ is*

$$(O - E) = 26 - 18.6 = 7.4$$

The expected cell count, 18.6, and difference $(O - E) = 7.4$ are listed in Table 9.12 as 18.6(7.4). The remaining expected cell counts and differences of Table 9.12 were computed in the same way.

Table 9.12 *Expected Cell Counts and Differences $(O - E)$ for the Data of Table 9.11*

| Initiation Ceremonies | Duration | |
	Up to a Year	One Year or More
Absent	18.6 (7.4)	17.4 (-7.4)
Present	10.4 (-7.4)	9.6 (7.4)

Since two of the expected cell frequencies are near five we will use Yates' correction to test the hypothesis of independence of the classifications. The calculations for this example are quite easy because the absolute value of all differences $(O - E)$ is 7.4. Hence

$$\chi_C^2 = \sum \frac{(|O - E| - .5)^2}{E}$$

$$= \frac{(7.4 - .5)^2}{18.6} + \frac{(7.4 - .5)^2}{17.4}$$

$$+ \frac{(7.4 - .5)^2}{10.4} + \frac{(7.4 - .5)^2}{9.6}$$

$$= 47.61 \left(\frac{1}{18.6} + \frac{1}{17.4} + \frac{1}{10.4} + \frac{1}{9.6} \right)$$

$$= 47.61(.312) = 14.85$$

The rejection region for $\alpha = .05$ *and* $(r - 1)(c - 1) = (2 - 1)(2 - 1) = 1$ *degree of freedom is 3.84. Since the observed value of* χ_C^2 *is greater than 3.84, we have sufficient evidence to indicate that the two classifications are dependent.*

In Chapter 8 we presented a two-sample test for comparing two proportions which assumed that the sample sizes were both 30 or more. In each sample we observed the number of successes in a fixed number of trials. The sample results could then be displayed in the form of Table 9.13. Note that in sample 1 we observed X_1 successes (and $n_1 - X_1$ failures) in n_1 trials. Similarly, for sample 2 we observed X_2 successes in the n_2 trials where we assume that n_1 and n_2 are both 30 or more.

What happens when the sample sizes are not large (30 or more)? We shall consider this problem in Section 9.4.

Table 9.13   *Summary of Data*

	Sample 1	Sample 2	Total
Success	X_1	X_2	$X_1 + X_2$
Failure	$n_1 - X_1$	$n_2 - X_2$	$n_1 + n_2 - X_1 - X_2$
Total	n_1	n_2	$n_1 + n_2$

EXERCISES

3   What do we mean by saying that the chi-square test for contingency tables is a test of independence?

4   What are the assumptions one should meet before computing a chi-square test of independence?

5   A preelection survey was conducted in three different districts to compare the fraction of voters favoring the incumbent governor. Random samples of 50 registered voters were polled in each of the districts. These results are presented in the table. Do these data present sufficient evidence to indicate that the fractions favoring the incumbent governor differ in the three districts? (Use $\alpha = .05$.)

District	1	2	3
Favor incumbent governor	19	14	26
Do not favor incumbent governor	31	36	24

6   Conduct a chi-square test of independence for the data of Table 9.1 (page 318). (Use $\alpha = .01$.) Interpret your results.

9.4   A TEST OF SIGNIFICANCE FOR CROSS-CLASSIFICATION DATA: FISHER'S EXACT TEST

*T*he data of Table 9.13 represent a cross-classification of data by sample and success or failure. *Fisher's exact test* provides a nonparametric test for analyzing count (nominal or ordinal) data when the sample sizes are small (recall that the chi-square test required that all expected cell counts be five or more). We assume that there are two independent random samples and each observation can be classified into one of two categories, which we shall call a success or failure.

We can compute the exact probability of observing X_1 successes in the first sample and X_2 successes in the second using Fisher's exact distribution, often referred to as the *hypergeometric probability distribution*. The exact probability of observing X_1 and X_2 successes from independent random samples of sizes n_1 and n_2, respectively, is given by

$$P(X_1, X_2) = \frac{n_1!\,n_2!\,(X_1 + X_2)!\,(n_1 + n_2 - X_1 - X_2)!}{(n_1 + n_2)!\,X_1!\,X_2!\,(n_1 - X_1)!\,(n_2 - X_2)!}$$

If we use the marginal totals of Table 9.13, we see that $P(X_1, X_2)$ is computed by taking the ratio of the products of the factorials for the four marginal totals to the product of the factorials of the individual cell frequencies and the total sample size, $n_1 + n_2$.

To illustrate the use consider the following example.

Example 9.4   *A sampling of 11 professors teaching "fine arts" in church-related colleges and another of 11 fine arts teachers in state-supported colleges were asked to classify their perception of departmental pressures to publish. The sample data are summarized in Table 9.14.*

Table 9.14   *Comparison of Faculty Perception of the Pressure to Publish*

Amount of Pressure to Publish	Type of College		Total
	Church-Related	State-Supported	
Quite a lot/some	0	4	4
Very little/none	11	7	18
Total	11	11	22

Determine the probability of observing 0 and 4 "successes" in the samples of size 11 and 11, respectively.

Solution   *Utilizing the marginal totals and cell frequencies we have*

$$P(0, 4) = \frac{11!\,11!\,4!\,18!}{22!\,0!\,4!\,11!\,7!}$$

(Recall that $0! = 1$.)

After much simplification and cancellation of terms,

$$P(0, 4) = \frac{11 \cdot 10 \cdot 9 \cdot 8}{22 \cdot 21 \cdot 20 \cdot 19} = .045$$

That is, the probability of observing $X_1 = 0$ and $X_2 = 4$ is .045.

Example 9.4 could have been stated as a statistical test of an hypothesis. It may be important to test the null hypothesis that the proportion of "successes" in population I is identical to the proportion of "successes" in population II against the alternative that one population has a higher fraction of successes. We shall illustrate the test with an example.

Example 9.5 *Suppose that the sample data of Example 9.4 were those given in Table 9.15. Test the hypothesis that the proportions of professors who feel that quite a bit of departmental pressure is brought to bear on publishing for the two types of colleges are identical against the one-sided alternative that there is a higher proportion who feel this way in state-supported colleges. (Use $\alpha = .05$.)*

Table 9.15 *Sample Data for the Faculty Comparison*

Amount of Pressure to Publish	Type of College		Total
	Church-Related	State-Supported	
Quite a lot/some	1	3	4
Very little/none	10	8	18
Total	11	11	22

Solution *Observed values of X_2 greater than X_1 will indicate possible rejection of the null hypothesis; hence we must compute the probability of observing the sample values for X_1 and X_2, but in addition we must compute values of X_1 and X_2 which are more extreme (more unlikely) while maintaining the same marginal total, $X_1 + X_2$. In our example we* *observed $X_1 = 1$, $X_2 = 3$, and $X_1 + X_2 = 4$, for which*

$$P(1, 3) = \frac{11!\,11!\,4!\,18!}{22!\,1!\,3!\,10!\,8!}$$

$$= .248$$

A more extreme set of values, X_1 and X_2, with the same marginal total, $X_1 + X_2 = 4$, would be $X_1 = 0$, $X_2 = 4$. From Example 9.4,

$$P(0, 4) = .045$$

The probability of observing $X_1 = 1$ and $X_2 = 3$ or worse is $P(1, 3) + P(0, 4) = .293$. Since this value is greater than $\alpha = .05$, we have insufficient evidence to reject the null hypothesis that these proportions differ for church-supported and state-supported colleges.

Summary of Fisher's Exact Test (for n_1 and n_2 less than 30)

Null Hypothesis: The fractions of successes in the two populations are identical.

Alternative Hypothesis: The fraction of successes in population II is greater than that for population I (a one-sided test).

Test Statistic: For the marginal total, $X_1 + X_2$, equal to some value k, compute $P(X_1, X_2)$ for the observed number of successes and other more extreme results for which $X_1 + X_2 = k$.

Rejection Region: Reject if the sum of the probabilities computed in the previous step is less than the specified value of α.

(Note: Either population can be labeled as I.)

Fisher's exact test is not always easy to apply, especially if one must compute several of the probabilities $P(X_1, X_2)$. There are tables of critical values which can be helpful in applying this test, but since they are so cumbersome, we omit them from our presentation. If you are faced with a problem requiring several difficult calculations of $P(X_1, X_2)$, you can refer to Siegel (1956) for the necessary critical values.

m
9.5   A TEST OF SIGNIFICANCE FOR CROSS-CLASSIFICATION DATA: THE McNEMAR TEST

any times cross-classification data arise from studies where each of several individuals is sampled at two different points in time to determine if a change in attitude, preference, view point, philosophy, or whatever has occurred between the time of the first and second test period. For example, an advertising firm in charge of the promotional work for a political candidate may sample 100 voters in a precinct to estimate the proportion of voters in favor of the candidate of interest. Then following an extensive advertising campaign, the firm might recontact the same 100 voters to ascertain their preference and, in particular, to see if a change in preference (for better or for worse) has taken place.

To determine if a significant change has taken place over time, we construct a two-way table in which we characterize an individual's response at each test period into one of two categories (see Table 9.16). For lack of better words, we shall call these categories "success" or "failure."

Table 9.16   *McNemar Data Summary*

	After	
Before	*Success*	*Failure*
Failure	Cell 1 (n_1)	Cell 2
Success	Cell 3	Cell 4 (n_4)

If we let n_1 be the number of observations in cell 1 and n_4 be the number of observations in cell 4, the total for cells 1 and 4, $n_1 + n_4$, represents the total number of persons whose response has changed in either direction. Under the null hypothesis that changes from failure to success (cell 1) or success to failure (cell 4) are equally likely, we would expect $(n_1 + n_4)/2$ changes to occur in the two directions. We can utilize this information in testing the significance of any observed changes.

The test statistic or decision maker for this test makes use of this information and again involves the quantities $(O - E)^2/E$. But unlike the chi-square test of independence, we only sum the terms for cells 1 and 4. Also, the expected cell count for both cells 1 and 4 is $(n_1 + n_4)/2$. When the null hypothesis of no change is true, this quantity

$$\chi_M^2 = \sum_{\text{cells 1 and 4}} \frac{(O - E)^2}{E}$$

follows a chi-square distribution with d.f. $= 1$.

A shortcut formula for χ_M^2, which incorporates the expected cell counts, $(n_1 + n_4)/2$ is

$$\chi_M^2 = \frac{(n_1 - n_4)^2}{n_1 + n_4}$$

We shall now summarize the test procedure.

McNemar Test of Change

Null Hypothesis: *A change in attitudes, etc., from failure to success or success to failure is equally likely.*

Alternative Hypothesis: *Changes in one direction are more likely than changes in another.*

Test Statistic:

$$\chi_M^2 = \sum_{\text{cells 1 and 4}} \frac{(O - E)^2}{E} = \frac{(n_1 - n_4)^2}{n_1 + n_4}$$

Rejection Region: *Reject H_0 if χ_M^2 exceeds the critical chi-square value with d.f. $= 1$ and $a = \alpha$.*

Example 9.6 *A random sample of 50 delegates was drawn from the party delegates chosen for the national convention. Prior to the convention 15 of these delegates favored unilateral withdrawal of all U.S. troops stationed in Western Europe as an aftermath of World War II, while 35 were opposed. After several hours of heated debate, the national party adopted a plank for their platform, univocally supporting unilateral withdrawal. The same 50 delegates were then polled again after the convention. Use the data of Table 9.17 to test for a significant change in opinion from the pre- to postconvention polls. (Use $\alpha = .05$.)*

Table 9.17   *Pre- and Postconvention Results*

	Postconvention		
Preconvention	*Against*	*For*	*Total*
For	0	15	15
Against	20	15	35
Total	20	30	50

Solution   *The observed frequencies are given in Table 9.17, and we can compute the value of our test statistic using the shortcut formula with $n_1 = 0$ and $n_4 = 15$. The quantity χ_M^2 is then*

$$\chi_M^2 = \sum_{\text{cells 1 and 4}} \frac{(O - E)^2}{E} = \frac{(n_1 - n_4)^2}{n_1 + n_4}$$

$$= \frac{(0 - 15)^2}{0 + 15} = \frac{225}{15} = 15$$

The critical χ^2 value with $a = .05$ and d.f. $= 1$ is 3.84 (see Table 3 of the Appendix). Since χ_M^2 exceeds 3.84, we reject the null hypothesis and conclude that a significant change of opinion has occurred with the party's convention.

9.6   SUMMARY

Chapter 9 concerns the classification of bivariate data in two-way tables and particularly with tests of an hypothesis of independence between the two variables. You should have the data arranged in a two-way table with the cells showing the frequency appropriate for each. Then, if the expected cell frequency is equal to 5 or more, for all cells, you can use a chi-square test for independence. When the expected cell frequency is less than 5 and you have a 2×2 table, you can use Fisher's test for independence.

For what types of data are the tests appropriate? The tests are appropriate for data measured on any scale of measurement: nominal, ordinal, interval, or ratio. We assume that each pair of bivariate measurements will fall in one and only one cell of the two-way table. It is further assumed that the bivariate data have been selected in a random manner from a fairly large population so that the classification of one bivariate pair of measurements will be unaffected by the classification of another.

Chapter 9 introduces the notion of association between two or more variables. Specific measures of association for nominal and ordinal data will be presented in Chapter 10 and for interval and ratio data in Chapter 11. Remember, the objective is to determine whether a relationship exists between two or more variables and if possible to provide a measure of the strength of that relationship. Prediction of the value of one variable, based on knowledge of another, is frequently the ultimate objective of the researcher.

QUESTIONS AND PROBLEMS

7   What are the objectives in cross-classifying two variables?

8   State the rules for constructing cross-classification tables.

9   The governor of each state was polled to determine his opinion concerning a particular domestic policy issue. At the same time, his party affiliation was recorded. The data are given here. If we assume that the 50 governors represent a random sample of political leaders throughout the nation, do the data present sufficient evidence to indicate a dependence between party affiliation and the opinion expressed on the domestic policy issue? (Use $\alpha = .05$.)

	Approve of Policy	*Do Not Approve*	*No Opinion*
Republican	18	5	5
Democrat	8	8	6

10   The data in Table 9.4 (page 320) show only a few cases of murder rates in the category 19–24. Combine categories 13–18 and 19–24 into one interval, 13–24, and test the hypothesis that murder rates are independent of geographic region. Interpret your findings. (Let $\alpha = .05$.)

11   When is Yates' correction for continuity appropriate?

12   Discuss the kinds of problems appropriate for using each of the following statistical tests.

(a) Chi-square test of independence
(b) Fisher's exact test
(c) McNemar test

13 Use the data given here to test the research hypothesis that physicians' concern for the need to communicate with lower-class people will increase after spending 4 weeks in a general hospital.

Concern Before Experience in General Hospital	After Experience in a General Hospital		
	High	Low	Total
Low	27	5	32
High	9	9	18

14 In a study of small groups, a researcher reported the results contained in the table. Do the data provide sufficient evidence to indicate that completion or noncompletion of a task is related to "type of leadership"? Test an hypothesis of "no relationship" using $\alpha = .05$.

Completion of Task	Group Leadership Style	
	Democratic	Authoritarian
Yes	6	2
No	1	4
Total	7	6

15 Use the data in the table to test the null hypothesis that the proportions of Protestants and Catholics who believe abortion is always permissible are equal. (Use Fisher's exact test with $\alpha = .05$.)

Believe Abortion Is Always Permissible	Religious Background	
	Catholic	Protestant
Yes	1	4
No	7	3

16 A study of changes in belief concerning "the permissibility of abortion at all times" was reported involving 100 Catholics who had undergone extensive study on the moral and ethical ramifications of the issue. Use the McNemar test of change of beliefs before and after the extensive study.

	After Study	
Before	*No*	*Yes*
Yes	0	20
No	50	30

17 If a researcher wished to test the null hypothesis that social class is independent of sex (Exercise 13, Chapter 2, page 31), would he have any difficulties in meeting the assumptions underlying the use of the chi-square test of independence? Which assumption(s)?

18 See Exercise 13, Chapter 2 (page 31). Combine the groups corresponding to "lower" and "lower-middle" and test whether the distribution of people by social class is independent of sex. (Use the chi-square test with $\alpha = .05$.)

19 A study to determine the effectiveness of a drug (serum) for arthritis resulted in the comparison of two groups, each consisting of 200 arthritic patients. One group was inoculated with the serum, the other received a placebo (an inoculation that appears to contain serum but actually is nonactive). After a period of time, each person in the study was asked to state whether his arthritic condition was improved. The results were as given in the table. Do these data present sufficient evidence to indicate that the serum was effective in improving the condition of arthritic patients? Test using chi square and set $\alpha = .05$.

	Treated	*Untreated*
Improved	117	74
Not improved	83	126

20 Convert the raw frequencies in Exercise 19 to proportions and compute the z test for the difference between two proportions. Are the conclusions comparable to those in Exercise 19 if we set $\alpha = .05$? Explain your findings.

21 A group of 306 people were interviewed to determine their opinion concerning a particular current American foreign-policy issue. At the same time their political affiliation was recorded. The data are as given. Do the data present sufficient evidence to indicate a dependence between party affiliation and the opinion expressed for the sampled population? Explain.

	Approve	*Disapprove*	*No Opinion*
Republicans	114	53	17
Democrats	87	27	8

22 A survey of student opinion concerning a resolution presented to the student council was studied to determine whether the resulting opinion was independent of fraternity and sorority affiliation. Two hundred students were interviewed, with the results shown. In a chi-square test of independence, how many degrees of freedom are there?

	Favor	*Opposed*	*Undecided*
Fraternity	37	16	5
Sorority	30	22	8
Unaffiliated	32	44	6

23 Use the data in Exercise 22 to determine if there is sufficient evidence to indicate that student opinion concerning the resolution is independent of status (i.e., fraternity, sorority, and unaffiliated). (Use $\alpha = .05$.)

REFERENCES

Anderson, T. R., and M. Zelditch. *A Basic Course in Statistics*, 2nd ed. New York: Holt, Rinehart and Winston, Inc., 1968. Chapter 12.

Blalock, H. M. *Social Statistics*, 2nd ed. New York: McGraw-Hill Book Company, 1972. Chapter 15.

Champion, D. J. *Basic Statistics for Social Research*. Scranton, Pa.: Chandler Publishing Company, 1970. Chapter 8.

Mendenhall, W., and L. Ott. *Understanding Statistics*. North Scituate, Mass.: Duxbury Press, 1972. Chapter 14.

Palumbo, D. J. *Statistics in Political and Behavioral Science*. New York: Appleton-Century-Crofts, 1969. Chapter 8.

Siegel, S. *Nonparametric Statistics for the Behavioral Sciences*. New York: McGraw-Hill Book Company, 1956. Chapter 6.

Weiss, R. S. *Statistics in Social Research*. New York: John Wiley & Sons, Inc., 1968. Chapter 14.

10

MEASURES OF ASSOCIATION: NOMINAL AND ORDINAL DATA

10.1 INTRODUCTION

Cross classification of data, presented in Chapter 9, is performed to determine whether two variables, representing the two directions of classification, are related. Hence Chapter 9 deals solely with the question of whether the two variables are dependent and presents statistical tests for the null hypothesis "the two variables are independent." The objective is to collect evidence to support the research hypothesis that "the two variables are dependent and therefore related." Chapter 10 carries the study of the relationship between two variables one step further. It is not sufficient to know only that two variables are related. Specifically, we desire measures of the strength of the relationship or association. This is because most often the ultimate goal in the social sciences is to predict the value of one variable based on knowledge of a second variable.

Chapter 10 presents the important concept of reduction in error and several related measures of association. The resultant measures of association, which are available for both nominal and ordinal data, are called proportional reduction in error statistics. These expressions will be explained in the following sections. It should be noted that three additional measures of association useful in the social sciences—the phi coefficient, the contingency coefficient, and Kendall's W—are presented but do not possess a proportional reduction in error interpretation.

10.2 A NOMINAL MEASURE OF ASSOCIATION: THE PHI COEFFICIENT

In Chapter 9 we discussed the use of the chi-square test of independence for cross-classification data. Following a significant chi-square test social scientists may also be interested in the strength of the dependence between two classifications. The phi coefficient, which can be obtained from the decision maker for the chi-square test of independence, is such a measure of the strength of the association between two classifications.

Phi Coefficient: Nominal Measure of Association

$$\phi = \sqrt{\frac{\chi^2}{n}}$$

where

$$\chi^2 = \sum \frac{(O - E)^2}{E}$$

and n is the total sample size.

The phi coefficient is widely used as a measure of association for nominal data cross-classified in a 2 × 2 table (2 rows and 2 columns) following a significant chi-square test of independence.

Example 10.1   *Clark and Larson conducted a study in 1972 titled "Mobility, Productivity, and Inbreeding at Small Colleges: A Comparative Study" which was concerned with the relationship between type of book publisher for productive faculty members and the type of college represented. Using the data of Table 10.1, run a chi-square test of independence and compute ϕ to measure the strength of the association between the two classifications.*

Table 10.1   *Type of Book Publisher for Productive Faculty Members by Type of College*

Type of College	Type of Publisher		Total
	Secular Press	*Religious Press*	
Church-supported	15	15	30
State-supported	25	0	25
Total	40	15	55

SOURCE: Stanley A. Clark and Richard F. Larson, "Mobility, Productivity, and Inbreeding at Small Colleges: A Comparative Study," *Sociology of Education, 45* (Fall 1972), p. 433; by permission.

Solution *Before computing χ^2 we must determine the expected cell frequencies for the data of Table 10.1. Following the procedure of Section 9.3, we obtain*

$$\text{expected cell count (church-supported secular)} = \frac{30(40)}{55} = 21.82$$

$$\text{expected cell count (church-supported religious)} = \frac{30(15)}{55} = 8.18$$

$$\text{expected cell count (state-supported secular)} = \frac{25(40)}{55} = 18.18$$

$$\text{expected cell count (state-supported religious)} = \frac{25(15)}{55} = 6.82$$

The quantity χ^2 is then

$$
\begin{aligned}
\chi^2 &= \sum \frac{(O - E)^2}{E} \\
&= \frac{(-6.82)^2}{21.82} + \frac{(6.82)^2}{8.18} + \frac{(6.82)^2}{18.18} + \frac{(-6.82)^2}{6.82} \\
&= \frac{46.51}{21.82} + \frac{46.51}{8.18} + \frac{46.51}{18.18} + \frac{46.51}{6.82} \\
&= 17.20
\end{aligned}
$$

The critical value for χ^2 with $\alpha = .05$ and d.f. $= 1$ is 3.84. Since the observed value of χ^2 exceeds 3.84, we reject the null hypothesis of independence and conclude that the classifications "type of college" and "type of publisher" are dependent. How great is the association? One measure of the strength of this association is given by ϕ. For our example,

$$\phi = \sqrt{\frac{\chi^2}{n}} = \sqrt{\frac{17.20}{55}} = .56$$

Statisticians and social scientists have generally agreed to develop measures of association for two variables which have unity as an upper limit (achieved when there is perfect positive or negative association) and zero as a lower limit (achieved when there is no association between the two variables). For Example 10.1, suppose that $n = 110$. If there were no association between the variables "type of college" and "type of publisher," the data might

Table 10.2   *No Association Between the Variables*
"Type of College" and "Type of Publisher"

| | Type of Publisher | | |
Type of College	Secular Press	Religious Press	Total
Church-supported	40	15	55
State-supported	40	15	55
Total	80	30	110

appear as shown in Table 10.2. It is obvious from this table that the variable "type of college" gives no information toward the prediction of the variable "type of publisher." For these data it is easy to verify that the phi coefficient equals zero since in each cell the observed cell frequency equals the expected cell frequency.

Table 10.3   *Perfect Association Between the Variables*
"Type of College" and "Type of Publisher"

| | Type of Publisher | | |
Type of College	Secular Press	Religious Press	Total
Church-supported	0	30	30
State-supported	80	0	80
Total	80	30	110

If the data of Example 10.1 appeared as in Table 10.3, then there would be a perfect association between the variables "type of college" and "type of publisher." Hence we could predict the type of publisher without error by knowing the type of college represented. It is easy to compute χ^2 for the data of Table 10.3 to see that

$$\chi^2 = 110$$

and hence

$$\phi = \sqrt{\frac{\chi^2}{n}} = \sqrt{\frac{110}{110}} = 1$$

Thus we have illustrated that for 2 × 2 tables the phi coefficient equals zero when there is no association between the two variables and equals 1 where there is a perfect association. Unfortunately for larger tables (more rows and columns) the upper limit of ϕ is greater than 1. For this reason, its use is limited to 2×2 tables.

10.3 A NOMINAL MEASURE OF ASSOCIATION: THE CONTINGENCY COEFFICIENT

The contingency coefficient, which is a modification of the phi coefficient, is a useful measure of association for cross-classification tables larger than a 2×2. The contingency coefficient is computed as follows:

The Contingency Coefficient

$$C = \sqrt{\frac{\phi^2}{1 + \phi^2}} \quad \text{or equivalently} \quad C = \sqrt{\frac{\chi^2}{n + \chi^2}}$$

Example 10.2 *A study of migrants and their family relations was conducted to ascertain whether the classification "degree of kinship participation" in the extended family was independent of a family's socioeconomic status. Use the data in Table 10.4 to run a chi-square test of independence and then compute C, the contingency coefficient. Expected cell values have been computed and placed in parentheses in Table 10.4.*

Table 10.4 *Study of Newcomer Adaptations*

Socioeconomic Status	Degree of Kinship Participation			Total
	Low	*Medium*	*High*	*Total*
Low	75 (88.69)	98 (102.48)	75 (56.83)	248
Medium	182 (184.53)	211 (213.23)	123 (118.24)	516
High	116 (99.78)	122 (115.29)	41 (63.93)	279
Total	373	431	239	1043

SOURCE: Felix M. Berardo, "Internal Migrants and Extended Family Relations—A Study of Newcomer Adaptation," unpublished Ph.D. dissertation, Florida State University, 1965, p. 122; by permission.

Solution   *Before computing C we should first run a test of independence of the two classifications to see if the data provide sufficient evidence to show that the two variables "socioeconomic status" and "degree of kinship participation" are dependent. Using the observed and expected cell frequencies from Table 10.4 we have*

$$\chi^2 = \sum \frac{(O - E)^2}{E}$$

$$= \frac{(-13.69)^2}{88.69} + \frac{(-4.48)^2}{102.48} + \frac{(18.17)^2}{56.83}$$

$$+ \frac{(-2.53)^2}{184.53} + \frac{(-2.23)^2}{213.23} + \frac{(4.76)^2}{118.24}$$

$$+ \frac{(16.22)^2}{99.78} + \frac{(6.71)^2}{115.29} + \frac{(-22.93)^2}{63.93}$$

$$= 19.61$$

Since the rejection region for $\alpha = .05$ (and hence $a = .05$) with d.f. $= (r - 1)(c - 1) = 2(2) = 4$ is 9.49, we conclude that the classifications are dependent. Now let us compute a measure of the strength of this dependence, the contingency coefficient.

Using $\chi^2 = 19.61$ we can compute the contingency coefficient as

$$C = \sqrt{\frac{\chi^2}{n + \chi^2}} = \sqrt{\frac{19.61}{1043 + 19.61}}$$

$$= \sqrt{.0184} = .136$$

The contingency coefficient is relatively easy to compute and satisfies the condition that it go to zero when there is no association between the classifications, but it does have some disadvantages as a measure of association. First, the contingency coefficient is always less than 1, even when the two classifications are completely dependent on one another. Second, note that C provides only an intuitive measure of the degree of association between two directions of classification. It is used after a chi-squared test of independence and hence is most frequently employed for data that satisfy the conditions required of a chi-square test. The expected values of a cell should be five or more before using C as a measure of association. Third, contingency coefficients for two different sets of data can only be compared if the two-way tables are of the same size (same number of rows and columns). This restriction is due to the fact that C can attain larger values for larger tables, so the range of possible values of C differs depending on the size of the table involved. Finally, it is difficult to compare the contingency coefficient with any of the other measures of association which will be presented later in this chapter.

Some of the disadvantages of C can be alleviated by adjusting the computed value of C depending on the maximum value C may attain. It can be shown that for a square two-way table (i.e., the number of rows equals the number of columns), the maximum value of C obtainable under perfect association is

$$C_{max} = \sqrt{\frac{r-1}{r}}$$

where r is the number of rows in the table. For $r = 2$,

$$C_{max} = \sqrt{1/2} = .707$$

If we now form a modified (adjusted) version of the contingency coefficient as

$$C_{adj} = \frac{C}{C_{max}}$$

the largest value C_{adj} may assume is 1.0, which occurs under perfect association when $C = C_{max}$. Applying this result to the data in Table 10.4 we see that with $r = 3$ rows

$$C_{max} = \sqrt{2/3} = .816$$

Hence the value of the adjusted value of C is

$$C_{adj} = \frac{C}{C_{max}} = \frac{.136}{.816} = .17$$

Using C_{adj}, we now have a measure of association that equals zero under no association and equals 1.0 when the two variables are perfectly associated. In spite of this, it is still difficult to compare C_{adj} for two tables with different numbers of rows. We turn now to some measures of association that are based on a concept called "reduction in error" and have fewer disadvantages than those previously discussed. All the procedures presented will be calculated in a similar manner and have a common means for interpretation. As a preface we need to discuss the concept of reduction in error.

EXERCISES

1   Using the data in Exercise 19, Chapter 9 (page 342), compute the phi coefficient.

2   Compute the phi coefficient for the following information:

	Male	*Female*
Democrat	20	10
Republican	10	20

3   What is the advantage of the contingency coefficient C over the phi coefficient?

4   Compute C_{max} for the data in Exercise 19, Chapter 9 (page 342). Interpret.

5   Determine C for the data in Exercise 19, Chapter 9 (page 342). What is the value of C_{adj}? Intepret.

6   What does C_{adj} tell us that C does not?

10.4   REDUCTION IN ERROR FOR MEASURES OF ASSOCIATION

*T*he concept "reduction in error" is based on the premise that we wish to predict the value of one variable based on knowledge of another. Specifically, assume that we can classify objects (people, relationships, beliefs, types of government, and so on) in a two-way table according to two directions of classification. Knowing that an object is in a particular category for one direction of classification (this method of classification is called the *independent variable*), we wish to predict the category in which the object will fall in the second direction of classification (called the *dependent variable*). For example, suppose that married couples are categorized according to the two directions of classification, "religious affiliation of the husband" and "religious affiliation of the wife," as shown in Table 10.5. If we wish to predict the religious affiliation of the wife when we know the religious affiliation of the husband, then "religious affiliation of the husband" is the independent variable, and the "religious affiliation of the wife" is the dependent variable. Hence the objective of the classification is prediction of the dependent variable, assuming that you know the value of the independent variable.

Table 10.5 *Two Directions of Classification*

Religious Affiliation of Husband	Religious Affiliation of Wife	
	Protestant	*Catholic*
Protestant Catholic		

The reasoning behind reduction in errors is based on two "rules of association." The first is intended to suggest "no association" between the variables (implying that the independent variable contributes no information for the prediction of the dependent variable). The second rule is intended to indicate "perfect association" between the two variables. Then given a set of elements classified in a two-way table, we calculate the number of elements misclassified (called number of errors of classification) for the "no association" rule and again for the "perfect association" rule. If the two variables in the

Table 10.6 *Perfect Association*
(a)

Affiliation of Husband	Affiliation of Wife		Total
	Protestant	*Catholic*	
Protestant Catholic	50 0	0 50	50 50
Total	50	50	100

(b)

Affiliation of Husband	Affiliation of Wife		Total
	Protestant	*Catholic*	
Protestant Catholic	0 50	50 0	50 50
Total	50	50	100

two-way table are really associated, the number of errors for the "perfect association" rule should be less than for the "no association" rule. The difference between these two numbers is called the reduction in error.

The "perfect association" rule is easy to define. Perfect association, meaning no prediction error, would imply that given a particular row, all the elements would fall in one (and only one) column. If the religious affiliation of husbands and wives were perfectly associated, the religious affiliations for a sample of 100 married couples would appear as shown in Table 10.6(a) or (b). Table 10.6(a) would result if husbands and wives always possessed the same religious affiliation. The data would fall as shown in Table 10.6(b) if husbands and wives always possessed different religious affiliations. For either case, knowing the husband's religious affiliations, you could predict, without error, the affiliation of the wife.

Similarly, perfect association for a 3 × 3 table (3 rows and 3 columns) might yield data as shown in Table 10.7(a) and (b). Note that for any given row in the tables all the couples fall in one (and only one) column.

Table 10.7   *Perfect Association*

(*a*)

Affiliation of Husband	*Affiliation of Wife*			*Total*
	Protestant	*Catholic*	*Other*	
Protestant	40	0	0	40
Catholic	0	35	0	35
Other	0	0	25	25
Total	40	35	25	100

(*b*)

Affiliation of Husband	*Affiliation of Wife*			*Total*
	Protestant	*Catholic*	*Other*	
Protestant	25	0	0	25
Catholic	0	0	25	25
Other	0	50	0	50
Total	25	50	25	100

A rule for classification that implies "no association" is more difficult to define. In fact, the definition of the "no association" rule is the major source of difference between the various measures of association based on reduction in error.

In real life we know that perfect association between the two directions of classification, religious affiliations of husband and wife, does not exist. Hence it is more likely that the sample data would appear as shown in Table 10.8. Now we can no longer infer a wife's religious affiliation by merely examining her husband's (or vice versa). We can state this more formally as follows. If our objective is to infer correctly the religious affiliation of a wife,

Table 10.8 *Less Than Perfect Association Between Religious Affiliation of Husband and Wife*

Affiliation of Husband	Affiliation of Wife		Total
	Protestant	Catholic	
Protestant	35	15	50
Catholic	15	35	50
Total	50	50	100

the perfect association rule "a wife's religious affiliation matches that of her husband" would work only for the case where there is a perfect association between the two classifications. If the association is less than perfect, application of this rule would introduce some error into the assignment of religious affiliation of wives.

Applying the perfect association rule to the data of Table 10.8, we would classify all 50 wives corresponding to the 50 Protestant husbands as Protestants. Note, however, that only 35 of the wives are Protestants; hence we have incurred 15 errors of classification. Similarly, by classifying all 50 wives of the 50 Catholic husbands as Catholic we would incur an additional 15 errors. The total number of misclassifications (errors) using the perfect association rule would be 30.

In spite of the errors, knowledge of the husband's affiliation does give some information concerning a wife's religious affiliation. For example, if we ignore the husband's affiliation and tried to infer a wife's affiliation, we could use the "no association" rule "all wives are Protestant." If we were to assume the column totals of Table 10.8 as fixed with 50 Catholic wives and

50 Protestant wives, use of this rule would result in 50 wives being misclassified. Hence in determining the degree of association between the classification of religious affiliations of husband and wife, it seems reasonable to compare the number of errors (misclassifications) resulting from inferring a wife's affiliation with and without the information concerning the husband's religious affiliation. We could also choose the no association rule "all wives are Catholic." For this example (since the numbers of Catholic and Protestant wives are equal), the number of misclassifications will again be 50.

Suppose that the numbers of Catholic and Protestant wives were not equal (that is, the column totals were not equal); how would we choose the better no association rule of the two? Certainly, we would select the rule that gives the smaller number of errors of misclassification. This will always be the column with the larger column total.

For the data of Table 10.8 the number of errors resulting from the no association rule, "all wives are Protestant," is 50 while utilizing the perfect association rule "a wife's religious affiliation is the same as her husband's" results in 30 errors. The reduction in errors obtained by incorporating information on the husband's religious affiliation is $50 - 30 = 20$. The proportional reduction in error in going from the first rule to the second is

$$\frac{50 - 30}{50} = \frac{20}{50} = .4$$

Thus there is a 40 percent reduction in error attributable to the second variable (religious affiliation of husband). The higher the proportional reduction in error, the higher the degree of association between the two variables (or classifications).

Proportional Reduction in Error

$$PRE = \frac{errors\ for\ ``no\ association\ rule"\ -\ errors\ for\ ``perfect\ association\ rule"}{errors\ for\ ``no\ association\ rule"}$$

If we applied this procedure to the data of Table 10.6, there would again be 50 errors in classifying wives by the rule "all wives are Protestant"; however, in using the rule "a wife's affiliation is the same as her husband's religious affiliation" we incur no errors. Hence the reduction in errors is 50 and the proportional reduction in error is

$$\frac{50 - 0}{50} = 1.00$$

In cases such as Table 10.6 where there is a perfect association between two variables, the percentage reduction in error going from a rule that ignores a second variable (e.g., husband's affiliation) to one that utilizes this information is 100.

Let us now extend this concept to a 3 × 3 table. Suppose that the religious affiliations of 100 husbands and wives were determined and summarized as in Table 10.9.

Table 10.9 *Less Than Perfect Association Between the Religious Affiliations of Husbands and Wives*

Affiliation of Husband	Affiliation of Wife			Total
	Protestant	*Catholic*	*Other*	
Protestant	40	7	3	50
Catholic	8	30	2	40
Other	2	3	5	10
Total	50	40	10	100

If we were to ignore the information of the husband's religious affiliation and try to infer the wife's, we could either identify all wives as Protestants, Catholics, or other. The best no association rule is one that would make the smallest number of errors. Fixing the column totals of Table 10.6 we would make the smallest number of errors (50) using the no association rule "all wives are Protestant" because the most wives (50) fall in the Protestant category. Incorporating the information of the husband's affiliation and using the perfect association "the wife's affiliation is the same as her husband's," we would have correctly classified 75 of the wives in Table 10.9 or, equivalently, incurred 25 errors. The reduction in errors going from the first rule to the second is $50 - 25 = 25$, and hence the proportional reduction in error obtained by using the information on the husband's affiliation is

$$\text{PRE} = \frac{50 - 25}{50} = .50$$

It is important at this point to summarize the procedures involved in using proportional reduction-in-error measures of association. Two rules and two errors will be defined.

Summary of the General Concept of Reduction in Error

Rule 1: The No Association Rule
A rule for classifying data according to one variable with knowledge only of the marginal totals for that variable. Information on a second variable is ignored.
E_1: *The number of errors of classification obtained using rule 1.*

Rule 2: The Perfect Association Rule
A rule for classifying data according to one variable that utilizes information on a second variable.
E_2: *The number of errors of classification obtained from rule 2.*

The proportional reduction in error is then

$$PRE = \frac{E_1 - E_2}{E_1}$$

Although rules for classifying data will change for various measures of association, the concept of the proportional reduction in error is the same. We proceed now with specific reduction-in-error measures of association.

7   Explain the concept "reduction in error" as applied to measures of association.

8   State the rule of "no association."

9   State the rule of "perfect association."

10   Express, in symbolic form, the formulas for any proportional reduction in error statistic.

10.5 A PROPORTIONAL REDUCTION IN
ERROR MEASURE OF ASSOCIATION
FOR NOMINAL SCALES: LAMBDA

or any cross classification of two variables measured on a nominal scale, we can compute two lambdas, one lambda designated by λ_r, when the row variable is the "dependent" variable, and the other, designated by λ_c, which is computed when the column variable is the "dependent" variable. It should be noted that λ_r and λ_c calculated from the same set of data are not necessarily identical or even similar in numerical value; however, the nature of the problem will usually dictate which variable is being predicted (the dependent variable) based on information from another variable (the independent variable). If this is not the case and it is impossible to designate an independent and a dependent variable, then we can calculate $(\lambda_r + \lambda_c)/2$, which provides a measure of mutual predictability between two variables.

The calculation of λ requires that we define rules 1 and 2, corresponding to "no association" and "perfect association," and the two errors of classification associated with these rules. To aid in understanding, we shall give these definitions for a specific example and then state them for the general case.

Example 10.3 *Fred L. Strodtbeck conducted a study to determine whether kinship lineage is predictive of which spouse will dominate family decisions. Two societies, one matrilineal and one patrilineal, were included in the study and families were classified as to which spouse dominated family decisions. Use the data of Table 10.10 to compute λ_r to measure the degree of association between sex of the decision maker and kinship lineage.*

Table 10.10 *Data from Strodtbeck's Study*

	Kinship Lineage		
Decision Maker	Matrilineal Navajo	Patrilineal Mormon	Total
Husband	34	42	76
Wife	46	29	75
Total C	80	71	151

SOURCE: Fred L. Strodtbeck, "Husband–Wife Interaction over Related Differences," *American Sociological Review, 16* (Aug. 1951), p. 472; by permission.

Solution   *Since we are trying to predict the sex of the decision maker within the family based on kinship lineage, the row variable is the dependent variable.*

: *(No Association Rule): Place all families in the row with the largest row total, namely row 1.*
: *The number of errors (misclassifications) utilizing this rule would be* 151 − 76 = 75.

: *(Perfect Association Rule): Place all 80 families of column 1 in the row with the largest cell frequency, namely row 2. The number of errors incurred in column 1 would be* 80 − 46 = 34. *Place all 71 families of column 2 in the row with the largest cell frequency, namely row 1. The number of errors incurred in column 2 would be* 71 − 42 = 29.
: *The number of errors (misclassifications) utilizing the perfect association rule would be* 34 + 29 = 63.

Hence λ_r *the proportional reduction in error due to the independent variable kinship lineage, is*

$$\lambda_r = \frac{E_1 - E_2}{E_1} = \frac{75 - 63}{75} = \frac{12}{75} = .16$$

We can now formulate the general rule for computing λ_r (when the row variable is the dependent variable).

Rules and Errors for Computing λ_r

Rule 1 (No Association Rule): Place all observations in the row with the largest row total.
E_1 : *The number of errors for rule 1 is* (n − *largest row total*).

Rule 2 (Perfect Association Rule): Place all observations of column 1 in the row with the largest cell frequency. Compute the number of errors for column 1 as (*column 1 total* − *largest cell frequency of column 1*). *Place all observations of column 2 in the row with the largest cell frequency. Compute the number of errors for column 2 as* (*column 2 total* − *largest cell frequency of column 2*), *etc.*
E_2 : *The number of errors for rule 2 is the sum of the errors for each column.*
The reduction in errors obtained by using information on the column variable to predict the row variable is $E_1 - E_2$ *and the* proportional reduction in error, λ_r, *is*

$$\lambda_r = \frac{E_1 - E_2}{E_1}$$

It should be noted that the procedure for computing λ_c (when the column variable is the dependent variable) is identical to that for computing λ_r with the roles of rows and columns reversed. Thus rule 1 would specify that all observations be placed in the column with the largest column total. Because of the similarity of λ_r and λ_c, we omit a formal restatement of the rules and errors.

To compute λ_c for our example (the data, Table 10.10), rule 1, the no association rule, would place all observations in the column with the largest column total (column 1). The number of errors incurred would be

$$E_1 = 151 - 80 = 71$$

Rule 2, the perfect association rule, would place all 76 families of row 1 in the column with the largest cell frequency (column 2) and the 75 families of row 2 in the column with the largest cell frequency (column 1). The number of errors incurred in row 1 would be 34. Similarly, the number of errors in row 2 would be 29. Hence

$$E_2 = 34 + 29 = 63$$

Then

$$\lambda_c = \frac{E_1 - E_2}{E_1} = \frac{71 - 63}{71} = \frac{8}{71} = .11$$

Several additional comments should be made concerning λ. First, λ can range between 0 and 1 (hence it is impossible to get a negative value of λ). The closer λ is to 1, the more relative reduction in error is achieved by shifting from rule 1 to rule 2. Since rule 2 utilizes a second variable for prediction purposes, the larger λ is, the greater the degree of association between the two variables (or classifications). Second, λ can be used for any two-way tables with two or more rows and two or more columns. Third, as stated previously, some researchers utilize the average of λ_r and λ_c when it is not clear which variable is the dependent variable and which is the independent. In such cases

$$\frac{\lambda_r + \lambda_c}{2}$$

provides a combined measure of association between the two variables which lies between 0 and 1. But $(\lambda_r + \lambda_c)/2$ cannot be interpreted as a proportional reduction in error statistic as we have done with the separate components,

λ_r and λ_c. For our example, a combined measure of association is

$$\frac{\lambda_r + \lambda_c}{2} = \frac{.16 + .11}{2} = .135$$

We turn now to several ordinal measures of association which are proportional reduction-in-error measures.

11 Define λ.

12 What is the difference between λ_r and λ_c?

13 Use the data of Exercise 14, Chapter 9 (page 341), to compute λ_r. Interpret.

14 Use the data of Exercise 14, Chapter 9 (page 341), to compute λ_c. Interpret.

15 Compute a measure of mutual predictability for the data in Exercise 14, Chapter 9 (page 341), assuming no independent variable.

16 What is the advantage of λ over C_{adj}?

G 10.6 A PROPORTIONAL REDUCTION IN ERROR MEASURE OF ASSOCIATION FOR ORDINAL SCALES: GAMMA

amma provides a measure of association between two variables measured on ordinal scales. In particular, it measures the predictability of orders of ranks associated with one variable from orders of ranks associated with a second variable. We shall explain the "no association" and "perfect association" rule for gamma as well as its computation by means of an example.

Seventy-seven individuals sampled in a community were ranked according to their annual income (low, medium, high) and the prestige of their occupation (low, medium, or high). These data are recorded in Table 10.11. The objective is to make inferences about the rankings for a dependent variable (annual income) based on knowledge of an independent variable (occupational prestige).

Table 10.11 *Income and Occupational Ranks*

Occupational Prestige	Annual Income		
	Low	Medium	High
Low	6	10	8
Medium	4	12	12
High	2	8	15

In order to fully understand the meaning and logic behind the use of gamma, we must think in terms of pairs of observations rather than single observations. Gamma attempts to answer the question: If we know the order of ranks on one variable, occupational prestige, for a pair of individuals, can we predict the order of ranks for that pair on a second variable, annual income? For example, if one individual of the pair has a higher occupational prestige ranking than the other, can we also predict that he will have a higher ranking on annual income? Note that we are not trying to predict the actual rank but rather we are trying to predict the rank of one individual on the dependent variable, annual income, relative to that of another individual. The no association rule will be used to predict the order of ranks for pairs of observations on one variable (the dependent variable) without reference to a second variable (the independent variable) while the perfect association rule utilizes information concerning the order of ranks of pairs of observations on the independent variable to predict the order or ranks for pairs of observations on the dependent variable.

The definition of error is different from that used for the nominal measures of association. An error is incurred if the predicted order of ranks of the dependent variable for a pair of observations differs from the actual order of ranks. For example, if we predict that one individual has a higher ranking on annual income than another when, in fact, the reverse is true. we have incurred an error. Thus for gamma, the term "error" denotes a wrong predicted order of ranks for a pair of observations rather than a misclassification of a single observation. We shall now consider the no association rule for our example.

If we have no information concerning the independent variable for a pair of observations that would be useful in predicting which individual of the pair has a higher annual income rank, we have no basis for a prediction rule. We resort to a random assignment of orderings of annual income ranks for pairs of individuals. For our example, we could form all possible pairs of individuals from the 77 persons sampled. (Since a mathematician could show that there would be 2926 such pairs, we shall not actually list the pairs.) We could then randomly label one individual within each pair as *A* and the other *B*. Our no association rule could then be stated as

---

Rule 1 (No Association Rule)

For each pair of individuals, we predict, on the dependent variable, that person A has a higher rank than person B.

(Note: The roles of A and B are arbitrary and could be switched.)

In our case the no association rule would be that for each pair, person *A* has a higher annual income rank than person *B*. If we ignore pairs where the ranks of the two observations are the same, it would seem reasonable to assume that our random assignment of *A* and *B* within a pair would lead to errors in 50 percent of the orderings (i.e., we would expect to predict the wrong order of ranks for 50 percent of the pairs). Indeed this is true. If we ignore tied ranks on either variable and let N_s denote the number of pairs of observations where the ordering is the same on both variables and N_r be the number of pairs where the ordering is reversed for the two variables, then $N_s + N_r$ is the total number of pairs without ties on either variable. The number of errors incurred using rule 1 is

$$E_1 = .5(N_s + N_r)$$

The computation of γ can be illustrated using the data of Table 10.11. For convenience, we have relisted the data in Table 10.12. Starting from the upper-left-hand corner of Table 10.12, if any of the six persons classified as low prestige, low income (i.e., low, low) were paired with any of the other 18 persons in the first row of Table 10.12, the two individuals of a pair would have the same prestige ranking and hence none of these pairs would go into the computation of N_s or N_r. Similarly, if any of the six persons classified (low, low) were paired with any of the remaining six individuals in the first column of Table 10.12, the two individuals would have a tied rank for the variable, annual income. Again, none of these pairs would go into the calculation of N_s or N_r. However, if any of these six individuals classified (low, low)

Table 10.12   *Data of Table 10.11*

Occupational Prestige	Annual Income		
	Low	Medium	High
Low	6	10	8
Medium	4	12	12
High	2	8	15

were paired with any of the other $(12 + 12 + 8 + 15 = 47)$ individuals not in either row 1 or column 1, the order of ranks for both variables, occupation prestige and annual income, would be the same. For example, if a person classified (low, low) were paired with one classified (medium, high), the second individual has a higher rank on both variables and hence the order of ranks is the same for both variables. There are

$$6(12 + 12 + 8 + 15) = 282$$

pairs that could be formed using one individual from the upper-left-hand cell of Table 10.12 and another from the 47 individuals that fall below and to the right of that cell.

Consider now the 10 individuals in the first row, second column, of Table 10.12. If any of these 10 individuals classified as (low, medium) is paired with one of the $(12 + 15 = 27)$ individuals that fall below and to the right of the first row, second column, of Table 10.12, we would obtain the same ordering of ranks for both variables. There would be

$$10(12 + 15) = 270$$

pairs formed in this way.

Moving across the first row of Table 10.12 to the (low, high) category, there are no cells to the right and below, so we could not pair an individual in this cell with any other to form a pair with the same ordering of ranks in both variables.

Moving to the second row, first column, of Table 10.12, we could form

$$4(8 + 15) = 92$$

pairs with one individual categorized (medium, low) and another from the categories to the right and below this cell. Each of these pairs would have the same ordering of ranks on both variables. Similarly, pairing any of the 12 individuals in the second row, second column, with one from cells to the right and below we obtain

$$12(15) = 180$$

more pairs with the same ordering of ranks on the two variables. Proceeding to the remaining cell in row 2 and the three cells in row 3, there are no more pairs which can be formed such that the ordering of ranks is the same for both variables.

We have now determined N_s, the number of pairs of observations where the ordering is the same on both variables:

$$N_s = 282 + 270 + 92 + 180 = 824$$

In a similar way, N_r, the number of pairs where the ordering of ranks is reversed for the two variables, can be computed by starting at the upper-right-hand corner of Table 10.12 and working to the left and below. For example, if any of the eight individuals classified as (low, high) were paired with any of the other $(4 + 12 + 2 + 8 = 26)$ individuals not in either row 1 or column 3, the order of ranks on the first variable would be the reverse of the order of ranks on the second variable. We might pair an individual classified (low, high) with one classified (medium, medium). Note here that ranking for the first individual is lower on the first variable but higher on the second variable. Hence the order of ranks is reversed.

Beginning at the upper-right-hand corner of Table 10.12 and working to the left and below we find

$$8(4 + 12 + 2 + 8) = 208$$
$$10(4 + 2) = 60$$
$$12(2 + 8) = 120$$

and

$$12(2) = 24 \text{ pairs}$$

Hence

$$N_r = 208 + 60 + 120 + 24 = 412$$

The number of errors using the no association rule is

$$E_1 = .5(N_s + N_r)$$
$$= .5(824 + 412) = .5(1236) = 618$$

We can now summarize the procedure for computing N_s and N_r:

Procedure for Computing N_s and N_r

1 Arrange the categories for both variables from the lowest to highest order in the cross classification of the data.

2 The number of pairs with the same ordering of ranks for the two variables, N_s, can be computed by beginning in the upper-left-hand corner of the table and moving across the rows multiplying the number of entries in a cell times the number of entries in the cells to the right and below. The sum of all these multiplications equals N_s.

3 Reverse procedure of step 2 to compute N_r. Beginning in the upper-right-hand corner of the table and moving across the rows, multiply the number of entries in each cell times the number of entries in all cells to the left and below. The sum of all these is N_r, the number of pairs having the reverse ordering of ranks for the two variables.

The computation of N_s for the data of Table 10.12 can be shown graphically as in Figure 10.1. Thus for each table in Figure 10.1 you compute the product of the number in the cross-hatched square times the sum of the numbers in the shaded squares. Notice how you start with the cross-hatched square in the upper left corner and move to the right across the first row,

Figure 10.1 *Graphic illustration of the computation of N_s*

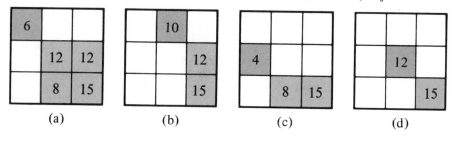

(a) (b) (c) (d)

stopping at the next-to-last square. This is because there are no cells below and to the right of the last square in the first row. This operation yields the products in Figure 10.1(a) and (b). Then you select the first element to the left in the second row and repeat the process to obtain Figure 10.1(c) and (d). The operation is stopped at the next-to-last row because no cells lie below the last row.

The graphic illustration of the computation of N_r for the same data produces figures with the shaded area diminishing as you move from top right to bottom left (see Figure 10.2). You will see that the computational procedure for N_r is the same as for N_s except that you move from right to left.

Figure 10.2   *Graphic illustration of the computation of N_r*

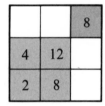 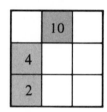 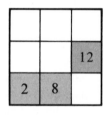 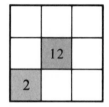

We can now state the perfect association rule which utilizes information on the orders of pairs for one variable (often called the independent variable) to predict the orders of ranks for pairs of observations on a second variable (the dependent variable).

Rule 2 (Perfect Association Rule)

If N_s is larger than N_r, predict the same *order for a pair of ranks on the dependent variable as was observed for the same pair on the independent variable. If N_r is greater than N_s, predict the* reverse *order for a pair of ranks on the dependent variable as was observed for the same pair on the independent variable. If $N_s = N_r$, predict the same order of ranks. The number of errors incurred using rule 2 will be either N_s or N_r, whichever is smaller. We designate this by $E_2 = $ minimum (N_s, N_r).*

For our example

$$E_2 = \text{minimum } (824,412) = 412$$

Having specified both rules and the number of errors incurred with each procedure, we can define γ, a proportional reduction in error measure, as

$$\gamma = \frac{E_1 - E_2}{E_1}$$

$$= \frac{.5(N_s + N_r) - \text{minimum}(N_s, N_r)}{.5(N_s + N_r)}$$

For our example the relative reduction in errors obtained by utilizing information on the independent variable, occupational prestige, to predict the order of ranks for pairs of observations on the dependent variable, annual income, is

$$\gamma = \frac{E_1 - E_2}{E_1}$$

$$= \frac{618 - 412}{618} = \frac{206}{618} = \frac{1}{3} = .33$$

Although γ can be readily computed using the previous formula, most social scientists use the following computational formulas.

Computation Formula for γ

If $N_s \geqslant N_r$,

$$\gamma = \frac{N_s - N_r}{N_s + N_r}$$

If $N_r > N_s$,

$$\gamma = -\frac{N_s - N_r}{N_s + N_r}$$

Note that had we used this computational formula in our annual-income example we would have obtained the same value for γ:

$$\gamma = \frac{N_s - N_r}{N_s + N_r} = \frac{824 - 412}{824 + 412} = \frac{412}{1236} = \frac{1}{3} = .33$$

The interpretation of γ is fairly simple in light of the concept of a proportional reduction-in-error measure of association. When there is a positive association between the orders for pairs of ranks for the two variables (N_s larger than N_r) the quantity $(N_s - N_r)/(N_s + N_r)$ will be positive. Similarly, when there is a negative association between the orders for pairs of ranks on the two variables (N_s less than N_r) the quantity $(N_s - N_r)/(N_s + N_r)$ will be negative. In either case γ will be the proportional reduction in error due to the independent variable.

Example 10.4 *A study of the relationship between an individual's attitude toward the legalization of marijuana and his overt behavior with reference to marijuana was conducted on a sample of 204 individuals. Use the data given in Table 10.13 to compute γ, a measure of the association between ranks of attitudes and ranks of behavior concerning marijuana.*

Table 10.13 *Attitudes and Overt Behavior Concerning Legalization of Marijuana*

Attitude Toward Legalization	Overt Behavior	
	Unfavorable	Favorable
Least favorable	57	15
Next least favorable	23	7
Next most favorable	10	20
Most favorable	15	57

SOURCE: S. L. Albrecht, M. L. De Fleur, and L. G. Warner, "Attitude–Behavior Relationships," *Pacific Sociological Review*, *15* (April 1972), p. 195, by permission.

Solution *The tables needed to compute N_s are shown in Figure 10.3.*

Figure 10.3 *Tables for computation of N_s, Example 10.4*

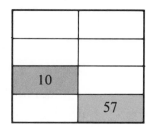

Then from Figure 10.3, multiplying the number in the cross-hatched cell by the sum of the numbers in the shaded cells we find

$$N_s = 57(7 + 20 + 57)$$
$$+ 23(20 + 57)$$
$$+ 10(57)$$
$$= 7129$$

Similarly, the tables needed to compute N_r are shown in Figure 10.4.

Figure 10.4 *Tables for computation of N_r, Example 10.4*

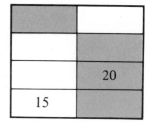

Then

$$N_r = 15(23 + 10 + 15)$$
$$+ 7(10 + 15)$$
$$+ 20(15)$$
$$= 1195$$

Thus for $N_s > N_r$,

$$\gamma = \frac{N_s - N_r}{N_s + N_r} = \frac{7129 - 1195}{7129 + 1195} = .71$$

Since $N_s > N_r$, the orderings of ranks for pairs on the two variables are positively associated. The magnitude of γ indicates that the error of prediction can be reduced by 71 percent if we use the orders of ranks for attitudes toward legalization of marijuana to predict orders of ranks for overt behavior with reference to marijuana.

As stated previously, γ is the proportional reduction in error obtained by using information on the order of one variable to predict the order of ranks for another variable. Gamma may vary between zero and 1.0. If $N_s > N_r$ there is a positive relationship between the orders of ranks for the two variables while for $N_r > N_s$ there is a negative relationship between orders of ranks for the two variables. A γ of 1.0 indicates that there are no errors using the perfect association rule, while a value of $\gamma = 0$ means there is no association between the orders of the independent and dependent variables.

17 State the no association rule for γ. How does the no association rule for γ differ from that for λ?

18 State the perfect association rule for γ.

19 Compute γ for the data in Exercise 5, Chapter 3 (page 56). Interpret.

20 When would you use γ in preference to λ?

21 Compute γ for the data in Table 9.6 (page 322). Interpret.

10.7 A PROPORTIONAL REDUCTION IN ERROR MEASURE OF ASSOCIATION FOR ORDINAL SCALES: SPEARMAN'S RHO

Although Spearman's rank-order correlation coefficient rho, ρ, can be viewed as a proportional reduction-in-error measure of association, it is more easily understood if we first examine its use and its computational formula and then present ρ as a proportional reduction-in-error measure.

In sociological studies it is often important to measure the strength of the relationship between two ordinal variables. One way to do this is to measure the strength of the relationship between the rank order of observations on one variable and the rank order of the same observations on a second variable. To illustrate, consider the following example. Ten incoming freshmen were first ranked (from 1 to 10) on their scores on the College Entrance Examinations. These same 10 students were ranked at the end of their freshman year according to grade-point averages (see Table 10.14). Of course it would be important for the admissions office to be able to predict the GPA ranks using the CEE ranks.

Table 10.14 *Ranks for 10 Students on the CEE and the Year-End Freshman GPA*

Student	CEE Ranks	GPA Ranks
A	1	1
B	3	2
C	2	3
D	5	4
E	7	6
F	8	8
G	9	7
H	10	9
I	6	10
J	4	5

One measure of the strength of the relationship between the sets of ranks is given by Spearman's rank-order correlation coefficient.

Spearman's Rank-Order Correlation Coefficient ρ

$$\rho = 1 - \frac{6\Sigma d^2}{n(n^2 - 1)}$$

where

$d = $ *difference between pairs of ranks*
$n = $ *total number of pairs of ranks*

The computing formula for Spearman's ρ given above is correct if there are no ties in the ranks. It is a good approximation to ρ if the number of ties is small relative to n, the number of pairs of ranks?

Example 10.5 *Use the data of Table 10.14 to compute ρ.*

Solution *Before computing ρ from Table 10.14 we need two additional columns, representing the differences and squares of differences in pairs of ranks. These columns are given in Table 10.15. For $n = 10$ pairs of ranks, the computed value of ρ is*

$$\rho = 1 - \frac{6\Sigma d^2}{n(n^2 - 1)}$$

$$= 1 - \frac{6(26)}{10(100 - 1)}$$

$$= 1 - .158 = .84$$

Table 10.15   *Differences in Pairs of Ranks*

Student	CEE Ranks	GPA Ranks	Differences (CEE − GPA)	d^2
A	1	1	0	0
B	3	2	1	1
C	2	3	−1	1
D	5	4	1	1
E	7	6	1	1
F	8	8	0	0
G	9	7	2	4
H	10	9	1	1
I	6	10	−4	16
J	4	5	−1	1
				$\Sigma\, d^2 = 26$

How do we interpret ρ as a measure of the strength of the association of the ranking for two variables? Spearman's correlation coefficient varies between -1.0 and 1.0. A value of $\rho = 1.0$ would indicate a perfect association between the ranking for the two variables. For our example, ρ would equal 1.0 if all differences were zero; that is, ranks on the CEE perfectly agreed with the ranks for the GPA. A value of $\rho = -1.0$ would again indicate a perfect relationship between the ranks, but in this case the relationship would be negative. For example, if a rank of 1 on the CEE was associated with a rank of 10 for the GPA, a rank of 2 with a 9, etc., ρ would equal -1.0.

As mentioned previously, ρ is a proportional reduction in error measure. We define rules 1 and 2 as follows:

Rule 1 (No Association Rule): Assign each observation of the dependent variable a rank of $(n + 1)/2$, the average of the ranks for n pairs.
Rule 2 (Perfect Association Rule): If r_I is the rank of the independent variable for an observation, then we predict r_d, the rank of the dependent variable, as

$$r_d = r_I(\rho) + \frac{n + 1}{2}(1 - \rho)$$

It can then be shown (see Mueller, Schuessler, and Costner, 1970) that the proportional reduction in error realized by utilizing rule 2 rather than rule 1 is ρ^2, the square of the Spearman rank correlation coefficient.

In contrast to the previously discussed measures of association for nominal or ordinal classifications, we can do more than merely quantify the degree of association using ρ. We can use the sample rank correlation coefficient, $\hat{\rho}$, as the decision maker in a statistical test of an hypothesis of "no association" ($\rho = 0$) between ranks of paired observation from two populations. For n equal to 10 or more the quantity

$$t = \hat{\rho} \sqrt{\frac{n-2}{1-\hat{\rho}^2}}$$

follows a Student's t distribution. A summary of the rank correlation test for ρ is as follows:

Statistical Test for ρ

Null Hypothesis: $\rho = 0$.
Alternative Hypothesis: ρ *is not equal to zero (two-tailed test).*
Test Statistic:

$$t = \hat{\rho}\sqrt{\frac{n-2}{1-\hat{\rho}^2}}$$

Rejection Region: **Reject the null hypothesis if the absolute value of** t **is greater than the tail-end value of** t **with** $a = \alpha/2$ **and** $d.f. = n - 2$.

$$\left[Note: \ \hat{\rho} = 1 - \frac{6\Sigma\, d^2}{n(n^2-1)} \right]$$

Example 10.6 *Use the data of Example 10.5 to test the null hypothesis that the correlation between the rank order of observations on the College Entrance Examinations and the rank order of observations on the year-end freshmen grade-point averages is zero (i.e., $\rho = 0$) against the alternative that there is a positive correlation between these two sets of ranks.*

Solution *The null hypothesis and alternative hypothesis are*

$$H_0: \rho = 0$$

$$H_a: \rho > 0$$

From a sample of n = 10 incoming freshmen, we obtain the corresponding ranks for the CEE and GPA (see Table 10.15). The sample rank-order correlation coefficient is, then, $\hat{\rho} = .84$, as we calculated it to be in Example 10.5. The decision maker for our test is

$$t = \hat{\rho}\sqrt{\frac{n-2}{1-\hat{\rho}^2}}$$

$$= (.84)\sqrt{\frac{10-2}{1-(.84)^2}}$$

$$= .84(5.21) = 4.38$$

For $\alpha = .05$ and d.f. $= n - 2 = 8$, the critical value of t (with a $= .05$ and d.f. $= 8$) is 1.86. Since the observed value of t, t $= 4.38$, exceeds 1.86, we reject the null hypothesis and conclude that ranks on the CEE and freshman GPA's are positively correlated.

In this section we discussed a measure of the association between two sets of rankings on n different items. We turn now to a measure of association among k (k greater than 2) different sets of ranking on the same n items. This measure, called Kendall's coefficient of concordance, would be useful in studies of the reliability of a set of judges or the degree of agreement between the rankings of social priorities by persons of different ethnic backgrounds.

10.8 A MEASURE OF ASSOCIATION
FOR ORDINAL DATA:
KENDALL'S *W*

*U*nlike several of the previous measures of association, Kendall's coefficient of concordance (Kendall's *W*) is not a proportional reduction in error measure of association. Kendall's *W* is used to measure the degree of similarity among two or more sets of ranks of levels for a quantitative variable measured at the ordinal level. For example, we might ask four people to rank 10 different types of jobs according to preference. If the people all have the same preference rankings, Kendall's *W* will be large; if they differ greatly in their rankings, *W* will be small. Kendall's *W* will take values from 0 to 1, 0 corresponding to no association (no common ranking) to 1 for identical rankings for all sets.

Table 10.16 *Occupational Prestige Rankings*

Occupation	Upper Class	Upper Middle Class	Middle Class	Lower Middle Class	Lower Class	R
Physician	1	1	1	1	1	5
Public school teacher	2	3	3	3	2	13
Priest or minister	3	2	2	2	4	13
Dentist	4	4	4	4	5	21
Lawyer	5	8	6	8	9	36
Pharmacist	6	7	9	6	7	35
Nurse in a hospital	7	6	5	5	3	26
Businessman	8	5	8	9.5	12	42.5
Banker	9	9	7	9.5	8	42.5
Policeman	10	10	10	7	6	43
Automobile mechanic	11	12	12	12	11	58
Plumber	12	11	13	11	10	57
Clerk in a store	13	13	11	13	13	63
Waiter or waitress	14	14	14	14	14	70

We can illustrate its use with the following example. Fourteen different occupations were ranked in terms of community prestige by five different people (one each from the upper class, upper-middle, middle, lower-middle, and lower class). These rankings are presented in Table 10.16. Note that for this example we have $k = 5$ different sets of rankings for the $n = 14$ occupations. Recall that when two observations are tied, they are assigned a rank equal to the average of the occupied ranks. For example, in the lower-middle-class ranking of Table 10.16, the occupations businessman and banker were judged to have the same occupational prestige. Both occupations were assigned a rank of 9.5, the average of the ranks, 9 and 10, which they occupied. In addition to the rankings in the table, we added another column, R, which gives the sum of the ranks associated with each occupation (row of the table). The sum of the five ranks for a dentist was 21.

We now define W as

$$W = \frac{12SS_R}{k^2 n(n^2 - 1)}$$

where

$$SS_R = \Sigma R^2 - \frac{(\Sigma R)^2}{n}$$

Example 10.7 *Use the data of Table 10.16 to compute W, Kendall's coefficient of concordance.*

Table 10.17 *Data from Table 10.16 Necessary for Computing SS_R*

Occupation	R (Row Totals)	R^2
Physician	5	25
Public school teacher	13	169
Priest or minister	13	169
Dentist	21	441
Lawyer	36	1296
Pharmacist	35	1225
Nurse in a hospital	26	626
Businessman	42.5	1806.25
Banker	42.5	1806.25
Policeman	43	1849
Automobile mechanic	58	3364
Plumber	57	3249
Clerk in a store	63	3969
Waiter or waitress	70	4900

$$\Sigma R = 525 \quad \Sigma R^2 = 24{,}944.5$$

Solution *Before computing W we must first calculate SS_R. An appropriate R and R^2 column have been listed in Table 10.17. Substituting into SS_R we have*

$$SS_R = \Sigma R^2 - \frac{(\Sigma R)^2}{n}$$

$$= 24{,}944.5 - \frac{(525)^2}{14} = 24{,}944.5 - 19{,}687.5$$

$$= 5257$$

Hence

$$W = \frac{12 SS_R}{k^2 n(n^2 - 1)} = \frac{12(5257)}{25(14)(195)} = \frac{63{,}084}{68{,}250}$$

$$= .92$$

For this example W = .90 expresses the degree of association or agreement among the five different prestige rankings of the 14 occupations. Since W = .90 is near the maximum value 1, fairly strong similarity exists among the five sets of rankings, corresponding to people from the five social classes.

As with Spearman's ρ, we can run a test of significance for an observed value of W. We envision a population of people for each social class grouping in which each person ranks 14 occupations. These rankings will differ among people within and between populations. However, there will be a pooled ranking for each population that represents the consensus of opinion for all people within that population. The research hypothesis that we wish to verify is that the pooled ranking of occupations is the same for all five social classes. Hence we shall wish to test the null hypothesis that there is no common ranking for the five populations against the alternative hypothesis that the rankings for the five populations are identical. The decision maker utilizes the observed value of W. In particular, for n, the number of items ranked, greater than 7, the quantity

$$k(n-1)W = \frac{12SS_R}{kn(n+1)}$$

follows a chi-square distribution with $n-1$ degrees of freedom when the null hypothesis is true. Hence for a given value of α, we reject the null hypothesis of independence of the classifications if $k(n-1)W$ exceeds the critical chi-square value with $a = \alpha$ and d.f. $= n-1$. This procedure is summarized below.

Test of Significance for Kendall's *W*

Null Hypothesis: **The k sets of ranks are independent.**
Alternative Hypothesis: **The k sets of ranks are dependent.**
Test Statistic:

$$k(n-1)W = \frac{12SS_R}{kn(n+1)}$$

Rejection Region: **Reject the null hypothesis if $k(n-1)W$ is greater than the critical value of chi square with $a = \alpha$, d.f. $= n-1$.**

(*Note: n, the number of items ranked, must be greater than 7.*)

Example 10.8   *Run a test of independence for the k = 5 sets of ranks in Example 10.7. (Use α = .05.)*

Solution   *The null hypothesis is that the 5 sets of ranks are independent and the alternative (research) hypothesis is that they are dependent. Recall the observed value of W was .92. Thus*

$$k(n - 1)W = 5(13)(.92) = 59.8$$

For a = .05 and d.f. = n − 1 = 13, the critical value of chi square is 22.36. Since 59.8 exceeds the critical value of chi square we conclude that there is a significance association among the sets of rankings.

One word of caution should be sounded. A significant value for W does not imply that the rankings are correct or true. We are judging the significance, not the validity, of the degree of association among the rankings.

EXERCISES

22   How does ρ differ from W?

23   A large corporation selects college graduates for employment, using both interviews and a psychological-achievement test. Interviews conducted at the home office of the company were far more expensive than the tests, which could be conducted on campus. Consequently, the personnel office was interested in determining whether the test scores were correlated with interview ratings and whether tests could be substituted for interviews. The idea was not to eliminate interviews but to reduce their number. To determine whether correlation was present, 10 prospects were ranked during interviews and tested. The paired scores are as shown. Calculate the Spearman rank correlation coefficient ρ. Rank 1 is assigned to the candidate judged to be the best.

Subject	Interview Rank	Test Score
1	8	74
2	5	81
3	10	66
4	3	83
5	6	66
6	1	94
7	4	96
8	7	70
9	9	61
10	2	86

24 Refer to Exercise 23. Do the data present sufficient evidence to indicate that the correlation between interview rankings and test scores is less than zero? If this evidence does exist, can we say that tests could be used to reduce the number of interviews?

25 A political scientist wished to examine the relationship of the voter image of a conservative political candidate and the distance between the residences of the voter and the candidate. Each of 12 voters rated the candidate on a scale of 1 to 20. The data are as shown. Calculate the Spearman rank correlation coefficient $\hat{\rho}$.

Voter	Rating	Distance
1	12	75
2	7	165
3	5	300
4	19	15
5	17	180
6	12	240
7	9	120
8	18	60
9	3	230
10	8	200
11	15	130
12	4	130

26 Refer to Exercise 25. Do these data provide sufficient evidence to indicate a negative correlation between rating and distance?

27 Is Kendall's W a proportional reduction in error measure of association?

28 Compute W on the sets of ranks pertaining to the preference of three judges (rank order) of attractive places to visit. Do the judges seem to agree on the qualities of an attractive city? (Use $\alpha = .05$.)

Cities	Judge A	Judge B	Judge C
Atlanta, Ga.	4	3	4
Chicago, Ill.	6	6	6
Detroit, Mich.	5	5	5
New York, N.Y.	1	1	2
Oakland, Calif.	3	2	1
St. Louis, Mo.	2	4	3
Philadelphia, Pa.	7	7	7

C 10.9   SUMMARY

hapter 10 presents measures of association for both nominal and ordinal data. Measures of association for nominal data included the ϕ coefficient, contingency coefficient C, and λ. Recall that ϕ was only appropriate for 2×2 tables and that the contingency coefficient C offers a measure that could be utilized for larger tables. Unfortunately, C is difficult to interpret because its maximum value increases as the size of the table increases. This limitation is overcome by λ, a proportional reduction-in-error statistic for nominal data that takes values in the interval 0 to 1. No relationship is implied if $\lambda = 0$; a perfect relationship exists if $\lambda = 1$.

Measures of association for ordinal data include gamma (γ), Spearman's ρ, and Kendall's W. All are proportional reduction-in-error statistics except for Kendall's W. Gamma utilizes the order of ranks for pairs of observations. The objective is to predict the order of ranks on the dependent variable based on knowledge about the independent variable. Spearman's ρ measures the strength of the relationship between the ranks of observations on one variable and the ranks of the same observations on a second variable. The objective of the researcher is to predict the rank of the dependent variable based on knowledge of the dependent variable (note the difference between the objectives for γ and ρ). Kendall's W extends this concept to more than two sets of rankings, giving a measure of the similarity of two rankings selected from two or more populations. Finally, note that both ρ and W can be employed as test statistics to test an hypothesis of "no association" between sets of rankings.

QUESTIONS AND PROBLEMS

29   As a measure of association, how does the phi coefficient differ from λ?

30   Using the data given, compute ϕ and then the contingency coefficient C. What is the value of C_{max} for these data? The value of C_{adj}? Interpret.

Type of Publisher by Type of College for Faculty Publications (Productive Faculty Only)

Type of College	Type of Publisher	
	Secular	Religious
Church-supported	64	29
State-supported	84	2

SOURCE: Stanley A. Clark and Richard F. Larson, "Mobility, Productivity, and Inbreeding at Small Colleges: A Comparative Study," *Sociology of Education, 45* (Fall 1972), p. 433.

31 In dealing with measures of association for nominal-level data, why might a social scientist prefer λ over other measures?

32 What is the primary advantage of using the contingency coefficient over ϕ?

33 Compute the value for C_{max} for 2×2, 3×3, 4×4, 5×5, and 6×6 tables. What do these values suggest about the nature of the contingency coefficient?

34 In general, and without reference to any specific measure of association, describe what is meant by the term "proportional reduction in error measure of association."

35 In computing λ, what is the no association rule? The perfect association rule?

36 How does λ_r differ from λ_c? Why would a social scientist be more interested in one as opposed to the other?

37 Compute and interpret λ_r using the data in the table. They come from a study by social psychologist David A. Ward in which he studied the relationship between occupations and orientation to modern life for the residents of Candelaria, Colombia.

A Study of Modernity Among 623 Residents of Candelaria, Colombia

Occupation	Orientation to Modern Life			Total
	High	Medium	Low	
Mental	148	27	8	183
Physical	253	105	82	440
Total	401	132	90	623

SOURCE: David A. Ward, "A Study of Modernity and Occupation in Candelaria, Colombia," unpublished paper; by permission.

38 Compute and interpret γ for the data given.

Relation Between Degree of Perceived Parental Interest and Self-esteem Among 461 High School Students

Self-esteem	Degree of Perceived Parental Interest			Total
	High	*Medium*	*Low*	
High	92	56	7	155
Medium	104	107	26	237
Low	15	44	10	69
Total	211	207	43	461

SOURCE: Billy L. Williams, "Self-esteem and Patterns of Group Memberships Among High School Adolescents," University of Florida, unpublished M.A. thesis, 1973; by permission.

39 Sociologist Marc Petrowsky conducted an investigation in the early 1970s in which he studied a number of sociology departments throughout the United States. The data contain the rank order of these departments by five variables. Measure the extent to which the departments are similarly ranked according to these variables. Is there evidence to indicate that the ranks on the five column headings are independent?

Department	Department Productivity Score	Size of Faculty	Productivity per Faculty	Productivity per Graduate	Number of Productive Graduates
Wisconsin	1	1	2	2	2
Washington (Seattle)	2	4.5	1	3	3
Stanford	3	7	4	6	6
Chicago	4	4.5	5	1	1
Vanderbilt	5	8.5	3	4	5
Missouri	6	2	8.5	7	7
Colorado	7	6	7	9	8.5
Pennsylvania	8	3	8.5	5	4
Notre Dame	9	10	6	8	8.5
New Hampshire	10	8.5	10	10	10

SOURCE: Marc Petrowsky, "Departmental Prestige and Scholarly Productivity: A Replication of Previous Research," University of Florida, unpublished M.A. thesis, 1971, pp. 69–77; by permission.

40 Consider the data in the table. Which statistical measure of association, λ or γ, would be more appropriate? Compute and interpret the appropriate measure.

Relationship Between Social Status of Father and Subject's Religiosity in Brazil

Religiosity	Social Status of Father	
	White Collar	Manual
Nonpracticing	43	149
Semipracticing	208	469
Practicing	134	173

SOURCE: E. Wilbur Bock and Suglyama Lutaka, "Social Status, Mobility and Premarital Pregnancy: A Case Study of Brazil," *Journal of Marriage and the Family, 32* (March 1970), p. 288; by permission.

41 Criminologist Don C. Gibbons investigated the reactions of San Francisco area adults to 20 criminal acts ranging from drunk driving to second-degree murder. Sociologist Kenneth J. Hodge repeated the study at the University of Florida. The crimes and their relative ranks, based on citizen reaction as to seriousness, are given in the table. Compute $\hat{\rho}$, the rank-order correlation coefficient. Test an hypothesis of no correlation between the crime rankings for the two cities. (Use $\alpha = .01$.)

Crime	San Francisco	University of Florida
Murder	1	3
Robbery	2	1
Burglary	3	2
Manslaughter	4	5
Rape	5	4
Embezzlement	6	6
Price rigging	7	7
Child molesting	8	8
Assault	9	14
Narcotics	10	10
Auto theft	11	11
Bad check	12	13
Misrepresenting advertising	13	9

(table continued overleaf)

Crime	San Francisco	University of Florida
Draft evasion	14	12
Exhibitionism	15	15
Homosexuality	16	18
Tax evasion	17	17
Marijuana	18	16
Statuatory rape	19	19
Drunk driving	20	20

SOURCE: Don C. B. Gibbons, "Crime and Punishment: A Study in Social Attitudes," *Social Forces, 48* (June 1969), p. 396; and Kenneth J. Hodge, "Student Conceptions and Reactions to Crime," University of Florida, unpublished M.A. thesis, 1971, p. 26; by permission.

REFERENCES

Blalock, H. M. *Social Statistics*, 2nd ed. New York: McGraw-Hill Book Company, 1972. Chapter 15.

Champion, D. J. *Basic Statistics for Social Research*. Scranton, Pa.: Chandler Publishing Company, 1970. Chapter 10.

Mueller, J. H., K. F. Schuessler, and H. L. Costner. *Statistical Reasoning in Sociology*, 2nd ed. Boston: Houghton Mifflin Company, 1970. Chapters 9 and 10.

Palumbo, D. J. *Statistics in Political and Behavioral Science*. New York: Appleton-Century-Crofts, 1969. Chapter 8.

Siegel, S. *Nonparametric Statistics for the Behavioral Sciences*. New York: McGraw-Hill Book Company, 1956. Chapter 9.

11

REGRESSION AND CORRELATION

11.1 INTRODUCTION

Chapter 11 extends our study of the relations between two variables measured on nominal or ordinal scales to the case where the data have been measured on interval or ratio scales. The chi-square test used in Chapter 9 was to determine "dependency" and, by implication, the strength of the relation between the two variables. Measures of the strength of a relationship, based on various proportional reduction-in-error statistics, such as lambda, gamma, and rho, were presented in Chapter 10. The objective was to establish whether two variables, measured on an appropriate scale—nominal or ordinal—were related and to measure the strength of the relationship. The social scientist's ultimate goal in many cases is to predict one variable based on information concerning another. This chapter considers exactly the same problems, with the same objectives and goals, as those discussed in Chapters 9 and 10 except that we now assume that the data for both variables have been measured on an interval or ratio scale.

For example, suppose that we wish to determine whether per capita income (X) in a state is related to the total expenditure per child per year in the state's public school system (Y). If we collect data on X and Y for a sample of states, we shall want to know whether the data provide sufficient evidence to indicate a relationship. We shall also want to know how strong is the relationship and be able to predict per child expenditures. If we think the per capita income in a state is increasing and will reach a level of $5500 in the near future, we might wish to predict the annual (state) expenditure per child that the state's legislature will approve for public education.

Often the social scientist seeks a more sophisticated model to predict one variable based on knowledge of the values of two or more independent variables. For example, prediction of expenditures per child per year for public education (Y) would likely be more accurate if we were to utilize, in addition to per capita income, knowledge of the assessed valuation of property throughout the state, the rate of economic inflation, the rate of growth of public school attendance, as well as many other variables. This multivariable model should be able to predict per child expenditures as well, or better, than if we used information solely on per capita income. This chapter, which confines its attention to a study of a straight-line (linear) relationship between two variables, illustrates concepts and techniques that have been developed for examining multivariable relationships. These multivariable techniques, which are beyond the scope of this text, are discussed in Blalock (1972) as well as other texts listed at the end of the chapter.

11.2 SCATTER DIAGRAMS
AND THE FREEHAND
REGRESSION LINE

efore specifying a measure of association between two quantitative variables measured on an interval or ratio scale, let us develop an
intuitive grasp of our problem by plotting data corresponding to measurements on two variables, *X* and *Y*, using a scatter diagram. To illustrate we
have taken 1970 census data for nine states to obtain information on per
capita income (*X*) and state expenditure per student for public education (*Y*).
In particular, we are interested in determining the strength of the relationship
between the two variables. The census data appear in Table 11.1.

Table 11.1 1970 *Census Data on per Capita Income and State
Expenditures on Education for Nine States*

State	Per Capita Income *X*	Public Education per Student Expenditure *Y*
	$	$
Arkansas	2520	534
California	4272	922
Colorado	3568	695
Michigan	3944	842
Mississippi	2194	476
North Carolina	2890	609
New York	4421	1237
Rhode Island	3779	904
South Dakota	3051	657

SOURCE: Bureau of the Census, *Statistical Abstract of the United States:
1970,* Washington, D.C., pp. 100, 320.

To aid in the interpretation of the relationship between the variables
"per capita income" and "state expenditure in education," it is useful to
plot the data of Table 11.1 in a scatter diagram. To do this we construct
horizontal and vertical axes of approximately equal length. Social scientists
generally agree that the independent variable (*X*) should be labeled along the
horizontal axis while the dependent variable (*Y*) is labeled along the vertical
axis. For our example it is convenient to label per capita income as the

independent variable and state expenditure per student as the dependent variable since we would expect the level of per capita income to influence expenditures on education. In some situations it is not possible to state that one variable depends on the other, so we would just make an arbitrary designation for our scatter diagram.

Having labeled the axes, we then draw scales along the axes in such a way that all measurements can easily be plotted along the appropriate scale. For our example per capita income ranges from $2192 to $4421 and state expenditure per student for public education ranges from $476 to $1237. Note that values on each of the variables fall within the scales chosen in Figure 11.1.

Having drawn, labeled, and scaled the axes of the scatter diagram, we plot the data of Table 11.1. Each dot on the scatter diagram represents the information concerning one state and can be obtained by plotting the state

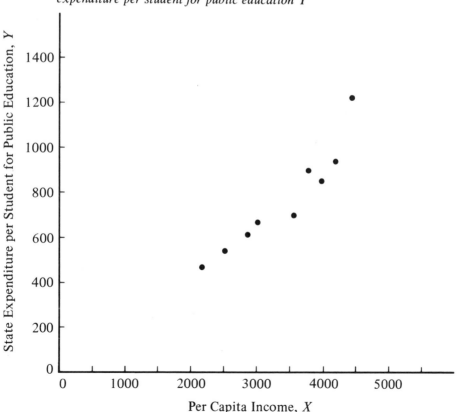

Figure 11.1   *Scatter diagram of per capita income X and state expenditure per student for public education Y*

educational expenditure per student (Y) versus the corresponding per capita income (X). It appears from Figure 11.1 that the state educational expenditure per student increases as the per capita income of a state increases. In fact, most dots of the scatter diagram lie on a straight line. We call a line running through the dots of a scatter diagram a trend line or regression line. In this case the trend line is a straight line and we say there is a linear relationship between X and Y [see Figure 11.2(a) and (b)]. However, not all regression lines are linear. In some cases we shall observe a trend line that is curved and then we would say there was a curvilinear relationship between X and Y [see Figure 11.2(c) and (d)].

Figure 11.2   *Different types of regression lines:* (a) *linear relationship between X and Y;* (b) *linear relationship between X and Y;* (c) *curvilinear relationship between X and Y;* (d) *curvilinear relationship between X and Y*

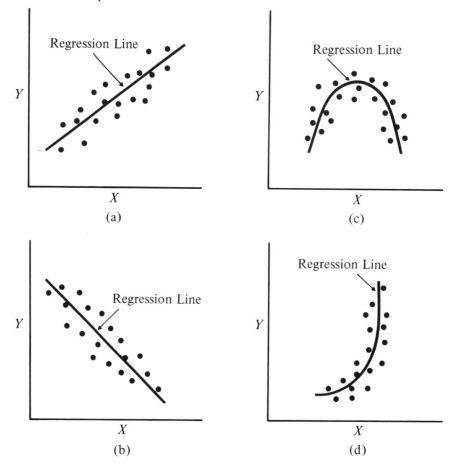

There are many methods for obtaining a regression line relating Y to X. The first is called an "eyeball fit" or freehand regression line and can be obtained by placing a ruler on the graph (Figure 11.1) and moving it about until it seems to pass through as many of the dots as possible (see Figure 11.3). The resulting line can be used to predict a state's educational expenditure based on the per capita income. To predict Y when $X = 3500$, refer to the graph and note that the coordinate for the point corresponding to $X = 3500$ is $Y = 800$ (see the arrows on Figure 11.3). We can predict Y from X using our regression line.

The freehand regression line in Figure 11.3 can be represented by an equation of the form

$$Y = a + bX$$

Only points (pairs of values for X, Y) that fall on the line will satisfy the equation. The two constants in the equation, a and b, determine the location

Figure 11.3 *Freehand regression line for the data of Table 11.1*

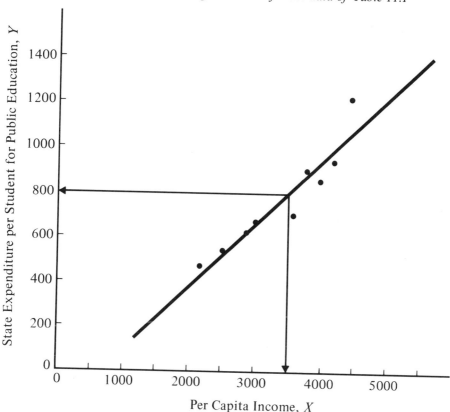

and slope of the line. a is the Y intercept, that is, the value of Y at the point where the line crosses the Y axis ($X = 0$); b is the slope of the regression line, that is, the increase in Y which corresponds to a one-unit increase in X (see Figure 11.4).

Figure 11.4 *Slope b and intercept a for the equation Y = a + bX*

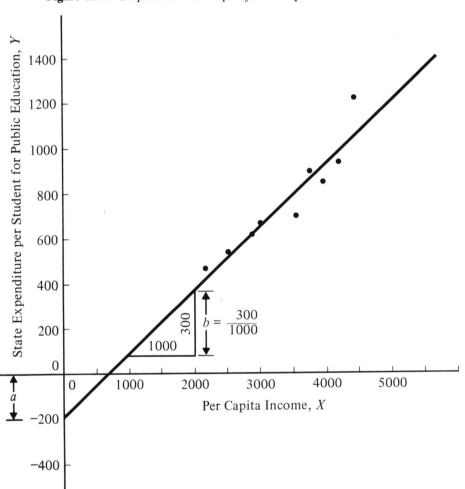

Looking at Figure 11.4, it appears that the Y intercept is approximately -200, so $a = -200$. The increase in Y for a 1000-unit increase in X appears to be 300. The slope is approximately $300/1000 = .3$ and the equation corresponding to the eyeball-fitted line or the freehand regression line of Figure 11.3 is $Y = -200 + .3X$. To predict Y when $X = 3500$ we substitute $X = 3500$

into the equation to obtain $Y = -200 + .3(3500) = 850$. We see that we can predict Y using either the graph, Figure 11.3, or the equivalent prediction equation, $Y = -200 + .3X$.

Although the freehand regression line provides us with a prediction equation; there could be many different prediction equations from different "eyeball" fits to the same data. What we seek is a precise procedure for determining the constants a and b in our prediction equation.

$$Y = a + bX$$

The procedure that we shall use for determining a and b is called the method of least squares.

EXERCISES

1   What is meant by a regression line?

2   What is meant by a scatter diagram?

3   Does a freehand regression line have any utility? Explain.

4   Plot a scatter diagram and draw a freehand regression line for the data in the table.

Religiosity Score	*Social Conscience Score*
7	9
4	8
3	3
12	15
8	13
1	4

5   What is meant by the Y intercept of a line?

6   Can you construct a scatter diagram for ordinal-level data? Explain.

11.3 METHOD OF LEAST SQUARES

*T*he statistical procedure for finding the best prediction equation is, in many respects, an objective way to obtain an eyeball fit to the points. For example, when we "eyeball" a line to a set of points we move the ruler until we think that we have minimized the distances from the points to the fitted line. We shall denote the predicted value of Y for a given value of X (obtained from the fitted line) as $\hat{Y}$ and the prediction equation as

$$\hat{Y} = a + bX$$

Figure 11.5 *Least-squares fit to the data in Table 11.1*

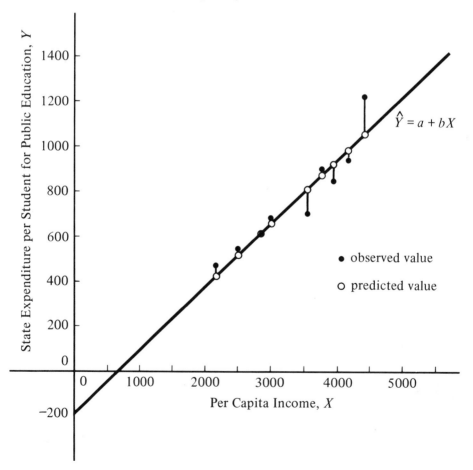

The vertical distance from a point to the prediction line represents the deviation of a point from the predicted value of Y (see Figure 11.5). Symbolically this deviation is $(Y - \hat{Y})$. To find the best prediction equation (regression line) we can work with deviations from the prediction line $\hat{Y}$. The method of least squares chooses the prediction that minimizes the sum of squares of the deviations of the observed from the predicted values of Y.

The sum of squares of deviations, denoted by SSE, can be written as

$$\text{SSE} = \Sigma\,(Y - \hat{Y})^2$$

where

$$\hat{Y} = a + bX$$

Substituting for $\hat{Y}$, we have

$$\text{SSE} = \Sigma\,[Y - (a + bX)]^2$$

The method of least squares chooses those values for a and b that make SSE a minimum. Derivation of these values is beyond the scope of this text, but they can be found using the formulas below.

Least-Squares Formulas for Computing a and b

$$b = \frac{S_{xy}}{S_{xx}}$$

and

$$a = \bar{Y} - b\bar{X}$$

where

$$S_{xx} = \Sigma\,X^2 - \frac{(\Sigma\,X)^2}{n}$$

and

$$S_{xy} = \Sigma\,XY - \frac{(\Sigma\,X)(\Sigma\,Y)}{n}$$

The use of these formulas for finding a and b and the least-squares line will be illustrated by an example.

Example 11.1 *Obtain the best-fitting prediction line for the educational expenditure data in Table 11.1 using the method of least squares.*

Solution *The calculation of a and b using the method of least squares is greatly simplified using Table 11.2.*

Table 11.2 *Calculation of a and b Using the Method of Least Squares for the Data of Table 11.1*

X	Y	X^2	XY	Y^2
2520	534	6,350,400	1,345,680	285,156
4272	922	18,249,984	3,938,784	850,084
3568	695	12,730,624	2,479,760	483,025
3944	842	15,555,136	3,320,848	708,964
2194	476	4,813,636	1,044,344	226,576
2890	609	8,352,100	1,760,010	370,881
4421	1237	19,545,241	5,468,777	1,530,169
3779	904	14,280,841	3,416,216	817,216
3051	657	9,308,601	2,004,507	431,649
30,639	6876	109,186,563	24,778,926	5,703,720

Substituting into the formulas we obtain

$$S_{xx} = \Sigma X^2 - \frac{(\Sigma X)^2}{n} = 109,186,563 - \frac{(30,639)^2}{9}$$

$$= 4,881,194$$

$$S_{xy} = \Sigma XY - \frac{(\Sigma X)(\Sigma Y)}{n} = 24,778,926 - \frac{(30,639)(6876)}{9}$$

$$= 1,370,730$$

$$\overline{Y} = \frac{\Sigma Y}{n} = \frac{6876}{9} = 764.0$$

and

$$\overline{X} = \frac{\Sigma X}{n} = \frac{30,639}{9} = 3404.33$$

Hence

$$b = \frac{S_{xy}}{S_{xx}} = \frac{1{,}370{,}730}{4{,}881{,}194} = .28$$

and

$$a = \bar{Y} - b\bar{X} = 764.0 - .28(3404.33)$$

$$= 764.0 - 953.21 = -189.21$$

The least-squares prediction equation relating state expenditure per student for public education $\bar{Y}$ to the corresponding state per capita income X is then

$$\hat{Y} = -189.21 + .28X$$

One important comment should be made concerning the least-squares prediction equation. The slope of the least-squares line, b, depends on the magnitude of the units of X and Y. Hence one cannot compare slopes of different regression lines unless the X's and Y's are recorded in the same units (such as inches, feet, meters, miles, etc.).

To avoid this difficulty we could work in the standardized units (z scores)

$$z_x = \frac{X - \bar{X}}{\sigma_x}$$

and

$$z_y = \frac{Y - \bar{Y}}{\sigma_y}$$

where

$$\sigma_x = \sqrt{\frac{S_{xx}}{n}}$$

$$\sigma_y = \sqrt{\frac{S_{yy}}{n}}$$

and

$$S_{yy} = \Sigma\, Y^2 - \frac{(\Sigma\, Y)^2}{n}$$

Rewriting our regression line in terms of z scores we have

$$\hat{z}_y = a_s + b_s z_x$$

where a_s is the intercept and b_s the slope of the regression line computed from standardized measurements z_x and z_y. Without proof we simply state that it can be shown that the least-squares value of a_s will always equal zero. Hence our regression line for standardized units would be

$$\hat{z}_y = b_s z_x$$

where b_s is computed as it was previously with X replaced by z_x and Y replaced by z_y. Thus

$$b_s = \frac{S_{z_x z_y}}{S_{z_x z_x}}$$

This formula simplifies to

$$b_s = \frac{\Sigma z_x z_y}{n}$$

Example 11.2 *Use the data of Table 11.1 to find the least-squares prediction equation*

$$\hat{z}_y = b_s z_x$$

Solution *The first step in solving this problem is to transform the X, Y data to z scores. Recall from Example 11.1 that we computed S_{xx}, $\overline{X}$, and $\overline{Y}$ to be*

$$S_{xx} = 4{,}881{,}194$$
$$\overline{X} = 3{,}404.33$$

and

$$\overline{Y} = 764.0$$

Similarly, we find that S_{yy} can be computed using the shortcut formula

$$S_{yy} = \Sigma Y^2 - \frac{(\Sigma Y)^2}{n}$$

Substituting from Table 11.2 we have

$$S_{yy} = 5{,}703{,}720 - \frac{(6876)^2}{9} = 450{,}456$$

The quantities σ_x and σ_y are then

$$\sigma_x = \sqrt{\frac{S_{xx}}{n}} = \sqrt{\frac{4{,}881{,}194}{9}}$$

$$= 736.45$$

and

$$\sigma_y = \sqrt{\frac{S_{yy}}{n}} = \sqrt{\frac{450{,}456}{9}}$$

$$= 223.72$$

We can now compute z scores for the X and Y of Table 11.3 using the formulas

$$z_x = \frac{X - \bar{X}}{\sigma_x} = \frac{X - 3404.33}{736.45}$$

$$z_y = \frac{Y - \bar{Y}}{\sigma_y} = \frac{Y - 764.0}{223.72}$$

Substituting each value of X and Y into the appropriate z formula we obtain the z scores listed in Table 11.3. Using the data of Table 11.3 we

Table 11.3 *Raw Scores, z Scores, and Cross Products for the Data of Table 11.1*

X	z_x	Y	z_y	$z_x z_y$
2520	− 1.2008	534	− 1.0281	1.2345
4272	1.1782	922	.7062	.8321
3568	.2222	695	− .3084	− .0685
3944	.7328	842	.3486	.2555
2192	− 1.6435	476	− 1.2873	2.1157
2890	− .6984	609	− .6928	.4839
4421	1.3805	1237	2.1142	2.9187
3779	.5088	904	.6258	.3184
3051	− .4798	657	− .4783	.2295

can proceed to compute b_s, the slope of the standardized least-squares regression line. Recall that

$$b_s = \frac{\Sigma z_x z_y}{n}$$

Substituting, we have

$$b_s = \frac{\Sigma z_x z_y}{n} = \frac{8.3198}{9} = .92$$

And the standardized least-squares prediction equation is

$$\hat{z}_y = b_s z_x = .92 z_x$$

Note we predict z_y for a given value of z_x by substituting into the prediction equation. For a standardized per capita income of $z_x = 1.20$, we would predict the standardized data expenditure per person for public education to be

$$\hat{z}_y = b_s z_x = .92(1.20) = 1.10$$

The quantity b_s, the slope of the standardized least-squares prediction equation, is a widely used measure of the strength of the linear relationship between two variables, X and Y. It is commonly called the Pearson product moment correlation coefficient and denoted by the letter r.

Studies of the relationship between two or more variables, measured on an interval or ratio scale, are numerous. Goode (1972) studied political activity and collective violence. He reports a correlation of .72 between the number of participants in collective violence and the number of arrests made. King and Hunt (1972) studied the relationship between frequency of church attendance and organizational activity ($r = .64$). In a third recent example, Kasarda (1972) studied the relationship between organizational components of central cities and the population size (log) of portions of SMSA's. To understand the meaning of this quantity we refer to the computational formula

$$r = \frac{\Sigma z_x z_y}{n}$$

Note that r is merely the mean of the cross products, $z_x z_y$. The terms "mean" and "moment" are synonymous in statistics; hence Karl Pearson, who first formulated this measure of association, labeled his statistic a product-moment correlation coefficient. The quantity r, the slope of the standardized regression line, represents the change in z_y for a one-unit change in z_x and can be interpreted much as b was in the regression line for Y and X. A positive value of r (positive slope) indicates that z_y increases with z_x [see **Figure 11.6(a)**].

Figure 11.6 *Interpretations of* r: (a) *positive-linear correlation;* (b) *negative-linear correlation;* (c) *no linear correlation*

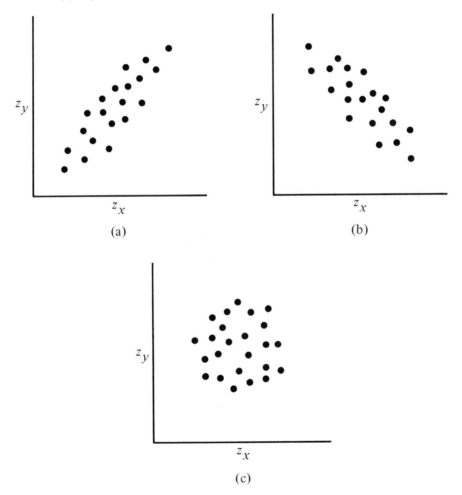

(a)

(b)

(c)

Similarly, a negative r (negative slope) implies that z_y decreases as z_x increases [see **Figure 11.6(b)**]. A value of $r = 0$ indicates no relationship between the standardized variables z_x and z_y [see **Figure 11.6(c)**].

We have now seen that the sign of r indicates the slope of the line $\hat{z}_y = r z_x$ and identifies the direction of the relationship between the original variables, X and Y (see Figure 11.7). A more precise interpretation to r will be given after we study the quantity r^2.

Figure 11.7 *Implications of possible values for r, the Pearson product moment correlation coefficient*

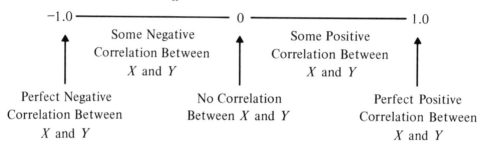

EXERCISES

7   Use the data in Exercise 4 (page 395) to compute a and b.

8   Use the data in Exercise 4 (page 395) to compute all the predicted Y values based on $a + bX$ and compare them with the observed values of Y. Plot the actual regression line.

9   Repeat Exercise 7 using standard scores.

10   Repeat Exercise 8 using standard scores.

11   Using the standard scores, compute r for the data in Exercise 4 (page 395).

11.4   THE COEFFICIENT OF DETERMINATION r^2

The coefficient of determination can be used to measure the strength of the relationship between two variables, X and Y. If we assume that Y and X are linearly related, we can calculate the least-squares regression line

$$\hat{Y} = a + bX$$

where

$$b = \frac{S_{xy}}{S_{xx}} \quad \text{and} \quad a = \bar{Y} - b\bar{X}$$

In addition, we found that the sum of squares of the deviations (often called the "sum of squares error") about the prediction line is given by

$$\text{SSE} = \sum (Y - \hat{Y})^2$$

Now suppose that you do not assume that X and Y are linearly related but rather that X contributes no information for predicting Y. Then you would drop the term bX out of the prediction equation and would predict Y to equal the sample mean,

$$\hat{Y} = \bar{Y}$$

This prediction line is shown in Figure 11.8(b) along with the vertical lines representing the error of prediction. Compare the magnitude of the deviations of the data points from the line

$$\hat{Y} = a + bX$$

Figure 11.8 *Two models fit to the same data*

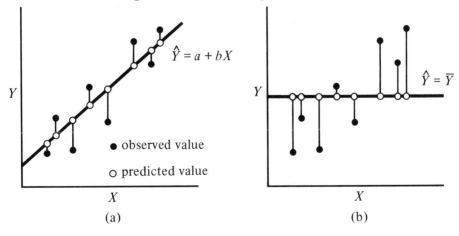

for Figure 11.8(a) with those about the line

$$\hat{Y} = \overline{Y}$$

of Figure 11.8(b). This subjective comparison can be expressed in a more objective way by comparing the sums of squares error for the two prediction equations.

The sum of square of deviations about the prediction equation $\hat{Y} = \overline{Y}$ is often called the total variation

$$\Sigma (Y - \hat{Y})^2 = \Sigma (Y - \overline{Y})^2 = S_{yy}$$

The reduction in the sum of squares error obtained by using the information on the independent variable X is then

$$S_{yy} - SSE$$

This quantity is called the explained variation, that is, that amount of the total variation in the Y values that can be explained by the X values. The proportional reduction in the error sum of squares is called the coefficient of determination r^2:

$$r^2 = \frac{\text{explained variation}}{\text{total variation}} = \frac{S_{yy} - SSE}{S_{yy}}$$

We can think of r^2 as being analogous to a proportional reduction in error statistic, as in Chapter 10. The difference here is that we compare the sum of squares of the errors for two prediction rules, whereas in Chapter 10 we compare the number of errors for two prediction rules. The "no association" rule for r^2 is

$$\hat{Y} = \overline{Y}$$

(i.e., X contributes no information for the prediction of Y). The corresponding "perfect association" rule is

$$\hat{Y} = a + bX$$

Substituting into the formula for r^2 we obtain the following computational formula for r^2 (and hence r):

Computational Formula for r^2 and r

$$r^2 = \frac{S_{xy}^2}{S_{xx}S_{yy}}$$

$$r = \sqrt{r^2} = \sqrt{\frac{S_{xy}^2}{S_{xx}S_{yy}}}$$

Recall that when using z scores we can compute r as

$$r = \frac{\Sigma z_x z_y}{n}$$

Example 11.3   *Use the data of Table 11.1 and the computational formula above to compute r^2 and r.*

Solution   *Recall that we previously computed S_{xy}, S_{xx}, and S_{yy} to be*

$$S_{xy} = 1{,}370{,}730$$
$$S_{xx} = 4{,}881{,}194$$

and

$$S_{yy} = 450{,}456$$

Substituting, we have

$$r^2 = \frac{S_{xy}^2}{S_{xx}S_{yy}} = \frac{(1{,}370{,}730)^2}{4{,}881{,}194(450{,}456)} = .85$$

and

$$r = \sqrt{.85} = .92$$

By incorporating the information on per capita income (X) into our prediction model, we have achieved an 85 percent reduction in the sum of squares error for the prediction equation, $\hat{Y} = \bar{Y}$. We note also that .92 is identically the same value as the one that we obtained for the Pearson product moment correlation coefficient using standardized scores. Hence we have two different ways of calculating r.

As we have seen, the coefficient of determination is a proportional reduction in error statistic which identifies the proportional drop in the sum of squares error obtained by using the model $\hat{Y} = a + bX$ rather than the model $\hat{Y} = \bar{Y}$. If the independent variable X contributes no information toward the prediction of Y, the coefficient of determination would be $r^2 = 0$. Similarly, if the incorporation of X into the model allows us to predict Y perfectly (with no error), then $r^2 = 1$.

We can now clear up some misinterpretations of the coefficient of linear correlation of Section 11.3. A coefficient of correlation equal to .5 does not mean that the strength of the relationship between Y and X is "halfway" between no correlation and perfect correlation. As we have stated, the fraction of the total variation in the Y values that is attributable to X is equal to r^2. If $r = .5$, the independent variable X is accounting for $r^2 = .25$ of the total variation in Y. A second point to note is that Y and X could be perfectly related in some way other than in a linear manner when $r = 0$ or some very small value (see Figure 11.9).

Figure 11.9   *Perfect curvilinear fit with r = 0*

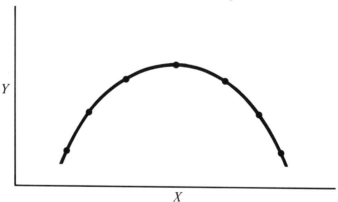

Finally, note that you cannot add correlations. If the simple linear correlations between Y and X_1, Y and X_2, and Y and X_3 are .10, .30, and .20, respectively, it *does not* follow that X_1, X_2, and X_3 account for $(.10)^2 + (.20)^2 + (.30)^2$ of the variability. In actuality these variables may be highly correlated and contribute the same information for the prediction of Y. The relationship between Y and several independent variables should not be studied by computing simple correlation coefficients for each of the independent variables. Rather, we should relate Y to X_1, X_2, and X_3 using a single multivariable model. This topic is beyond the scope of this text. The interested student should see Blalock (1972).

To summarize, the coefficient of variation r^2 determines the amount of the total variabilities in the Y values which can be accounted for using the model $\hat{Y} = a + bX$. In contrast, the Pearson product moment correlation coefficient r is the slope of the regression line $\hat{z}_y = rz_x$ and can be used for prediction purposes and for information concerning how X and Y vary. Thus the sign of r designates whether X and Y are positively related.

EXERCISES

12 What is the value of r^2 for the data in Exercise 4? (*Hint* : See answer to Exercise 11, page 404). Interpret your result.

13 Explain the differences in interpretation between r and r^2.

14 State the "perfect association" rule for r^2.

15 Is r^2 meaningful if one of the variables is measured on an ordinal scale and the other on a ratio?

m **11.5 COMPUTING FORMULAS
FOR r**

any formulas are available for calculating r.

Computational Formulas for r

Ungrouped data, raw scores

$$r = \sqrt{\frac{S_{xy}^2}{S_{xx}S_{yy}}}$$

where

$$S_{xy} = \Sigma XY - \frac{(\Sigma X)(\Sigma Y)}{n}$$

$$S_{xx} = \Sigma X^2 - \frac{(\Sigma X)^2}{n}$$

and

$$S_{yy} = \Sigma\, Y^2 - \frac{(\Sigma\, Y)^2}{n}$$

Ungrouped data, z scores

$$r = \frac{\Sigma\, z_x z_y}{n}$$

Grouped data, raw scores

$$r = \frac{n\Sigma f_{xy}X_c Y_c - (\Sigma f_x X_c)(\Sigma f_y Y_c)}{\sqrt{\left[n\Sigma f_x X_c^2 - (\Sigma f_x X_c)^2\right]\left[n\Sigma f_y Y_c^2 - (\Sigma f_y Y_c)^2\right]}}$$

where X_c and Y_c are coded values obtained from grouped data, as we discussed in Section 4.5.

Example 11.4 *A study was conducted to determine the strength of the relationship between the fertility ratio and the average number of persons per household for SMSA's. In particular, can the fertility ratio of an SMSA be predicted from knowledge of the number of persons per household for that area? Data for 243 SMSA's have been cross-classified by persons per household and fertility ratio in Table 11.4. Use the observed data to obtain r, the Pearson product moment correlation coefficient.*

Solution *Using the X values in Table 11.4, we shall look at the class intervals and their corresponding frequencies f_x. Based on inspection alone, we select the class interval that we think is likely to contain the mean, $\bar{X}$. The midpoint of that interval is defined to be m_x. For our example choose $m_x = 3.145$, the midpoint of the sixth interval on X. (Recall that the choice of m_x is not critical; just pick a class interval near the center of the X distribution.) Now subtract m_x from each midpoint value, X. We can further reduce the difficulty of computation by dividing each of the differences, $X - m_x$, by the interval width of the X's, i_x. For $i_x = .1$, the coded midpoints are then computed as*

$$X_c = \frac{X - m_x}{i_x} = \frac{X - 3.145}{.1}$$

Those values appear in the left-hand column of Table 11.5.

Table 11.4 *Cross-Classification of Persons per Households and Fertility Ratios for 243 SMSA's*

Persons per Household Class Interval	Class Midpoint X	Fertility Ratios* 255–269 262	270–284 277	285–299 292	300–314 307	315–329 322	330–344 337	345–359 352	360–374 367	375–389 382	390–404 397	405–419 412	420–434 427	435–449 442	450–464 457	465–479 472	Total
2.60–2.69	2.645					1											1
2.70–2.79	2.745			1		1		1									3
2.80–2.89	2.845	1		1	3	2	2	1									10
2.90–2.99	2.945		2	1	3	5	10	1	1								23
3.00–3.09	3.045		1	3	10	10	13	2									62
3.10–3.19	3.145			2	3	7	8	11	8	5							67
3.20–3.29	3.245					3	7	22	12	9	4						38
3.30–3.39	3.345						1	9	8	6	4	1					17
3.40–3.49	3.445							4	2	6	1	2	1				12
3.50–3.59	3.545								1	2	3	3	1	1			4
3.60–3.69	3.645										2		1	2			3
3.70–3.79	3.745																0
3.80–3.89	3.845														1		0
3.90–3.99	3.945															1	1
4.00–4.09	4.045													1			1
4.10–4.19	4.145														2	2	1
Total		1	3	7	19	29	41	50	33	28	16	6	3	3	2	2	243

*The first numbers represent the class interval; the next number is the class midpoint, Y.

Table 11.5 Computations for Example 11.4*

							Y_c												
X_c	-6	-5	-4	-3	-2	-1	0	1	2	3	4	5	6	7	8	f_x	f_xX_c	$f_xX_c^2$	$f_{xy}X_cY_c$
-5					1 (10)											1	-5	25	10
-4			1 (16)		1 (8)		1 (0)									3	-12	48	24
-3	1 (18)			2 (18)	3 (18)	2 (6)	2 (0)									10	-30	90	60
-2		2 (20)	1 (8)	4 (24)	5 (20)	7 (14)	4 (0)									23	-46	92	86
-1		1 (5)	3 (12)	7 (21)	11 (22)	15 (15)	15 (0)	6 (-8)	5 (-10)	1 (-4)	1 (-4)					62	-62	62	58
0				3 (0)	7 (0)	9 (0)	22 (0)	9 (0)	9 (0)	4 (0)						67	0	0	0
1					3 (-6)	8 (-8)	12 (0)	9 (0)	6 (12)	4 (12)		1 (6)				38	38	38	25
2				1 (-6)	4 (-7)	2 (-4)	4 (0)	2 (6)	3 (24)	1 (10)	2 (16)		2 (36)			17	34	68	58
3			1 (-2)	1 (-6)	2 (12)	1 (3)	3 (0)	2 (27)	2 (36)	3 (36)	1 (15)	1 (20)		1 (56)		12	36	108	129
4								1 (5)	2 (24)	2 (30)					1 (32)	4	16	64	76
5						1 (5)					1 (20)					3	15	75	35
6																0	0	0	0
7																0	0	0	0
8						1 (8)									1 (72)	1	8	64	56
9																1	9	81	72
10															1 (70)	1	10	100	70
f_y	1	3	7	19	29	41	50	33	28	16	6	3	3	2	2	243	11	915	759
f_yY_c	-6	-15	-28	-57	-58	-41	0	33	56	48	24	15	18	14	16	19			Total
$f_yY_c^2$	36	75	112	171	116	41	0	33	112	144	96	75	108	98	128	1345			

Total

*Numbers inside the parentheses are $f_{xy}X_cY_c$; $\Sigma f_{xy}X_cY_c = 759$.

In a similar way we code the Y values of Table 11. 4 using the formula

$$Y_c = \frac{Y - m_y}{i_y} = \frac{Y - 352}{15}$$

The values appear in the first row of Table 11.5. The row and column entries $f_x, f_x X_c, f_x X_c^2, f_y, f_y Y_c$, and f_y, Y_c^2 are computed in the usual way and appropriate totals, Σf_x. $\Sigma f_x X_c$, $\Sigma f_x X_c^2$, Σf_y, $\Sigma f_y Y_c$, and $\Sigma f_y Y_c^2$, are given at the bottom of Table 11.5. The only entries that might cause some difficulty are those for the last column, labeled $f_{xy} X_c Y_c$. For a given row, we take the coded midpoint X_c for that row and multiply it times all cell frequencies f_{xy} and their class midpoints Y_c for that row. The sum of these values is the entry for the column $f_{xy} X_c Y_c$. For example, in the second row of Table 11.5, the coded class midpoint is -4. Multiplying $X_c = -4$ times all cell frequencies in the first row and their class midpoints Y_c we have

$$f_{xy} X_c Y_c = -4(-4) + (-4)(-2) + (-4)(0) = 24$$

This entry appears in the last column of Table 11.5. The sum of the entries in this column is

$$\Sigma f_{xy} X_c Y_c = 759$$

Using the appropriate summary data of Table 11.5 and the fact that

$$\Sigma f_x = \Sigma f_y = n$$

we can compute r as

$$r = \frac{n\Sigma f_{xy} X_c Y_c - (\Sigma f_x X_c)(\Sigma f_y Y_c)}{\sqrt{[n\Sigma f_x X_c^2 - (\Sigma f_x X_c)^2][n\Sigma f_y Y_c^2 - (\Sigma f_y Y_c)^2]}}$$

$$= \frac{243(759) - (11)(19)}{\sqrt{[243(915) - (11)^2][243(1345) - (19)^2]}}$$

$$= \frac{184,228}{\sqrt{(222,224)(326,474)}} = \frac{184,228}{269,351.74} = .68$$

Thus the independent variable, number of children per household, accounts for $(.68)^2 = .46$ of the total variability in the dependent variable, fertility ratio.

11.6   A TEST OF SIGNIFICANCE CONCERNING *r*

In the material presented thus far in Chapter 11 we have used *r* (and r^2) in strictly a descriptive sense under the assumption that we were *not* sampling from a larger body of data. In situations where our data represent a random sample from a larger body of data, the population, the Pearson product moment correlation $\hat{r}$, will be a sample correlation coefficient, which is an estimate of the population correlation coefficient *r*. We may be interested in a test of the null hypothesis $H_0 : r = 0$ (i.e., there is no linear correlation between *X* and *Y*). This test is summarized below.

Test of an Hypothesis Concerning *r*, the Population Correlation Coefficient

Null Hypothesis : $r = 0$

Alternative Hypothesis : $r \neq 0$ *(two-tailed test)*

Test Statistic :

$$t = \hat{r}\sqrt{\frac{n-2}{1-\hat{r}^2}}$$

Rejection Region : *For a specified value of* α, *reject the null hypothesis if the absolute value of t exceeds the tail-end value with* $a = \alpha/2$ *and d.f.* $= n - 2$.

Example 11.5   *Suppose that the per capita income and state public education expenditure data of Table 11.1 represented a random sample of nine states drawn from the population of 50 states. The research hypothesis for this study is that an increase in a state's expenditures per student on public education (Y) accompanies an increase in per capita income (X). Use the sample data to test the null hypothesis $r = 0$ (i.e., there is no linear correlation between X and Y).*

Solution   *The alternative hypothesis for our test is $H_a : r > 0$ since we expect that there is a positive correlation between per capita income and state expenditure on education. The null hypothesis is then $H_0 : r = 0$.*

In Example 11.2 we computed the sample correlation coefficient
to be $\hat{r} = .92$. Hence the test statistic is

$$t = \hat{r}\sqrt{\frac{n-2}{1-\hat{r}^2}} = .92\sqrt{\frac{7}{1-(.92)^2}}$$
$$= .92\sqrt{45.57} = .92(6.75) = 6.21$$

*For n = 9 and α = .05, we shall reject if the observed value of t is greater
than the tabulated value with d.f. = n − 2 = 7 and a = .05. From Table
2 of the Appendix, the critical value of t is t = 1.895. Since the observed
value of t is greater than 1.895, we reject the null hypothesis and conclude
that r is greater than zero. Thus as per capita income increases through
salary increases, state educators can begin to predict the money that will
be available for public schools.*

Figure 11.10 *Various shapes for scatter diagrams: (a) homoscedastic;
(b) heteroscedastic gourd-shaped; (c) heteroscedastic dumbbell-shaped*

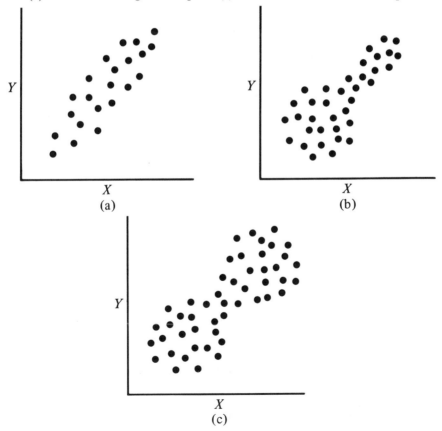

Several comments should be made concerning the assumptions for using the *t* test just described. First, the scatter of *Y* values for corresponding *X* values, known as the scedasticity, should be uniform across values of *X* [see Figure 11.10(a)]. If this relationship holds, we say that the *Y* values are homoscedastic; if not, the *Y* values are heteroscedastic. This condition can be easily checked using the scatter diagram. Several kinds of heteroscedasticity are illustrated in Figure 11.10(b) and (c). Second, although a test of the null hypothesis $H_0 : r = 0$ may turn out to be nonsignificant, this may not mean that the two variables *X* and *Y* are unrelated. Rather it implies that we have insufficient evidence to indicate that *X* and *Y* are linearly related.

EXERCISES

16 Explain the concept of homoscedasticity.

17 If $\hat{r} = +.80$ and $n = 100$, is *r* significantly different from zero? (Use $\alpha = .01$.)

18 Compute $\hat{r}$ for the following data on 12 high school students:

Grade-Point Average	IQ
2.1	116
2.2	129
3.1	123
2.3	121
3.4	131
2.9	134
2.9	126
2.7	122
2.1	114
1.7	109
3.3	132
3.5	140

19 Refer to Exercise 18. Is *r* significantly different from zero? (Use $\alpha = .05$.) Explain.

20 Refer to Exercise 18. Find the coefficient of determination and interpret it for this set of data.

21 What are some of the factors which might account for the fact that $\hat{r}$ is not unity for Exercise 18?

T   11.7   SUMMARY

his chapter presents techniques for studying the relationship between two variables measured on an interval or a ratio scale. The scatter diagram provides a graphic method for viewing the relationship between two variables. A regression line, computed from sample data and graphed on the scatter diagram, yields further graphic evidence of this relationship and permits the prediction of the dependent variable Y based on knowledge of the independent variable X.

A numerical measure of the association between Y and X was provided by the Pearson product moment correlation coefficient r, which was, in fact, the slope of the computed regression line when using standardized scores for the variables X and Y. We noted that r ranged between -1 and 1 with values near 0 implying little or no relationship between Y and X. Values near -1 or 1 indicated a strong relationship. The sign of r was also significant: positive values imply that Y increases as X increases; negative values indicate that Y decreases as X increases. A test of the hypothesis of "no correlation" was presented in Section 11.6.

The coefficient of determination r^2, a proportional reduction in error statistic, provides another, and perhaps more meaningful, measure of the strength of the relationship between Y and X. Based on the notion of prediction and the proportional reduction in error computed from "no association" and "perfect association" models, r^2 gives the proportion of the total variability in the Y values that can be attributed to a linear relationship between Y and X. The logical basis for the interpretation of r^2 is identical to that encountered in Chapter 10 except that the magnitudes as well as the numbers of errors are taken into account when dealing with interval and ratio data.

Finally, we note that the methodology of this chapter, concerned with the relation between two variables X and Y, can be extended to study the relationship between a dependent variable Y and a set of independent variables. Multivariable techniques for studying the relationships between sociological variables and for developing multivariable predictions will achieve a prominent and important role in social science investigations in future years.

QUESTIONS AND PROBLEMS

22   Why is it useful to construct a scatter diagram before computing Pearson's r?

23   What is a regression line?

24 What do we know about the limits on r?

25 What do we mean by the expression "b is the slope of the regression line"?

26 What is the method of least squares?

27 How does r as the slope of the regression line differ from b as the slope of the regression line?

28 Why is a researcher more likely to be interested in r as opposed to b?

29 What is the difference between an r of $-.80$ and an r of $+.80$?

30 What is the coefficient of determination?

31 Social adjustment and perceived self-image tests were administered to $n = 6$ ex-drug addicts. Compute the least-squares regression line for the data. Predict the social adjustment score for an ex-addict who scores 29 on the perceived self-image test.

Perceived Self-image Score	Social Adjustment Score
X	Y
35	55
23	37
42	61
18	28
31	52
45	70

32 Refer to Exercise 31. Compute r and r^2. Interpret your results.

33 To compute r, the data should be homoscedastic, linear, and normal. What do these three assumptions involve?

34 Use the data on the number of physicians and nurses per 100,000 U.S. population [see Exercise 40, Chapter 3 (page 89)], and let X denote the number of physicians and Y the number of nurses. Plot the data on a scatter diagram and draw a freehand regression line.

35 Write the equation of the line obtained from your freehand regression line in Exercise 34.

36 Use the method of least squares to obtain a prediction equation relating actual number of physicians X, to number of nurses (by year) Y, for the data in Exercise 34.

37 Compute r for the data contained in Exercise 36.

38 Compute r^2 for the data contained in Exercise 36.

39 Refer to Exercise 34. Test the hypothesis that the number of nurses in a year is linearly related to the number of physicians in a year. (Use α = .05.)

REFERENCES

Anderson, T. R., and M. Zelditch. *A Basic Course in Statistics*, 2nd ed. New York: Holt, Rinehart and Winston, Inc., 1968. Chapter 12.

Blalock, H. M. *Social Statistics*, 2nd ed. New York: McGraw-Hill Book Company, 1972. Chapters 17 and 18.

Champion, D. J. *Basic Statistics for Social Research*. Scranton, Pa.: Chandler Publishing Company, 1970. Chapter 10.

Goode, J., "Presidential Address: The Place of Force in Human Society." *American Sociological Review*, 37 (Oct. 1972), p. 530.

Kasarda, J. D., "The Theory of Ecological Expansion: An Empirical Test." *Social Forces*, 51 (Dec. 1972), pp. 165–175.

King, M. B., and R. A. Hunt, "Measuring the Religious Variable: Replication." *Journal for the Scientific Study of Religion*, 11 (Sept. 1972), p. 244.

Mendenhall, W. *Introduction to Linear Models and the Design and Analysis of Experiments*. Belmont, Calif.: Wadsworth Publishing Company, Inc., 1968. Chapters 6 and 7.

Mendenhall, W., and L. Ott. *Understanding Statistics*. North Scituate, Mass.: Duxbury Press, 1972. Chapter 10.

Mueller, J. H., K. F. Schuessler, and H. L. Costner. *Statistical Reasoning in Sociology*, 2nd ed. Boston: Houghton Mifflin Company, 1970. Chapter 11.

Palumbo, D. J. *Statistics in Political and Behavioral Science*. New York: Appleton-Century-Crofts, 1969. Chapter 9.

12

DIFFERENCES AMONG MORE THAN TWO POPULATION MEANS

12.1 INTRODUCTION

ethods for comparing two population means, based on random samples of interval or ratio data, were presented in Chapter 8. Very often the two-sample problem is a simplification of what is encountered in real life. That is, frequently we wish to compare more than two population means.

For example, suppose that we wish to compare the mean incomes of steelworkers for three different ethnic groups, say black, white, and Spanish Americans, in a certain steel city. Independent random samples of steel workers would be selected from each of the three ethnic groups (the three populations). One would have to consult the personnel files of the steel companies in the city, list steelworkers in each ethnic group, and select a random sample from each. On the basis of the three sample means, we wish to know whether the population mean incomes differ and, if so, by how much. Note that the sample means most likely will differ, but this does not imply a difference in mean income for the three ethnic groups. Even if the population mean incomes were identical, the sample means most probably would differ. Then how do you decide whether the differences among the sample means are large enough to imply a difference among the corresponding population means? We shall answer this question using a technique known as an analysis of variance.

12.2 THE LOGIC BEHIND AN ANALYSIS OF VARIANCE

hy we call the method "an analysis of variance" can be seen easily with an example. Assume that we wish to compare three population means based on samples of five observations each selected from their respective populations. Suppose the data for the three samples appear as shown in Table 12.1. Do the data present sufficient evidence to indicate a difference among the three population means? A brief visual analysis of the data in Table 12.1 might lead us to a rapid, intuitive "yes." A glance at each of the three samples indicates very little variation within the populations. That is, the variation in the measurements within each sample is very small. In contrast, the spread or variation among the sample means is so large in

Table 12.1 *Comparison of Three Sample Means (Small Amount of Within-Sample Variation)*

Samples from Population		
1	*2*	*3*
29.0	25.1	20.1
29.2	25.0	20.0
29.1	25.0	19.9
28.9	24.9	19.8
28.8	25.0	20.2
$\bar{X}_1 = 29.0$	$\bar{X}_2 = 25.0$	$\bar{X}_3 = 20.0$

comparison to the within-sample variation that we intuitively conclude that a real difference exists among the population means.

How your intuition works when a larger within-sample variation is present can be illustrated by viewing the data in Table 12.2. Now the variation within samples is quite large. In comparison to the within-sample variation, the spread or variation among the sample means is not so great that it could not have occurred due to chance.

The variations in the observations for the two sets of data, Tables 12.1 and 12.2, are shown graphically in Figure 12.1. The strong evidence to indicate

Table 12.2 *Comparison of Three Sample Means (Large Amount of Within-Sample Variation)*

Samples from Population		
1	*2*	*3*
29.0	33.1	15.2
14.2	5.4	39.3
45.1	17.3	14.8
48.9	42.0	25.5
7.8	27.2	5.2
$\bar{X}_1 = 29.0$	$\bar{X}_2 = 25.0$	$\bar{X}_3 = 20.0$

Figure 12.1   *Dot diagrams for the data of (a) Table 12.1 and (b) Table 12.2:* ○, *measurement from sample 1;* ●, *measurement from sample 2;* ☐, *measurement from sample 3*

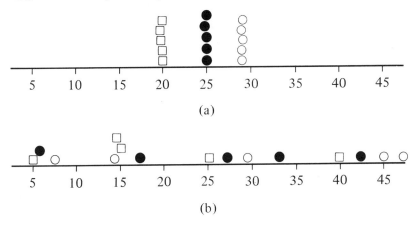

(a)

(b)

a difference in population means for the data of Table 12.1 is apparent in Figure 12.1(a). The lack of evidence is indicated by the overlapping of data points for the samples in Figure 12.1(b).

The data of Tables 12.1 and 12.2 (and Figure 12.1) should indicate very clearly what we mean by "an analysis of variance." All differences in sample means are judged statistically significant (or not) by comparing them with a measure of the random variation within the population data.

12.3   A TEST OF AN HYPOTHESIS CONCERNING MORE THAN TWO POPULATION MEANS: AN EXAMPLE OF AN ANALYSIS OF VARIANCE

J n Chapter 8 we presented a method for testing the equality of two population means. We hypothesized two normal populations (I and II) with means denoted by μ_1 and μ_2, respectively. To test the null hypothesis that $\mu_1 = \mu_2$, independent random samples of size n_1 and n_2 were drawn from the two populations. The sample data were then used to compute the value of the test statistic

$$t = \frac{\bar{X}_1 - \bar{X}_2}{s\sqrt{1/n_1 + 1/n_2}}$$

where

$$s^2 = \frac{(n_1 - 1)s_1^2 + (n_2 - 1)s_2^2}{(n_1 - 1) + (n_2 - 1)}$$

$$= \frac{(n_1 - 1)s_1^2 + (n_2 - 1)s_2^2}{n_1 + n_2 - 2}$$

is the pooled estimator of the common population variance σ^2. The rejection region for a specified value of α, the probability of a type I error, was then found using Table 2 of the Appendix.

Suppose now that we wish to extend this method to test the equality of more than two population means. The test procedure described above applies only to two means and therefore is inappropriate. Hence we shall employ a more general method of data analysis known as the analysis of variance. We shall illustrate its use with the following example.

Students from five different campuses throughout the country were surveyed to determine their attitudes toward industrial pollution. Each student sampled was asked a specified number of questions and then given a total interval-level score for the interview. Suppose that nine students are surveyed at each of the five campuses and we wish to compare the responses among the campuses. In particular, suppose that we wish to examine the average student score for each of the five campuses.

We shall label the set of all test scores that could have been obtained from campus I as population I and will assume that this population possesses a mean μ_1. A random sample of nine measurements (scores) is obtained from this population to monitor student attitudes toward pollution. The set of all scores that could have been obtained from students on campus II will be labeled population II (which has a mean μ_2). The data from a random sample of nine scores are obtained from this population. Similarly, μ_3, μ_4, and μ_5 represent the means of the populations for scores from campuses

Table 12.3   *Summary of the Sample Results for Five Populations*

	Population				
	I	*II*	*III*	*IV*	*V*
Sample mean	$\overline{X}_1$	$\overline{X}_2$	$\overline{X}_3$	$\overline{X}_4$	$\overline{X}_5$
Sample variance	s_1^2	s_2^2	s_3^2	s_4^2	s_5^2

III, IV, and V, respectively. We also obtain random samples of nine student scores from these populations.

From each of these five samples we calculate a sample mean and variance. The sample results can then be summarized as shown in Table 12.3.

If we are interested in testing the equality of the population means (i.e., $\mu_1 = \mu_2 = \mu_3 = \mu_4 = \mu_5$), we might be tempted to run all possible pairwise comparisons of two population means. Hence if we assume the five distributions are approximately normal with a common variance σ^2, we could run the $10\,t$ tests comparing two means as listed in Table 12.4 (see Section 8.8).

Table 12.4 *All Possible Null Hypotheses for Comparing Two Means from Five Populations*

$\mu_1 = \mu_2$	$\mu_2 = \mu_4$
$\mu_1 = \mu_3$	$\mu_2 = \mu_5$
$\mu_1 = \mu_4$	$\mu_3 = \mu_4$
$\mu_1 = \mu_5$	$\mu_3 = \mu_5$
$\mu_2 = \mu_3$	$\mu_4 = \mu_5$

One obvious disadvantage to this test procedure is that it is tedious and time consuming. But the more important and less apparent disadvantage of running multiple t tests to compare means is that the probability of falsely rejecting at least one of the hypotheses increases as the number of t tests increases. Thus, although we may have the probability of a type I error fixed at $\alpha = .05$ for each individual test, the probability of falsely rejecting on at least one of these tests is larger than .05. In other words, the combined probability of a type I error for the set of 10 hypotheses would be larger than the value .05 set for each individual test. Indeed, it could be as large as .40.

What we need is a single test of the hypothesis "all five population means are equal," which will be less tedious than the individual t tests and can be performed with a specified probability of a type I error (say .05). First we assume that the five sets of measurements are normally distributed with means given by μ_1, μ_2, μ_3, μ_4, and μ_5 and a common variance σ^2. Consider the quantity

$$s_w^2 = \frac{(n_1 - 1)s_1^2 + (n_2 - 1)s_2^2 + (n_3 - 1)s_3^2 + (n_4 - 1)s_4^2 + (n_5 - 1)s_5^2}{(n_1 - 1) + (n_2 - 1) + (n_3 - 1) + (n_4 - 1) + (n_5 - 1)}$$

$$= \frac{(n_1 - 1)s_1^2 + (n_2 - 1)s_2^2 + (n_3 - 1)s_3^2 + (n_4 - 1)s_4^2 + (n_5 - 1)s_5^2}{n_1 + n_2 + n_3 + n_4 + n_5 - 5}$$

Note this quantity is merely an extension of

$$s^2 = \frac{(n_1 - 1)s_1^2 + (n_2 - 1)s_2^2}{n_1 + n_2 - 2}$$

which is used as the estimator of the common variance for two populations in a test of the hypothesis $\mu_1 = \mu_2$ (Section 8.8). Thus s_w^2 represents a combined estimate of the common variance σ^2, and measures the variability of the observations "within" the five populations. (The subscript w refers to the within-population variability.)

Next we consider a quantity that measures the variability "between" or "among" the sample means. If the null hypothesis ($\mu_1 = \mu_2 = \mu_3 = \mu_4 = \mu_5$) is true, the populations are identical with mean μ and variance σ^2. Drawing single samples from the five populations is then equivalent to drawing five different samples from the same population. What kind of variation might be expected for these sample means? If the variation is too great, we would reject the hypothesis that $\mu_1 = \mu_2 = \mu_3 = \mu_4 = \mu_5$.

To assess the variation from sample mean to sample mean, we need to know the distribution of the mean of a sample of nine observations in repeated sampling. Each sample mean will have a probability distribution with the same mean, μ, and variance, $\sigma^2/9$. Since we have drawn five samples of nine observations each, we can estimate the population variance of the sample means, $\sigma^2/9$, using the formula

$$\begin{array}{c} \text{sample variance} \\ \text{(of the means)} \end{array} = \frac{\Sigma(\overline{X})^2 - (\Sigma\overline{X})^2/5}{5 - 1}$$

Note that we merely consider the $\overline{X}$'s as a sample of five observations and calculate their variance. This quantity estimates $\sigma^2/9$ and hence

$$s_B^2 = 9[\text{sample variance (of the means)}]$$

estimates σ^2. (The subscript B designates a measure of the variability among the sample means for the five populations.)

Under the null hypothesis that all five population means are identical, we have two estimates of σ^2, s_w^2 and s_B^2. Suppose that the ratio

$$\frac{s_B^2}{s_w^2}$$

is used as a test statistic to test the hypothesis that $\mu_1 = \mu_2 = \mu_3 = \mu_4 = \mu_5$. What is the distribution of this quantity if we were to repeat the experiment over and over again each time calculating s_B^2 and s_w^2?

When independent random samples are drawn from normal populations with equal variances, the quantity s_B^2/s_w^2 possesses a probability distribution in repeated sampling that is known to statisticians as an F distribution. The equation for this probability distribution is omitted, but we note several of its important properties:

1   The F distribution, unlike the normal distribution or t distribution, is nonsymmetrical (see Figure 12.2).

2   There are many F distributions (and hence many shapes). We can specify a particular one by designating the degrees of freedom associated with s_B^2 and s_w^2. We designate these quantities as d.f.$_1$ and d.f.$_2$, respectively.

3   Tail-end values for the F distribution are tabulated and appear in Tables 4 and 5 of the Appendix.

Figure 12.2   *Distribution of s_B^2/s_w^2*

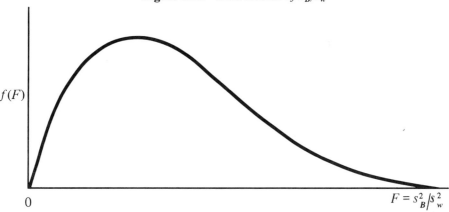

$f(F)$

0

$F = s_B^2 / s_w^2$

Table 4 records the upper-tail value of F, which has an area equal to .05 to its right (see Figure 12.3). The degrees of freedom for s_B^2, designated by d.f.$_1$, are indicated across the top of the table, while the degrees of freedom for s_w^2, designated by d.f.$_2$, appear in the first column on the left. For d.f.$_1 = 8$ and d.f.$_2 = 10$, the tabulated value is 3.07; that is, only 5 percent of the measurements for s_B^2/s_w^2 in repeated sampling from an F distribution with d.f.$_1 = 8$ and d.f.$_2 = 10$ degrees of freedom will exceed 3.07 (see Figure 12.3). Similarly, an entry in Table 5 records an upper-tail value of the F distribution that has an area equal to .01 to its right. The .01 tail-end value of the F distribution with d.f.$_1 = 5$ and d.f.$_2 = 3$ degrees of freedom is 28.24.

Figure 12.3   *Critical value for the F distribution d.f.$_1$ = 8, d.f.$_2$ = 10*

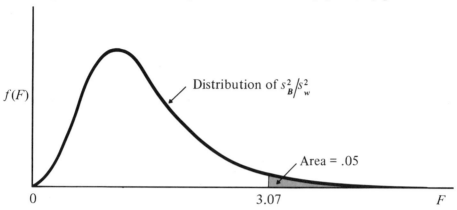

For our example, s_B^2/s_w^2 follows an F distribution with degrees of freedom which can be shown to be d.f.$_1$ = 4 for s_B^2 and d.f.$_2$ = 40 for s_w^2. The proof of these remarks is beyond the scope of this text. However, we make use of this result for testing the null hypothesis, $\mu_1 = \mu_2 = \mu_3 = \mu_4 = \mu_5$.

The decision maker used to test equality of the population means is

$$F = \frac{s_B^2}{s_w^2}$$

When the null hypothesis is true, both s_B^2 and s_w^2 estimate σ^2 and F would be expected to assume a value near $F = 1$. When the hypothesis of equality is false, s_B^2 will tend to be larger than s_w^2, owing to the differences among the

Figure 12.4   *Critical Value of F for α = .05, d.f.$_1$ = 4, d.f.$_2$ = 40*

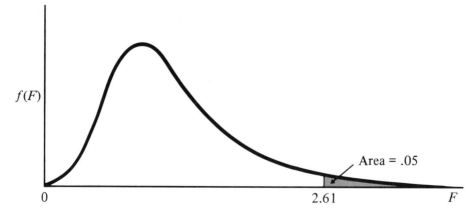

population means. Hence we shall reject the null hypothesis in the upper tail of the distribution of $F = s_B^2/s_w^2$. For α, the probability of a type I error equal to .05 or .01, we can locate the rejection region for this one-tailed test using Table 4 or 5 of the Appendix with d.f.$_1$ = 4 and d.f.$_2$ = 40. For $\alpha = .05$ the critical value of $F = s_B^2/s_w^2$ is 2.61 (see Figure 12.4). If the calculated value of F falls in the rejection region, we conclude that not all five population means are identical.

This procedure can be generalized with only slight modifications in the formulas to test the quality of k (where k is an integer equal to or greater than two) population means from normal populations with a common variance σ^2. Random samples of size $n_1, n_2, \ldots, n_k$ are drawn from the respective populations. We then compute the sample means and variances. The null hypothesis $\mu_1 = \mu_2 = \cdots = \mu_k$ is tested against the alternative that at least one of the population means is different from the others. The test procedure is given next.

Analysis of Variance for Testing the Equality of k Population Means

Null Hypothesis : $\mu_1 = \mu_2 = \cdots = \mu_k$.

Alternative Hypothesis : At least one of the population means is different from the others.

Test Statistic :

$$F = \frac{s_B^2}{s_w^2}$$

where

$$s_w^2 = \frac{(n_1 - 1)s_1^2 + (n_2 - 1)s_2^2 + \cdots + (n_k - 1)s_k^2}{n_1 + n_2 + \cdots + n_k - k}$$

and

$$s_B^2 = \frac{\sum n_i \bar{X}_i^2 - (\sum n_i \bar{X}_i)^2/n}{k - 1}$$

Rejection Region : Reject if F is greater than the tabulated value for $a = \alpha$, $d.f._1 = (k - 1)$, and $d.f._2 = (n_1 + n_2 + \cdots + n_k - k)$. See Tables 4 and 5 for F values corresponding to $a = .05$ and $a = .01$, respectively. (Note: $n = n_1 + n_2 + \cdots + n_k$.)

Recall that $k = 5$ and $n_1 = n_2 = \cdots = n_k = 9$ for our previous example. The rejection region in general can be determined for $\alpha = .05$ or $.01$ using Tables 4 and 5 of the Appendix with $\text{d.f.}_1 = (k - 1)$ and $\text{d.f.}_2 = n_1 + n_2 + \cdots + n_k - k$.

Example 12.1 *A group of social psychologists was interested in studying the effect of anxiety on learning as measured by student performance on a series of tests. Using a prestudy test, 27 students were classified into one of three anxiety groups. Group I students were those who scored extremely low on a scale measuring anxiety. Those placed in group III were students who scored extremely high on the anxiety scale. The remaining students were placed in group II. The results of the prestudy anxiety test indicated that 6 students were in group I, 12 in group II, and 9 in group II. Following the prestudy assignment of students to groups, the same battery of tests was given to each of the 27 students. The sample mean and variance of the battery test scores (based on a total of 100 points) have been summarized for each group in Table 12.5. Use the sample data to test the hypothesis that the average test score for low-, middle-, and high-anxiety students is identical (i.e., that anxiety has no effect on a student's performance on this battery of tests).*

Table 12.5 *Summary of Battery Test Scores*

Group I	Group II	Group III
$n_1 = 6$	$n_2 = 12$	$n_3 = 9$
$\overline{X}_1 = 88$	$\overline{X}_2 = 82$	$\overline{X}_3 = 78$
$s_1^2 = 10.1$	$s_2^2 = 14.8$	$s_3^2 = 13.9$

Solution *We first hypothesize a population for each anxiety group corresponding to all possible battery test scores for students who could have been included in the study. We shall assume that the measurements in each population are approximately normally distributed with the mean of population I equal to μ_1, the mean of II, μ_2, and the mean of III, μ_3. In addition, we will assume that the populations have a common variance σ^2. From these populations, random samples of size $n_1 = 6$, $n_2 = 12$, and $n_3 = 9$ students were obtained and assigned to the respective groups. To test the null hypothesis of equality of the population means, $\mu_1 = \mu_2 = \mu_3$, we must first compute s_w^2 and s_B^2. We can calculate s_w^2 directly from the* *sample data.*

$$s_w^2 = \frac{(n_1 - 1)s_1^2 + (n_2 - 1)s_2^2 + (n_3 - 1)s_3^2}{n_1 + n_2 + n_3 - 3}$$

$$= \frac{5(10.1) + 11(14.8) + 8(13.9)}{6 + 12 + 9 - 3}$$

$$= \frac{324.5}{24} = 13.52$$

Before obtaining s_B^2, we must first compute $\Sigma\, n_i\overline{X}_i$ and $\Sigma\, n_i\overline{X}_i^2$. From Table 12.5,

$$\Sigma\, n_i\overline{X}_i = 6(88) + 12(82) + 9(78) = 2214$$

and

$$\Sigma\, n_i\overline{X}_i^2 = 6(88)^2 + 12(82)^2 + 9(78)^2 = 181{,}908$$

Hence for $k = 3$ and $n = n_1 + n_2 + n_3 = 27$,

$$s_B^2 = \frac{\Sigma\, n_i\overline{X}_i^2 - (\Sigma\, n_i\overline{X}_i)^2/n}{k - 1}$$

$$= \frac{181{,}908 - (2214)^2/27}{2} = \frac{181{,}908 - 181{,}548}{2} = 180$$

The test statistic for the null hypothesis $\mu_1 = \mu_2 = \mu_3$ is

$$F = \frac{s_B^2}{s_w^2} = \frac{180.0}{13.52} = 13.31$$

Using the probability of a type I error, $\alpha = .01$, we can locate the upper-tail rejection region for this one-tailed test using Table 5 of the Appendix with

$$d.f._1 = k - 1 = 2$$

and

$$d.f._2 = n_1 + n_2 + n_3 - 3 = 24$$

This value is 5.61. Since the observed value of F is greater than 5.61, we reject the hypothesis of equality of the population means (i.e., at least one of the means is different from the rest). Although we do not know exactly where these differences lie, it is rather obvious from the sample data that anxiety has a detrimental effect on a student's performance in the battery of tests.

EXERCISES

1   State the assumptions underlying an analysis of variance concerning k population means.

2   Distinguish between s_w^2 and s_B^2 in testing the equality of k population means.

3   What is the F ratio and how does it function in an analysis of variance?

4   Six small groups were randomly selected from each of three populations. Group tension was measured and the results are shown in the table. Perform an analysis of variance for the experiment to determine if there are differences in the mean group-tension scores among the three populations. (Use $\alpha = .05$.)

A	B	C
4.2	5.6	3.2
1.1	5.1	2.5
3.7	4.4	2.9
2.6	4.2	3.6
2.1	4.2	3.2
3.7	5.1	4.1

5   The duration of fights between grade-school children was studied at five different schools. A sample of nine fights at each school was selected. A summary of the resulting durations of fights is recorded for each of the five schools. Run an analysis of variance to test the hypothesis that the mean durations of fights at the five schools, $\mu_1, \mu_2, \ldots, \mu_s$, are equal. (Use $\alpha = .05$.)

A	B	C	D	E
$n_1 = 9$	$n_2 = 9$	$n_3 = 9$	$n_4 = 9$	$n_5 = 9$
$\overline{X}_1 = 3.2$	$\overline{X}_2 = 3.8$	$\overline{X}_3 = 4.1$	$\overline{X}_4 = 4.0$	$\overline{X}_5 = 3.7$
$s_1^2 = .46$	$s_2^2 = .51$	$s_3^2 = .45$	$s_4^2 = .20$	$s_5^2 = .28$

J
12.4   KRUSKAL–WALLIS
ONE-WAY ANALYSIS OF
VARIANCE BY RANKS

n this section we shall present a one-way analysis of variance procedure for use on data measured on an ordinal scale. Unlike the procedure of Section 12.3, we do not require that the data for each of the samples be drawn from a normal population with common variance σ^2. We merely assume that the sample data are obtained from a population with a continuous frequency distribution. Although the Kruskal–Wallis procedure was developed for use with ordinal data, it provides a useful alternative to the analysis-of-variance procedure of Section 12.3, where the normality assumption or the equality of variance assumption (often called the assumption of homoscedasticity) does not hold. In such situations, the data from an interval or ratio scale can be converted to ordinal data and then analyzed using the Kruskal–Wallis procedure.

Under the null hypothesis we shall assume that each of k samples was drawn from identical populations. The observations from all k samples then are combined and jointly ranked, with the largest observation receiving the rank 1, the second largest receiving the rank 2, and the smallest receiving the rank n. If a tie occurs between two or more measurements, each measurement is assigned a rank equal to the mean of the occupied ranks.

Having jointly ranked the measurements from the k samples, we compute the sum of the ranks for each sample. The Kruskal–Wallis test procedure then asks the question: Are the differences among the sums of ranks for the k samples too large to have occurred if the samples were drawn from identical populations? To answer this question we utilize the following notation. Let

S_1 be the sum of the ranks for the n_1 observations in sample 1

S_2 be the sum of the ranks for the n_2 observations in sample 2

$\vdots$

S_k be the sum of the ranks for the n_k observations in sample k

Then our test statistic is

$$H = \frac{12}{n(n+1)}\left(\frac{S_1^2}{n_1} + \frac{S_2^2}{n_2} + \cdots + \frac{S_k^2}{n_k}\right) - 3(n+1)$$

Under the null hypothesis that the k samples were drawn from identical populations, the statistic H has a distribution that can be approximated by a chi-square distribution with d.f. $= k - 1$. This approximation should be reasonably good provided the sample sizes $n_1, n_2, \ldots, n_k$ are all greater than 5.

We can summarize the Kruskal–Wallis analysis of variance by ranks as follows.

Summary of the Kruskal–Wallis One-Way Analysis of Variance by Ranks

Null Hypothesis: All k samples were drawn from identical populations.

Alternative Hypothesis: At least one of the k samples was drawn from a population that was not identical to the other populations.

Test Statistic:

$$H = \frac{12}{n(n+1)} \left(\frac{S_1^2}{n_1} + \frac{S_2^2}{n_2} + \cdots + \frac{S_k^2}{n_k} \right) - 3(n+1)$$

where $S_1, S_2, \ldots, S_k$ are the sums of the ranks and $n_1, n_2, \ldots, n_k$ are the sample sizes for populations $1, 2, \ldots, k$, respectively.

Rejection Region: For a specified value of α, reject the null hypothesis if H exceeds the critical value of chi square for $a = \alpha$, d.f. $= k - 1$.

Table 12.6 *Mental Illness Knowledge Scores for the Clergymen*

Methodist	Catholic	Pentecostal
32	32	28
30	32	21
30	26	15
29	26	15
26	22	14
23	20	14
20	19	14
19	16	11
18	14	9
12	14	8

Example 12.2 *Three random samples of clergymen were drawn, one containing 10 Methodist ministers, the second containing 10 Catholic priests, and the third containing 10 Pentecostal ministers. Each of the clergymen was then examined using a test to measure his knowledge about causes of mental illness. These test scores are listed in Table 12.6. Use the data in the table to determine if the three groups of clergymen differ with respect to their knowledge about the causes of mental illness. (Use $\alpha = .05$.)*

Solution *The null and research hypothesis for this example can be stated as*

H_0: *There is no difference among the three groups with respect to knowledge about the causes of mental illness (i.e., the samples of scores were drawn from identical populations).*

H_a: *At least one of the three groups of clergymen differs from the others with respect to knowledge about causes of mental illness.*

Before computing H we must first jointly rank the 30 test scores from highest to lowest. From Table 12.6 we see that 32 is the highest test score, and three clergymen achieved this score. These three men occupy the ranks 1, 2, and 3. Each is assigned the average of the first three ranks, namely 2. The second highest score, 30, was obtained by two clergymen. Each of these scores is assigned the rank 4.5, the average of the fourth and fifth ranks occupied by these scores. In a similar way we can assign the remaining ranks to test scores. Table 12.7 lists the 30 test scores and

Table 12.7 *Test Scores and Ranks for the Clergymen Study*

Methodist	Catholic	Pentecostal
32 (2)	32 (2)	28 (7)
30 (4.5)	32 (2)	21 (13)
30 (4.5)	26 (9)	15 (20.5)
29 (6)	26 (9)	15 (20.5)
26 (9)	22 (12)	14 (24)
23 (11)	20 (14.5)	14 (24)
20 (14.5)	19 (16.5)	14 (24)
19 (16.5)	16 (19)	11 (28)
18 (18)	14 (24)	9 (29)
12 (27)	14 (24)	8 (30)
$n_1 = 10, S_1 = 113$	$n_2 = 10, S_2 = 132$	$n_3 = 10, S_3 = 220$

associated ranks (in parentheses). Note from the table that the sums of the ranks for the three groups are 113, 132, and 220, respectively. Hence the computed value of H is

$$H = \frac{12}{30(30 + 1)}\left[\frac{(113)^2}{10} + \frac{(132)^2}{10} + \frac{(220)^2}{10}\right] - 3(30 + 1)$$

$$= \frac{12}{930}(1276.9 + 1742.4 + 4840.0) - 93$$

$$= 8.4$$

The critical value of chi square with a = .05 and d.f. = k − 1 = 2 can be found using Table 3 of the Appendix. This value is 5.99. Since the observed value of H is greater than 5.99, we reject the null hypothesis and conclude that at least one of the clergy groups has more knowledge about causes of mental illness than the other two groups. In fact, the data suggest that Penecostal ministers are less knowledgeable in this area.

EXERCISES

6   Why would a social scientist be particularly interested in Kruskal–Wallis' one-way analysis of variance by ranks? Explain in detail.

7   Treat the data in Exercise 4 (page 432) at the ordinal level and compute the one-way analysis of variance by ranks. (Use $\alpha = .05$.)

8   Random samples of six people were selected from four minority ethnic groups and asked to score, on a 0 to 30 scale, how well they have been socially accepted in their community. Run an analysis of variance—one-way analysis of variance by ranks—on the data in the table. (Use $\alpha = .05$.)

Group A	Group B	Group C	Group D
10	18	15	22
9	16	13	24
14	14	12	21
11	15	14	23
12	16	13	22
8	19	11	26

C

12.5 SUMMARY

hapter 12 considers the problem of comparing more than two population means based on independent random samples of interval (or ratio) data from each population. To test the hypothesis of "no difference in population mean," we employ an analysis of variance. That is, we compare the variation in sample means with the variability within samples using an F statistic. If the F statistic is larger than expected (by comparing with the critical value in the F tables), we have evidence to indicate a difference in population means. Note that while an analysis of variance is usually applied when comparing three or more population means, it is also appropriate for the special two-sample case.

An analysis of variance is appropriate when the data for each population are normally distributed and when the populations possess identical variances. That is, the variation of the data within populations is identical. We shall not usually know whether these assumptions hold, but this is not a major drawback if the variation is similar for the populations. When you have doubts that this is true, or when the data are measured only at an ordinal level of measurement, we must adopt an alternative method. Then a suitable test statistic is the Kruskal–Wallis H.

Chapter 12 presents only a most elementary introduction to an analysis of variance. This technique can be employed to investigate the differences among population means when the data are linked in the manner of the paired-difference experiment of Chapter 8. It can also be used to explore the effect of two or more independent variables on a response. The interested reader will find information on these applications of the analysis of variance in the books cited at the end of the chapter.

QUESTIONS AND PROBLEMS

9 Refer to Exercise 5 (page 432). Suppose that three fights from schools A and B, two from school C, and one each from D and E were eliminated because they

A	B	C	D	E
$n_1 = 6$	$n_2 = 6$	$n_3 = 7$	$n_4 = 8$	$n_5 = 8$
$\overline{X}_1 = 3.2$	$\overline{X}_2 = 3.8$	$\overline{X}_3 = 4.1$	$\overline{X}_4 = 4.0$	$\overline{X}_5 = 3.7$
$s_1^2 = .36$	$s_2^2 = .53$	$s_3^2 = .40$	$s_4^2 = .31$	$s_5^2 = .35$

were staged. The sample means remained constant, but all the sample sizes changed as well as the standard deviations. Run an analysis of variance to test the equality of means, using the revised data given here. (Use $\alpha = .05$.)

10   A clinical psychologist wished to compare three methods for reducing hostility levels in university students. A certain psychological test (HLT) was used to measure the degree of hostility. High scores on this test were taken to indicate great hostility. Twenty-one students obtaining high, nearly equal scores were used in the experiment. Eight were selected at random from among the 21 problem cases and treated by method *A*. Seven were taken at random from the remaining 13 students and treated by method *B*. The other 6 students were treated by method *C*. All treatments continued throughout a semester. Each student was given the HLT test at the end of the semester, with the results shown. Do these data present sufficient evidence to indicate a difference in mean student response for the three methods after treatment? (Use $\alpha = .01$.)

Scores on the HLT Test

Method A	Method B	Method C
95	78	74
80	70	63
92	81	76
87	74	70
83	73	71
78	72	72
89	75	
86		

11   Refer to Exercise 10. Use the Kruskal–Wallis one-way analysis of variance. Do you obtain the same conclusions? (Use $\alpha = .05$.)

12   Three methods of instruction in group-encounter techniques were to be compared with respect to the mean level of group interaction. A total of 19 group leaders participated in the study; 6 were randomly assigned to technique 1, 6 to technique 2, and 7 to technique 3. After a one-week training period, all leaders were assigned to an encounter group and, after a 4-hour session, were scored on their ability to

Method 1	Method 2	Method 3
82	71	91
80	29	93
81	78	84
83	74	90
84	75	88
79	36	96
		92

achieve meaningful group interaction. Use the data in the table and run an analysis of variance to test the hypothesis that all three instructional techniques have achieved the same average level of group interaction. (Use $\alpha = .05$.)

13 Repeat Exercise 12 using the Kruskal–Wallis one-way analysis of variance. (Use $\alpha = .05$.)

14 When would the Kruskal–Wallis one-way analysis of variance be more appropriate than the analysis of variance of Section 12.3?

15 Set up a study in which an analysis of variance would be the appropriate statistical method of analysis.

16 If a researcher has ratio-level data, would he ever consider using the Kruskal–Wallis one-way analysis of variance on these data? Explain.

17 A sociologist randomly draws five samples of six cities each from regional areas. Within each city, he computes an index of consumer confidence concerning the government's ability to curb inflation. Considering the following data, carry out an analysis-of-variance test among the five sample means. (Use $\alpha = .05$.)

West	Southwest	North Central	East	South
12.3	19.2	34.8	19.2	31.9
15.9	20.5	29.3	21.8	42.5
11.7	18.4	31.4	17.5	29.6
14.8	9.7	18.4	24.3	37.1
25.7	21.5	26.3	18.7	33.3
16.2	22.6	30.5	25.3	38.8

18 Using the same data in Exercise 17, carry out the Kruskal–Wallis analysis of variance by ranks. How do these results compare with those in Exercise 17?

REFERENCES

Blalock, H. M. *Social Statistics*, 2nd ed. New York: McGraw-Hill Book Company, 1972. Chapter 16.

Dixon, W. J., and F. J. Massey. *Introduction to Statistical Analysis*, 3rd ed. New York: McGraw-Hill Book Company, 1969. Chapter 10.

Mendenhall, W., and L. Ott. *Understanding Statistics*. North Scituate, Mass.: Duxbury Press, 1972. Chapter 13.

Palumbo, D. J. *Statistics in Political and Behavioral Science*. New York: Appleton-Century-Crofts, 1969. Chapter 11.

Siegel, S. *Nonparametric Statistics for the Behavioral Sciences*. New York: McGraw-Hill Book Company, 1956. Chapter 8.

13

SUMMARY

The preceding chapters portray statistics as a theory of information used by social scientists to describe large masses of data and to make inferences about social phenomena based on information contained in samples. Thus the text divides into two portions, one aimed at data description, Chapters 3, 4, and 10, and the others concerned with statistical inference, Chapters 6, 7, 8, 9, 11, and 12. These two major topics are related in the sense that one must know how to describe data in order to phrase an inference about a population based on sample data. This explains why chapters dealing with data description usually precede those concerned with statistical inference.

Each chapter deals with some aspect of data description or statistical inference and follows in a logical connective sequence. We attempt to explain how and why social scientists describe data and how they take the next step and use sample data to make inferences about data generated by some social process. We hope to answer the following questions: What is statistics? What is its purpose in the social sciences? How does it accomplish its objective? If we have answered these questions to your satisfaction, if each chapter and section seem to fulfill a purpose and to complete a portion of the picture, we have in some measure accomplished our objective.

Chapter 1 presents statistics as a scientific tool (1) for describing large masses of sociological data, and (2) for making inferences, predictions, or decisions concerning a population of measurements based upon information contained in a sample. Statistics is employed in the evolutionary process known as the scientific method—which, in essence, is the observation of nature—in order that we may form inferences or theories concerning the structure of nature and test these theories against repeated observation. The major contributions of statistics to this procedure are clearly evident. Thus statistical descriptive procedures aid in the formation of social theories. Statistical inference is used later in the investigation to weigh the evidence in support of formulated theories. Since anyone can make inferences, the great value of statistical inference in the sciences is that it tells us how reliable an inference is.

Chapter 2 concerns a most important aspect of statistics that must precede a discussion of either data description or statistical inference: the problem of measurement and the acquisition of data. We learned about two classes of variables, quantitative and qualitative, and the four scales of measurement (nominal, ordinal, interval, and ratio) used to measure them. In particular, we noted that the scale of measurement we would employ to collect data on a social science variable will depend on its type as well as availability of information on the variable. For instance, we might like to measure a quantitative variable on an interval scale, but the best that can be done in a practical situation is to use an ordinal scale of measurement, that is, rank the data in order of magnitudes. To summarize, Chapter 2 deals with the first step in data collection, measurement, and points to the need for statistical data description and inferential techniques applicable to data acquired on the four scales of measurement. You will find these techniques described in subsequent chapters.

The method of phrasing an inference—that is, the way in which we describe a set of measurements—is presented in Chapters 3 and 4 in terms of a frequency distribution and the associated numerical descriptive measures. In particular, we note that the frequency distribution is subject to a probabilistic interpretation and that numerical descriptive measures are more suitable for inferential purposes because we can more easily associate a measure of goodness with them. Finally, a secondary but extremely important result of our study of numerical descriptive measures involved the notion of variation, its measurement in terms of a standard deviation, and its interpretation using the empirical rule. While concerned with describing a set of measurements, namely the population, we provided the basis for a description of the sampling distributions of estimators and a two-standard deviation bound on an error of estimation, encountered in Chapter 6. This basis also helped to locate regions of contradiction, called rejection regions, for the statistical tests of hypotheses in Chapter 7.

The mechanism involved in making an inference or a decision concerning a population of voter responses is introduced in Chapter 5. We hypothesized a precinct of voters in which 50 percent (or more) favored candidate Jones; we then randomly drew a sample of 20 voters from the precinct. Observing none in favor of Jones, we concluded that the observed sample was highly improbable if Jones were really a winner, and we therefore rejected our hypothesis. We were led to reject our hypothesis, not because the sample results were impossible but because they were highly improbable if the initial (null) hypothesis were true. This example showed us that probability is the mechanism employed in making statistical decisions and it explained why we need to know the probability distribution for the random variable measured in an experiment. Study of the normal random variable in Chapter 5 centers about its use in approximating the binomial probability distribution as well as the distributions of many other important statistics computed from sample data.

Chapters 6 and 7 formally discussed statistical inference, estimation and tests of hypotheses, and the methods of measuring the goodness of inferences generated by these two inferential procedures. The methodology applicable to inferences made from single samples, presented in Chapters 6 and 7, was extended to methods appropriate for two samples and the comparison of two populations in Chapter 8.

Describing or making inferences about data generated by a single social science variable, the topic of Chapters 3 through 8, is extended in Chapters 9, 10, and 11 to two or more social science variables. Specifically, we may take measurements on two or more social science variables to investigate their relationship in order to gain a better understanding of some social phenomenon. We may ultimately wish to observe one variable in a social situation and predict the value of another. Chapter 9 presents statistical tests to determine whether two variables measured on a nominal scale are related. The chi-square test of independence is applicable to data measured on any of the four scales of measurement. For the case where a relationship appears to exist between bivariate data measured on either a nominal or an ordinal scale, Chapter 10 presents a number of measures of association (most based on the percentage reduction in error concept) to measure the strength of the relationship. Included were the phi coefficient, the contingency coefficient, lambda, gamma, Spearman's rho, and Kendall's W. Chapter 11 applies specifically to a study of the linear relationship between two social science variables that have been measured on ratio or interval scales. We learned how to test the hypothesis of "no linear relationship" between the two variables, and how to use the simple coefficient of correlation r and the coefficient of determination r^2 to describe the strength of a relationship when evidence of a relationship was shown to exist.

The remaining chapter in the text, Chapter 12, extends the comparison of two treatment means, covered in Chapter 8, to a comparison of the means of three or more populations, using an analysis of variance. Chapter 12 most naturally follows directly after Chapter 8, but it is deferred until after a discussion of bivariate statistical methods because there is rarely time to include this topic in a beginning social statistics course.

Now that you have waded through part or all of the preceding 12 chapters, we might ask you: What is statistics? How does it work? What are some simple but practical applications to problems that occur in the social sciences or, for that matter, in everyday life? And, most important, of what value is statistics to you? A synopsis of the text shows statistics to be a theory of information with data description and inference as its goals. In descriptive statistics, we seek graphical or numerical descriptive methods that will rapidly convey the characteristics of a mass of social science data to a reader. In inferential statistics, a sample is selected from a population, and an inference is made based on information contained in the sample. How good is the inference? The ability to answer this question is the major contribution of statistical inference to the social sciences.

APPENDIX

Table 1 *Normal-Curve Areas*

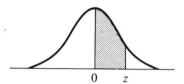

0   z

z	.00	.01	.02	.03	.04	.05	.06	.07	.08	.09
0.0	.0000	.0040	.0080	.0120	.0160	.0199	.0239	.0279	.0319	.0359
0.1	.0398	.0438	.0478	.0517	.0557	.0596	.0636	.0675	.0714	.0753
0.2	.0793	.0832	.0871	.0910	.0948	.0987	.1026	.1064	.1103	.1141
0.3	.1179	.1217	.1255	.1293	.1331	.1368	.1406	.1443	.1480	.1517
0.4	.1554	.1591	.1628	.1664	.1700	.1736	.1772	.1808	.1844	.1879
0.5	.1915	.1950	.1985	.2019	.2054	.2088	.2123	.2157	.2190	.2224
0.6	.2257	.2291	.2324	.2357	.2389	.2422	.2454	.2486	.2517	.2549
0.7	.2580	.2611	.2642	.2673	.2704	.2734	.2764	.2794	.2823	.2852
0.8	.2881	.2910	.2939	.2967	.2995	.3023	.3051	.3078	.3106	.3133
0.9	.3159	.3186	.3212	.3238	.3264	.3289	.3315	.3340	.3365	.3389
1.0	.3413	.3438	.3461	.3485	.3508	.3531	.3554	.3577	.3599	.3621
1.1	.3643	.3665	.3686	.3708	.3729	.3749	.3770	.3790	.3810	.3830
1.2	.3849	.3869	.3888	.3907	.3925	.3944	.3962	.3980	.3997	.4015
1.3	.4032	.4049	.4066	.4082	.4099	.4115	.4131	.4147	.4162	.4177
1.4	.4192	.4207	.4222	.4236	.4251	.4265	.4279	.4292	.4306	.4319
1.5	.4332	.4345	.4357	.4370	.4382	.4394	.4406	.4418	.4429	.4441
1.6	.4452	.4463	.4474	.4484	.4495	.4505	.4515	.4525	.4535	.4545
1.7	.4554	.4564	.4573	.4582	.4591	.4599	.4608	.4616	.4625	.4633
1.8	.4641	.4649	.4656	.4664	.4671	.4678	.4686	.4693	.4699	.4706
1.9	.4713	.4719	.4726	.4732	.4738	.4744	.4750	.4756	.4761	.4767
2.0	.4772	.4778	.4783	.4788	.4793	.4798	.4803	.4808	.4812	.4817
2.1	.4821	.4826	.4830	.4834	.4838	.4842	.4846	.4850	.4854	.4857
2.2	.4861	.4864	.4868	.4871	.4875	.4878	.4881	.4884	.4887	.4890
2.3	.4893	.4896	.4898	.4901	.4904	.4906	.4909	.4911	.4913	.4916
2.4	.4918	.4920	.4922	.4925	.4927	.4929	.4931	.4932	.4934	.4936
2.5	.4938	.4940	.4941	.4943	.4945	.4946	.4948	.4949	.4951	.4952
2.6	.4953	.4955	.4956	.4957	.4959	.4960	.4961	.4962	.4963	.4964
2.7	.4965	.4966	.4967	.4968	.4969	.4970	.4971	.4972	.4973	.4974
2.8	.4974	.4975	.4976	.4977	.4977	.4978	.4979	.4979	.4980	.4981
2.9	.4981	.4982	.4982	.4983	.4984	.4984	.4985	.4985	.4986	.4986
3.0	.4987	.4987	.4987	.4988	.4988	.4989	.4989	.4989	.4990	.4990

This table is abridged from Table I of *Statistical Tables and Formulas*, by A. Hald (New York: John Wiley & Sons, Inc., 1952). Reproduced by permission of A. Hald and the publishers, John Wiley & Sons, Inc.

Table 2   *Percentage Points of the t distribution*

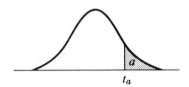

d.f.	a = .10	a = .05	a = .025	a = .010	a = .005
1	3.078	6.314	12.706	31.821	63.657
2	1.886	2.920	4.303	6.965	9.925
3	1.638	2.353	3.182	4.541	5.841
4	1.533	2.132	2.776	3.747	4.604
5	1.476	2.015	2.571	3.365	4.032
6	1.440	1.943	2.447	3.143	3.707
7	1.415	1.895	2.365	2.998	3.499
8	1.397	1.860	2.306	2.896	3.355
9	1.383	1.833	2.262	2.821	3.250
10	1.372	1.812	2.228	2.764	3.169
11	1.363	1.796	2.201	2.718	3.106
12	1.356	1.782	2.179	2.681	3.055
13	1.350	1.771	2.160	2.650	3.012
14	1.345	1.761	2.145	2.624	2.977
15	1.341	1.753	2.131	2.602	2.947
16	1.337	1.746	2.120	2.583	2.921
17	1.333	1.740	2.110	2.567	2.898
18	1.330	1.734	2.101	2.552	2.878
19	1.328	1.729	2.093	2.539	2.861
20	1.325	1.725	2.086	2.528	2.845
21	1.323	1.721	2.080	2.518	2.831
22	1.321	1.717	2.074	2.508	2.819
23	1.319	1.714	2.069	2.500	2.807
24	1.318	1.711	2.064	2.492	2.797
25	1.316	1.708	2.060	2.485	2.787
26	1.315	1.706	2.056	2.479	2.779
27	1.314	1.703	2.052	2.473	2.771
28	1.313	1.701	2.048	2.467	2.763
29	1.311	1.699	2.045	2.462	2.756
inf.	1.282	1.645	1.960	2.326	2.576

From "Table of Percentage Points of the *t*-distribution." Computed by Maxine Merrington, *Biometrika*, Vol. 32 (1941), p. 300. Reproduced by permission of the *Biometrika* Trustees.

Table 3 *Percentage Points of the Chi-Square Distribution*

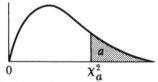

d.f.	$a = .995$	$a = .990$	$a = .975$	$a = .950$	$a = .900$
1	0.0000393	0.0001571	0.0009821	0.0039321	0.0157908
2	0.0100251	0.0201007	0.0506356	0.102587	0.210720
3	0.0717212	0.114832	0.215795	0.351846	0.584375
4	0.206990	0.297110	0.484419	0.710721	1.063623
5	0.411740	0.554300	0.831211	1.145476	1.61031
6	0.675727	0.872085	1.237347	1.63539	2.20413
7	0.989265	1.239043	1.68987	2.16735	2.83311
8	1.344419	1.646482	2.17973	2.73264	3.48954
9	1.734926	2.087912	2.70039	3.32511	4.16816
10	2.15585	2.55821	3.24697	3.94030	4.86518
11	2.60321	3.05347	3.81575	4.57481	5.57779
12	3.07382	3.57056	4.40379	5.22603	6.30380
13	3.56503	4.10691	5.00874	5.89186	7.04150
14	4.07468	4.66043	5.62872	6.57063	7.78953
15	4.60094	5.22935	6.26214	7.26094	8.54675
16	5.14224	5.81221	6.90766	7.96164	9.31223
17	5.69724	6.40776	7.56418	8.67176	10.0852
18	6.26481	7.01491	8.23075	9.39046	10.8649
19	6.84398	7.63273	8.90655	10.1170	11.6509
20	7.43386	8.26040	9.59083	10.8508	12.4426
21	8.03366	8.89720	10.28293	11.5913	13.2396
22	8.64272	9.54249	10.9823	12.3380	14.0415
23	9.26042	10.19567	11.6885	13.0905	14.8479
24	9.88623	10.8564	12.4011	13.8484	15.6587
25	10.5197	11.5240	13.1197	14.6114	16.4734
26	11.1603	12.1981	13.8439	15.3791	17.2919
27	11.8076	12.8786	14.5733	16.1513	18.1138
28	12.4613	13.5648	15.3079	16.9279	18.9392
29	13.1211	14.2565	16.0471	17.7083	19.7677
30	13.7867	14.9535	16.7908	18.4926	20.5992
40	20.7065	22.1643	24.4331	26.5093	29.0505
50	27.9907	29.7067	32.3574	34.7642	37.6886
60	35.5346	37.4848	40.4817	43.1879	46.4589
70	43.2752	45.4418	48.7576	51.7393	55.3290
80	51.1720	53.5400	57.1532	60.3915	64.2778
90	59.1963	61.7541	65.6466	69.1260	73.2912
100	67.3276	70.0648	74.2219	77.9295	82.3581

Table 3 (*continued*)

$a = .10$	$a = .05$	$a = .025$	$a = .010$	$a = .005$	d.f.
2.70554	3.84146	5.02389	6.63490	7.87944	1
4.60517	5.99147	7.37776	9.21034	10.5966	2
6.25139	7.81473	9.34840	11.3449	12.8381	3
7.77944	9.48773	11.1433	13.2767	14.8602	4
9.23635	11.0705	12.8325	15.0863	16.7496	5
10.6446	12.5916	14.4494	16.8119	18.5476	6
12.0170	14.0671	16.0128	18.4753	20.2777	7
13.3616	15.5073	17.5346	20.0902	21.9550	8
14.6837	16.9190	19.0228	21.6660	23.5893	9
15.9871	18.3070	20.4831	23.2093	25.1882	10
17.2750	19.6751	21.9200	24.7250	26.7569	11
18.5494	21.0261	23.3367	26.2170	28.2995	12
19.8119	22.3621	24.7356	27.6883	29.8194	13
21.0642	23.6848	26.1190	29.1413	31.3193	14
22.3072	24.9958	27.4884	30.5779	32.8013	15
23.5418	26.2962	28.8454	31.9999	34.2672	16
24.7690	27.5871	30.1910	33.4087	35.7185	17
25.9894	28.8693	31.5264	34.8053	37.1564	18
27.2036	30.1435	32.8523	36.1908	38.5822	19
28.4120	31.4104	34.1696	37.5662	39.9968	20
29.6151	32.6705	35.4789	38.9321	41.4010	21
30.8133	33.9244	36.7807	40.2894	42.7956	22
32.0069	35.1725	38.0757	41.6384	44.1813	23
33.1963	36.4151	39.3641	42.9798	45.5585	24
34.3816	37.6525	40.6465	44.3141	46.9278	25
35.5631	38.8852	41.9232	45.6417	48.2899	26
36.7412	40.1133	43.1944	46.9630	49.6449	27
37.9159	41.3372	44.4607	48.2782	50.9933	28
39.0875	42.5569	45.7222	49.5879	52.3356	29
40.2560	43.7729	46.9792	50.8922	53.6720	30
51.8050	55.7585	59.3417	63.6907	66.7659	40
63.1671	67.5048	71.4202	76.1539	79.4900	50
74.3970	79.0819	83.2976	88.3794	91.9517	60
85.5271	90.5312	95.0231	100.425	104.215	70
96.5782	101.879	106.629	112.329	116.321	80
107.565	113.145	118.136	124.116	128.299	90
118.498	124.342	129.561	135.807	140.169	100

From " Tables of the Percentage Points of the χ^2-Distribution." *Biometrika*, Vol. 32 (1941). pp. 188–189, by Catherine M. Thompson. Reproduced by permission of the *Biometrika* Trustees.

Table 4   *Percentage Points of the F Distribution, a = .05*

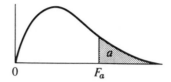

Degrees of Freedom   (a = .05)

d.f.₁ / d.f.₂	1	2	3	4	5	6	7	8	9
1	161.4	199.5	215.7	224.6	230.2	234.0	236.8	238.9	240.5
2	18.51	19.00	19.16	19.25	19.30	19.33	19.35	19.37	19.38
3	10.13	9.55	9.28	9.12	9.01	8.94	8.89	8.85	8.81
4	7.71	6.94	6.59	6.39	6.26	6.16	6.09	6.04	6.00
5	6.61	5.79	5.41	5.19	5.05	4.95	4.88	4.82	4.77
6	5.99	5.14	4.76	4.53	4.39	4.28	4.21	4.15	4.10
7	5.59	4.74	4.35	4.12	3.97	3.87	3.79	3.73	3.68
8	5.32	4.46	4.07	3.84	3.69	3.58	3.50	3.44	3.39
9	5.12	4.26	3.86	3.63	3.48	3.37	3.29	3.23	3.18
10	4.96	4.10	3.71	3.48	3.33	3.22	3.14	3.07	3.02
11	4.84	3.98	3.59	3.36	3.20	3.09	3.01	2.95	2.90
12	4.75	3.89	3.49	3.26	3.11	3.00	2.91	2.85	2.80
13	4.67	3.81	3.41	3.18	3.03	2.92	2.83	2.77	2.71
14	4.60	3.74	3.34	3.11	2.96	2.85	2.76	2.70	2.65
15	4.54	3.68	3.29	3.06	2.90	2.79	2.71	2.64	2.59
16	4.49	3.63	3.24	3.01	2.85	2.74	2.66	2.59	2.54
17	4.45	3.59	3.20	2.96	2.81	2.70	2.61	2.55	2.49
18	4.41	3.55	3.16	2.93	2.77	2.66	2.58	2.51	2.46
19	4.38	3.52	3.13	2.90	2.74	2.63	2.54	2.48	2.42
20	4.35	3.49	3.10	2.87	2.71	2.60	2.51	2.45	2.39
21	4.32	3.47	3.07	2.84	2.68	2.57	2.49	2.42	2.37
22	4.30	3.44	3.05	2.82	2.66	2.55	2.46	2.40	2.34
23	4.28	3.42	3.03	2.80	2.64	2.53	2.44	2.37	2.32
24	4.26	3.40	3.01	2.78	2.62	2.51	2.42	2.36	2.30
25	4.24	3.39	2.99	2.76	2.60	2.49	2.40	2.34	2.28
26	4.23	3.37	2.98	2.74	2.59	2.47	2.39	2.32	2.27
27	4.21	3.35	2.96	2.73	2.57	2.46	2.37	2.31	2.25
28	4.20	3.34	2.95	2.71	2.56	2.45	2.36	2.29	2.24
29	4.18	3.33	2.93	2.70	2.55	2.43	2.35	2.28	2.22
30	4.17	3.32	2.92	2.69	2.53	2.42	2.33	2.27	2.21
40	4.08	3.23	2.84	2.61	2.45	2.34	2.25	2.18	2.12
60	4.00	3.15	2.76	2.53	2.37	2.25	2.17	2.10	2.04
120	3.92	3.07	2.68	2.45	2.29	2.17	2.09	2.02	1.96
∞	3.84	3.00	2.60	2.37	2.21	2.10	2.01	1.94	1.88

Table 4   (*continued*)

10	12	15	20	24	30	40	60	120	∞	d.f.₁ / d.f.₂
241.9	243.9	245.9	248.0	249.1	250.1	251.1	252.2	253.3	254.3	1
19.40	19.41	19.43	19.45	19.45	19.46	19.47	19.48	19.49	19.50	2
8.79	8.74	8.70	8.66	8.64	8.62	8.59	8.57	8.55	8.53	3
5.96	5.91	5.86	5.80	5.77	5.75	5.72	5.69	5.66	5.63	4
4.74	4.68	4.62	4.56	4.53	4.50	4.46	4.43	4.40	4.36	5
4.06	4.00	3.94	3.87	3.84	3.81	3.77	3.74	3.70	3.67	6
3.64	3.57	3.51	3.44	3.41	3.38	3.34	3.30	3.27	3.23	7
3.35	3.28	3.22	3.15	3.12	3.08	3.04	3.01	2.97	2.93	8
3.14	3.07	3.01	2.94	2.90	2.86	2.83	2.79	2.75	2.71	9
2.98	2.91	2.85	2.77	2.74	2.70	2.66	2.62	2.58	2.54	10
2.85	2.79	2.72	2.65	2.61	2.57	2.53	2.49	2.45	2.40	11
2.75	2.69	2.62	2.54	2.51	2.47	2.43	2.38	2.34	2.30	12
2.67	2.60	2.53	2.46	2.42	2.38	2.34	2.30	2.25	2.21	13
2.60	2.53	2.46	2.39	2.35	2.31	2.27	2.22	2.18	2.13	14
2.54	2.48	2.40	2.33	2.29	2.25	2.20	2.16	2.11	2.07	15
2.49	2.42	2.35	2.28	2.24	2.19	2.15	2.11	2.06	2.01	16
2.45	2.38	2.31	2.23	2.19	2.15	2.10	2.06	2.01	1.96	17
2.41	2.34	2.27	2.19	2.15	2.11	2.06	2.02	1.97	1.92	18
2.38	2.31	2.23	2.16	2.11	2.07	2.03	1.98	1.93	1.88	19
2.35	2.28	2.20	2.12	2.08	2.04	1.99	1.95	1.90	1.84	20
2.32	2.25	2.18	2.10	2.05	2.01	1.96	1.92	1.87	1.81	21
2.30	2.23	2.15	2.07	2.03	1.98	1.94	1.89	1.84	1.78	22
2.27	2.20	2.13	2.05	2.01	1.96	1.91	1.86	1.81	1.76	23
2.25	2.18	2.11	2.03	1.98	1.94	1.89	1.84	1.79	1.73	24
2.24	2.16	2.09	2.01	1.96	1.92	1.87	1.82	1.77	1.71	25
2.22	2.15	2.07	1.99	1.95	1.90	1.85	1.80	1.75	1.69	26
2.20	2.13	2.06	1.97	1.93	1.88	1.84	1.79	1.73	1.67	27
2.19	2.12	2.04	1.96	1.91	1.87	1.82	1.77	1.71	1.65	28
2.18	2.10	2.03	1.94	1.90	1.85	1.81	1.75	1.70	1.64	29
2.16	2.09	2.01	1.93	1.89	1.84	1.79	1.74	1.68	1.62	30
2.08	2.00	1.92	1.84	1.79	1.74	1.69	1.64	1.58	1.51	40
1.99	1.92	1.84	1.75	1.70	1.65	1.59	1.53	1.47	1.39	60
1.91	1.83	1.75	1.66	1.61	1.55	1.50	1.43	1.35	1.25	120
1.83	1.75	1.67	1.57	1.52	1.46	1.39	1.32	1.22	1.00	∞

Table 5 *Percentage Points of the F Distribution, a = .01*

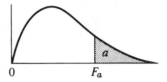

Degrees of Freedom $(a = .01)$

d.f.$_1$ d.f.$_2$	1	2	3	4	5	6	7	8	9
1	4052	4999.5	5403	5625	5764	5859	5928	5982	6022
2	98.50	99.00	99.17	99.25	99.30	99.33	99.36	99.37	99.39
3	34.12	30.82	29.46	28.71	28.24	27.91	27.67	27.49	27.35
4	21.20	18.00	16.69	15.98	15.52	15.21	14.98	14.80	14.66
5	16.26	13.27	12.06	11.39	10.97	10.67	10.46	10.29	10.16
6	13.75	10.92	9.78	9.15	8.75	8.47	8.26	8.10	7.98
7	12.25	9.55	8.45	7.85	7.46	7.19	6.99	6.84	6.72
8	11.26	8.65	7.59	7.01	6.63	6.37	6.18	6.03	5.91
9	10.56	8.02	6.99	6.42	6.06	5.80	5.61	5.47	5.35
10	10.04	7.56	6.55	5.99	5.64	5.39	5.20	5.06	4.94
11	9.65	7.21	6.22	5.67	5.32	5.07	4.89	4.74	4.63
12	9.33	6.93	5.95	5.41	5.06	4.82	4.64	4.50	4.39
13	9.07	6.70	5.74	5.21	4.86	4.62	4.44	4.30	4.19
14	8.86	6.51	5.56	5.04	4.69	4.46	4.28	4.14	4.03
15	8.68	6.36	5.42	4.89	4.56	4.32	4.14	4.00	3.89
16	8.53	6.23	5.29	4.77	4.44	4.20	4.03	3.89	3.78
17	8.40	6.11	5.18	4.67	4.34	4.10	3.93	3.79	3.68
18	8.29	6.01	5.09	4.58	4.25	4.01	3.84	3.71	3.60
19	8.18	5.93	5.01	4.50	4.17	3.94	3.77	3.63	3.52
20	8.10	5.85	4.94	4.43	4.10	3.87	3.70	3.56	3.46
21	8.02	5.78	4.87	4.37	4.04	3.81	3.64	3.51	3.40
22	7.95	5.72	4.82	4.31	3.99	3.76	3.59	3.45	3.35
23	7.88	5.66	4.76	4.26	3.94	3.71	3.54	3.41	3.30
24	7.82	5.61	4.72	4.22	3.90	3.67	3.50	3.36	3.26
25	7.77	5.57	4.68	4.18	3.85	3.63	3.46	3.32	3.22
26	7.72	5.53	4.64	4.14	3.82	3.59	3.42	3.29	3.18
27	7.68	5.49	4.60	4.11	3.78	3.56	3.39	3.26	3.15
28	7.64	5.45	4.57	4.07	3.75	3.53	3.36	3.23	3.12
29	7.60	5.42	4.54	4.04	3.73	3.50	3.33	3.20	3.09
30	7.56	5.39	4.51	4.02	3.70	3.47	3.30	3.17	3.07
40	7.31	5.18	4.31	3.83	3.51	3.29	3.12	2.99	2.89
60	7.08	4.98	4.13	3.65	3.34	3.12	2.95	2.82	2.72
120	6.85	4.79	3.95	3.48	3.17	2.96	2.79	2.66	2.56
∞	6.63	4.61	3.78	3.32	3.02	2.80	2.64	2.51	2.41

Table 5 *(continued)*

10	12	15	20	24	30	40	60	\120	∞	d.f.$_1$ / d.f.$_2$
6056	6106	6157	6209	6235	6261	6287	6313	6339	6366	1
99.40	99.42	99.43	99.45	99.46	99.47	99.47	99.48	99.49	99.50	2
27.23	27.05	26.87	26.69	26.60	26.50	26.41	26.32	26.22	26.13	3
14.55	14.37	14.20	14.02	13.93	13.84	13.75	13.65	13.56	13.46	4
10.05	9.89	9.72	9.55	9.47	9.38	9.29	9.20	9.11	9.02	5
7.87	7.72	7.56	7.40	7.31	7.23	7.14	7.06	6.97	6.88	6
6.62	6.47	6.31	6.16	6.07	5.99	5.91	5.82	5.74	5.65	7
5.81	5.67	5.52	5.36	5.28	5.20	5.12	5.03	4.95	4.86	8
5.26	5.11	4.96	4.81	4.73	4.65	4.57	4.48	4.40	4.31	9
4.85	4.71	4.56	4.41	4.33	4.25	4.17	4.08	4.00	3.91	10
4.54	4.40	4.25	4.10	4.02	3.94	3.86	3.78	3.69	3.60	11
4.30	4.16	4.01	3.86	3.78	3.70	3.62	3.54	3.45	3.36	12
4.10	3.96	3.82	3.66	3.59	3.51	3.43	3.34	3.25	3.17	13
3.94	3.80	3.66	3.51	3.43	3.35	3.27	3.18	3.09	3.00	14
3.80	3.67	3.52	3.37	3.29	3.21	3.13	3.05	2.96	2.87	15
3.69	3.55	3.41	3.26	3.18	3.10	3.02	2.93	2.84	2.75	16
3.59	3.46	3.31	3.16	3.08	3.00	2.92	2.83	2.75	2.65	17
3.51	3.37	3.23	3.08	3.00	2.92	2.84	2.75	2.66	2.57	18
3.43	3.30	3.15	3.00	2.92	2.84	2.76	2.67	2.58	2.49	19
3.37	3.23	3.09	2.94	2.86	2.78	2.69	2.61	2.52	2.42	20
3.31	3.17	3.03	2.88	2.80	2.72	2.64	2.55	2.46	2.36	21
3.26	3.12	2.98	2.83	2.75	2.67	2.58	2.50	2.40	2.31	22
3.21	3.07	2.93	2.78	2.70	2.62	2.54	2.45	2.35	2.26	23
3.17	3.03	2.89	2.74	2.66	2.58	2.49	2.40	2.31	2.21	24
3.13	2.99	2.85	2.70	2.62	2.54	2.45	2.36	2.27	2.17	25
3.09	2.96	2.81	2.66	2.58	2.50	2.42	2.33	2.23	2.13	26
3.06	2.93	2.78	2.63	2.55	2.47	2.38	2.29	2.20	2.10	27
3.03	2.90	2.75	2.60	2.52	2.44	2.35	2.26	2.17	2.06	28
3.00	2.87	2.73	2.57	2.49	2.41	2.33	2.23	2.14	2.03	29
2.98	2.84	2.70	2.55	2.47	2.39	2.30	2.21	2.11	2.01	30
2.80	2.66	2.52	2.37	2.29	2.20	2.11	2.02	1.92	1.80	40
2.63	2.50	2.35	2.20	2.12	2.03	1.94	1.84	1.73	1.60	60
2.47	2.34	2.19	2.03	1.95	1.86	1.76	1.66	1.53	1.38	120
2.32	2.18	2.04	1.88	1.79	1.70	1.59	1.47	1.32	1.00	∞

Table 6 *Squares and Square Roots*

n	n^2	$\sqrt{n}$	$\sqrt{10n}$	n	n^2	$\sqrt{n}$	$\sqrt{10n}$
				35	1 225	5.916 08	18.70829
1	1	1.000 00	3.162 28	36	1 296	6.000 00	18.97367
2	4	1.414 21	4.472 14	37	1 369	6.082 76	19.23538
3	9	1.732 05	5.477 23	38	1 444	6.164 41	19.49359
4	16	2.000 00	6.324 56	39	1 521	6.245 00	19.74842
5	25	2.236 07	7.071 07	40	1 600	6.324 56	20.00000
6	36	2.449 49	7.745 97	41	1 681	6.403 12	20.24846
7	49	2.645 75	8.366 60	42	1 764	6.480 74	20.49390
8	64	2.828 43	8.944 27	43	1 849	6.557 44	20.73644
9	81	3.000 00	9.486 83	44	1 936	6.633 25	20.97618
10	100	3.162 28	10.00000	45	2 025	6.708 20	21.21320
11	121	3.316 63	10.48809	46	2 116	6.782 33	21.44761
12	144	3.464 10	10.95445	47	2 209	6.855 66	21.67948
13	169	3.605 55	11.40175	48	2 304	6.928 20	21.90890
14	196	3.741 66	11.83216	49	2 401	7.000 00	22.13594
15	225	3.872 98	12.24745	50	2 500	7.071 07	22.36068
16	256	4.000 00	12.64911	51	2 601	7.141 43	22.58318
17	289	4.123 11	13.03840	52	2 704	7.211 10	22.80351
18	324	4.242 64	13.41641	53	2 809	7.280 11	23.02173
19	361	4.358 90	13.78405	54	2 916	7.348 47	23.23790
20	400	4.472 14	14.14214	55	3 025	7.416 20	23.45208
21	441	4.582 58	14.49138	56	3 136	7.483 32	23.66432
22	484	4.690 42	14.83240	57	3 249	7.549 83	23.87467
23	529	4.795 83	15.16575	58	3 364	7.615 77	24.08319
24	576	4.898 98	15.49193	59	3 481	7.618 15	24.28992
25	625	5.000 00	15.81139	60	3 600	7.745 97	24.49490
26	676	5.099 02	16.12452	61	3 721	7.810 25	24.69818
27	729	5.196 15	16.43168	62	3 844	7.874 01	24.89980
28	784	5.291 50	16.73320	63	3 969	7.937 25	25.09980
29	841	5.385 17	17.02939	64	4 096	8.000 00	25.29822
30	900	5.477 23	17.32051	65	4 225	8.062 26	25.49510
31	961	5.567 76	17.60682	66	4 356	8.124 04	25.69047
32	1 024	5.656 85	17.88854	67	4 489	8.185 35	25.88436
33	1 089	5.744 56	18.16590	68	4 624	8.246 21	26.07681
34	1 156	5.830 95	18.43909	69	4 761	8.306 62	26.26785

Table 6   (*continued*)

n	n^2	$\sqrt{n}$	$\sqrt{10n}$	n	n^2	$\sqrt{n}$	$\sqrt{10n}$
70	4 900	8.366 60	26.45751	105	11 025	10.24695	32.40370
71	5 041	8.426 15	26.64583	106	11 236	10.29563	32.55764
72	5 184	8.485 28	26.83282	107	11 449	10.34408	32.71085
73	5 329	8.544 00	27.01851	108	11 664	10.39230	32.86335
74	5 476	8.602 33	27.20294	109	11 881	10.44031	33.01515
75	5 625	8.660 25	27.38613	110	12 100	10.48809	33.16625
76	5 776	8.717 80	27.56810	111	12 321	10.53565	33.31666
77	5 929	8.774 96	27.74887	112	12 544	10.58301	33.46640
78	6 084	8.831 76	27.92848	113	12 769	10.63015	33.61547
79	6 241	8.888 19	28.10694	114	12 996	10.67708	33.76389
80	6 400	8.944 27	28.28427	115	13 225	10.72381	33.91165
81	6 561	9.000 00	28.46050	116	13 456	10.77033	34.05877
82	6 724	9.055 39	28.63564	117	13 689	10.81665	34.20526
83	6 889	9.110 43	28.80972	118	13 924	10.86278	34.35113
84	7 056	9.165 15	28.98275	119	14 161	10.90871	34.49638
85	7 225	9.219 54	29.15476	120	14 400	10.95445	34.64102
86	7 396	9.273 62	29.32576	121	14 641	11.00000	34.78505
87	7 569	9.327 38	29.49576	122	14 884	11.04536	34.92850
88	7 744	9.380 83	29.66479	123	15 129	11.09054	35.07136
89	7 921	9.433 98	29.83287	124	15 376	11.13553	35.21363
90	8 100	9.486 83	30.00000	125	15 625	11.18034	35.35534
91	8 281	9.539 39	30.16621	126	15 876	11.22497	35.49648
92	8 464	9.591 66	30.33150	127	16 129	11.26943	35.63706
93	8 649	9.643 65	30.49590	128	16 384	11.31371	35.77709
94	8 836	9.695 36	30.65942	129	16 641	11.35782	35.91657
95	9 025	9.746 79	30.82207	130	16 900	11.40175	36.05551
96	9 216	9.797 96	30.98387	131	17 161	11.44552	36.19392
97	9 409	9.848 86	31.14482	132	17 424	11.48913	36.33180
98	9 604	9.899 50	31.30495	133	17 689	11.53256	36.46917
99	9 801	9.949 87	31.46427	134	17 956	11.57584	36.60601
100	10 000	10.00000	31.62278	135	18 225	11.61895	36.74235
101	10 201	10.04998	31.78050	136	18 496	11.66190	36.87818
102	10 404	10.09950	31.93744	137	18 769	11.70470	37.01351
103	10 609	10.14889	32.09361	138	19 044	11.74734	37.14835
104	10 816	10.19804	32.24903	139	19 321	11.78983	37.28270

Table 6 (*continued*)

n	n^2	$\sqrt{n}$	$\sqrt{10n}$	n	n^2	$\sqrt{n}$	$\sqrt{10n}$
140	19 600	11.83216	37.41657	175	30 625	13.22876	41.83300
141	19 881	11.87434	37.54997	176	30 976	13.26650	41.95235
142	20 164	11.91638	37.68289	177	31 329	13.30413	42.07137
143	20 449	11.95826	37.81534	178	31 684	13.34166	42.19005
144	20 736	12.00000	37.94733	179	32 041	13.37909	42.30829
145	21 025	12.04159	38.07887	180	32 400	13.41641	42.42641
146	21 316	12.08305	38.20995	181	32 761	13.45362	42.54409
147	21 609	12.12436	38.34058	182	33 124	13.49074	42.66146
148	21 904	12.16553	38.47077	183	33 489	13.52775	42.77850
149	22 201	12.20656	38.60052	184	33 856	13.56466	42.89522
150	22 500	12.24745	38.72983	185	34 225	13.60147	43.01163
151	22 801	12.28821	38.85872	186	34 596	13.63818	43.12772
152	23 104	12.32883	38.98718	187	34 969	13.67479	43.24350
153	23 409	12.36932	39.11521	188	35 344	13.71131	43.35897
154	23 716	12.40967	39.24283	189	35 721	13.74773	43.47413
155	24 025	12.44990	39.37004	190	36 100	13.78405	43.58899
156	24 336	12.49000	39.49684	191	36 481	13.82027	43.70355
157	24 649	12.52996	39.62323	192	36 864	13.85641	43.81780
158	24 964	12.56981	39.74921	193	37 249	13.89244	43.93177
159	25 281	12.60952	39.87480	194	37 636	13.92839	44.04543
160	25 600	12.64911	40.00000	195	38 025	13.96424	44.15880
161	25 921	12.68858	40.12481	196	38 416	14.00000	44.27189
162	26 244	12.72792	40.24922	197	38 809	14.03567	44.38468
163	26 569	12.76715	40.37326	198	39 204	14.07125	44.49719
164	26 806	12.80625	40.49691	199	39 601	14.10674	44.60942
165	27 225	12.84523	40.62019	200	40 000	14.14214	44.72136
166	27 556	12.88410	40.74310	201	40 401	14.17745	44.83302
167	27 889	12.92285	40.86563	202	40 804	14.21267	44.94441
168	28 224	12.96148	40.98780	203	41 209	14.24781	45.05552
169	28 561	13.00000	41.10961	204	41 616	14.28286	45.16636
170	28 900	13.03840	41.23106	205	42 025	14.31782	45.27693
171	29 241	13.07670	41.35215	206	42 436	14.35270	45.38722
172	29 584	13.11488	41.47288	207	42 849	14.38749	45.49725
173	29 929	13.15295	41.59327	208	43 264	14.42221	45.60702
174	30 276	13.19091	41.71331	209	43 681	14.45683	45.71652

Table 6 (*continued*)

n	n^2	$\sqrt{n}$	$\sqrt{10n}$	n	n^2	$\sqrt{n}$	$\sqrt{10n}$
210	44 100	14.49138	45.82576	245	60 025	15.65248	49.49747
211	44 521	14.52584	45.93474	246	60 516	15.68439	49.59839
212	44 944	14.56022	46.04346	247	61 009	15.71623	49.69909
213	45 369	14.59452	46.15192	248	61 504	15.74902	49.79960
214	45 796	14.62874	46.26013	249	62 001	15.77973	49.89990
215	46 225	14.66288	46.36809	250	62 500	15.81139	50.00000
216	46 656	14.69694	46.47580	251	63 001	15.84298	50.09990
217	47 089	14.73092	46.58326	252	63 504	15.87451	50.19960
218	47 524	14.76482	46.69047	253	64 009	15.90597	50.29911
219	47 961	14.79865	46.79744	254	64 516	15.93738	50.39841
220	48 400	14.83240	46.90416	255	65 025	15.96872	50.49752
221	48 841	14.86607	47.01064	256	65 536	16.00000	50.59644
222	49 284	14.89966	47.11688	257	66 049	16.03122	50.69517
223	49 729	14.93318	47.22288	258	66 564	16.06238	50.79370
224	50 176	14.96663	47.32864	259	67 081	16.09348	50.89204
225	50 625	15.00000	47.43416	260	67 600	16.12452	50.99020
226	51 076	15.03330	47.53946	261	68 121	16.15549	51.08816
227	51 529	15.06652	47.64452	262	68 644	16.18641	51.18594
228	51 984	15.09967	47.74935	263	69 169	16.21727	51.28353
229	52 441	15.13275	47.85394	264	69 696	16.24808	51.38093
230	52 900	15.16575	47.95832	265	70 225	16.27882	51.47815
231	53 361	15.19868	48.06246	266	70 756	16.30951	51.57519
232	53 824	15.23155	48.16638	267	71 289	16.34013	51.67204
233	54 289	15.26434	48.27007	268	71 824	16.37071	51.76872
234	54 756	15.29706	48.37355	269	72 361	16.40122	51.86521
235	55 225	15.32971	48.47680	270	72 900	16.43168	51.96152
236	55 696	15.36229	48.57983	271	73 441	16.46208	52.05766
237	56 169	15.39480	48.68265	272	73 984	16.49242	52.15362
238	56 644	15.42725	48.78524	273	74 529	16.52271	52.24940
239	57 121	15.45962	48.88763	274	75 076	16.55295	52.34501
240	57 600	15.49193	48.98979	275	75 625	16.58312	52.44044
241	58 081	15.52417	49.09175	276	76 176	16.61235	52.53570
242	58 564	15.55635	49.19350	277	76 729	16.64332	52.63079
243	59 049	15.58846	49.29503	278	77 284	16.67333	52.72571
244	59 536	15.62050	49.39636	279	77 841	16.70329	52.82045

Table 6 (*continued*)

n	n^2	$\sqrt{n}$	$\sqrt{10n}$	n	n^2	$\sqrt{n}$	$\sqrt{10n}$
280	78 400	16.73320	52.91503	315	99 225	17.74824	56.12486
281	78 961	16.76305	53.00943	316	99 856	17.77639	56.21388
282	79 524	16.79286	53.10367	317	100 489	17.80449	56.30275
283	80 089	16.82260	53.19774	318	101 124	17.83255	56.39149
284	80 656	16.85230	53.29165	319	101 761	17.86057	56.48008
285	81 225	16.88194	53.38539	320	102 400	17.88854	56.56854
286	81 796	16.91153	53.47897	321	103 041	17.91647	56.65686
287	82 369	16.94107	53.57238	322	103 684	17.94436	56.74504
288	82 944	16.97056	53.66563	323	104 329	17.97220	56.83309
289	83 521	17.00000	53.75872	324	104 976	18.00000	56.92100
290	84 100	17.02939	53.85165	325	105 625	18.02776	57.00877
291	84 681	17.05872	53.94442	326	106 276	18.05547	57.09641
292	85 264	17.08801	54.03702	327	106 929	18.08314	57.18391
293	85 849	17.11724	54.12947	328	107 584	18.11077	57.27128
294	86 436	17.14643	54.22177	329	108 241	18.13836	57.35852
295	87 025	17.17556	54.31390	330	108 900	18.16590	57.44563
296	87 616	17.20465	54.40588	331	109 561	18.19341	57.53260
297	88 209	17.23369	54.49771	332	110 224	18.22087	57.61944
298	88 804	17.26268	54.58938	333	110 889	18.24829	57.70615
299	89 401	17.29162	54.68089	334	111 556	18.27567	57.79273
300	90 000	17.32051	54.77226	335	112 225	18.30301	57.87918
301	90 601	17.34935	54.86347	336	112 896	18.33030	57.96551
302	91 204	17.37815	54.95453	337	113 569	18.35756	58.05170
303	91 809	17.40690	55.04544	338	114 244	18.38478	58.13777
304	92 416	17.43560	55.13620	339	114 921	18.41195	58.22371
305	93 025	17.46425	55.22681	340	115 600	18.43909	58.30952
306	93 636	17.49286	55.31727	341	116 281	18.46619	58.39521
307	94 249	17.52142	55.40758	342	116 964	18.49324	58.48077
308	94 864	17.54993	55.49775	343	117 649	18.52026	58.56620
309	95 481	17.57840	55.58777	344	118 336	18.54724	58.65151
310	96 100	17.60682	55.67764	345	119 025	18.57418	58.73670
311	96 721	17.63519	55.76737	346	119 716	18.60108	58.82176
312	97 344	17.66352	55.85696	347	120 409	18.62794	58.90671
313	97 969	17.69181	55.94640	348	121 104	18.65476	58.99152
314	98 596	17.72005	56.03570	349	121 801	18.68154	59.07622

Table 6 (*continued*)

n	n^2	$\sqrt{n}$	$\sqrt{10n}$	n	n^2	$\sqrt{n}$	$\sqrt{10n}$
350	122 500	18.70829	59.16080	385	148 225	19.62142	62.04837
351	123 201	18.73499	59.24525	386	148 996	19.64688	62.12890
352	123 904	18.76166	59.32959	387	149 769	19.67232	62.20932
353	124 609	18.78829	59.41380	388	150 544	19.69772	62.28965
354	125 316	18.81489	59.49790	389	151 321	19.72308	62.36986
355	126 025	18.84144	59.58188	390	152 100	19.74842	62.44998
356	126 736	18.86796	59.66574	391	152 881	19.77372	62.52999
357	127 449	18.89444	59.74948	392	153 664	19.79899	62.60990
358	128 164	18.92089	59.83310	393	154 449	19.82423	62.68971
359	128 881	18.94730	59.91661	394	155 236	19.84943	62.76942
360	129 600	18.97367	60.00000	395	156 025	19.87461	62.84903
361	130 321	19.00000	60.08328	396	156 816	19.89975	62.92853
362	131 044	19.02630	60.16644	397	157 609	19.92486	63.00794
363	131 769	19.05256	60.24948	398	158 404	19.94994	63.08724
364	132 496	19.07878	60.33241	399	159 201	19.97498	63.16645
365	133 225	19.10497	60.41523	400	160 000	20.00000	63.24555
366	133 956	19.13113	60.49793	401	160 801	20.02498	63.32456
367	134 689	19.15724	60.58052	402	161 604	20.04994	63.40347
368	135 424	19.18333	60.66300	403	162 409	20.07486	63.48228
369	136 161	19.20937	60.74537	404	163 216	20.09975	63.56099
370	136 900	19.23538	60.82763	405	164 025	20.12461	63.63961
371	137 641	19.26136	60.90977	406	164 836	20.14944	63.71813
372	138 384	19.28730	60.99180	407	165 649	20.17424	63.79655
373	139 129	19.31321	61.07373	408	166 464	20.19901	63.87488
374	139 876	19.33908	61.15554	409	167 281	20.22375	63.95311
375	140 625	19.36492	61.23724	410	168 100	20.24864	64.03124
376	141 376	19.39072	61.31884	411	168 921	20.27313	64.10928
377	142 129	19.41649	61.40033	412	169 744	20.29778	64.18723
378	142 884	19.44222	61.48170	413	170 569	20.32240	64.26508
379	143 641	19.46792	61.56298	414	171 396	20.34699	64.34283
380	144 400	19.49359	61.64414	415	172 225	20.37155	64.42049
381	145 161	19.51922	61.72520	416	173 056	20.39608	64.49806
382	145 924	19.54482	61.80615	417	173 889	20.42058	64.57554
383	146 689	19.57039	61.88699	418	174 724	20.44505	64.65292
384	147 456	19.59592	61.96773	419	175 561	20.46949	64.73021

Table 6 (*continued*)

n	n²	√n	√10n	n	n²	√n	√10n
420	176 400	20.49390	64.80741	455	207 025	21.33073	67.45369
421	177 241	20.51828	64.88451	456	207 936	21.35416	67.52777
422	178 084	20.54264	64.96153	457	208 849	21.37756	67.60178
423	178 929	20.56696	65.03845	458	209 764	21.40093	67.67570
424	179 776	20.59126	65.11528	459	210 681	21.42429	67.74954
425	180 625	20.61553	65.19202	460	211 600	21.44761	67.82330
426	181 476	20.63977	65.26868	461	212 521	21.47091	67.89698
427	182 329	20.66398	65.34524	462	213 444	21.49419	67.97058
428	183 184	20.68816	65.42171	463	214 369	21.51743	68.04410
429	184 041	20.71232	65.49809	464	215 296	21.54066	68.11755
430	184 900	20.73644	65.57439	465	216 225	21.56386	68.19091
431	185 761	20.76054	65.65059	466	217 156	21.58703	68.26419
432	186 624	20.78461	65.72671	467	218 089	21.61018	68.33740
433	187 489	20.80865	65.80274	468	219 024	21.63331	68.41053
434	188 356	20.83267	65.87868	469	219 961	21.65641	68.48957
435	189 225	20.85665	65.95453	470	220 900	21.67948	68.55655
436	190 096	20.88061	66.03030	471	221 841	21.70253	68.62944
437	190 969	20.90454	66.10598	472	222 784	21.72556	68.70226
438	191 844	20.92845	66.18157	473	223 729	21.74856	68.77500
439	192 721	20.95233	66.25708	474	224 676	21.77154	68.84766
440	193 600	20.97618	66.33250	475	225 625	21.79449	68.92024
441	194 481	21.00000	66.40783	476	226 576	21.81742	68.99275
442	195 364	21.02380	66.48308	477	227 529	21.84033	69.06519
443	196 249	21.04757	66.55825	478	228 484	21.86321	69.13754
444	197 136	21.07131	66.63332	479	229 441	21.88607	69.20983
445	198 025	21.09502	66.70832	480	230 400	21.90890	69.28203
446	198 916	21.11871	66.78323	481	231 361	21.93171	69.35416
447	199 809	21.14237	66.85806	482	232 324	21.95450	69.42622
448	200 704	21.16601	66.93280	483	233 289	21.97726	69.49820
449	201 601	21.18962	67.00746	484	234 256	22.00000	69.57011
450	202 500	21.21320	67.08204	485	235 225	22.02272	69.64194
451	203 401	21.23676	67.15653	486	236 196	22.04541	69.71370
452	204 304	21.26029	67.23095	487	237 169	22.06808	69.78539
453	205 209	21.28380	67.30527	488	238 144	22.09072	69.85700
454	206 116	21.30728	67.37952	489	239 121	22.11334	69.92853

Table 6   (*continued*)

n	n^2	$\sqrt{n}$	$\sqrt{10n}$	n	n^2	$\sqrt{n}$	$\sqrt{10n}$
490	240 100	22.13594	70.00000	525	275 625	22.91288	72.45688
491	241 081	22.15852	70.07139	526	276 676	22.93469	72.52586
492	242 064	22.18107	70.14271	527	277 729	22.95648	72.59477
493	243 049	22.20360	70.21396	528	278 784	22.97825	72.66361
494	244 036	22.22611	70.28513	529	279 841	23.00000	72.73239
495	245 025	22.24860	70.35624	530	280 900	23.02173	72.80110
496	246 016	22.27106	70.42727	531	281 961	23.04344	72.86975
497	247 009	22.29350	70.49823	532	283 024	23.06513	72.93833
498	248 004	22.31591	70.56912	533	284 089	23.08679	73.00685
499	249 001	22.33831	70.63993	534	285 156	23.10844	73.07530
500	250 000	22.36068	70.71068	535	286 225	23.13007	73.14369
501	251 001	22.38303	70.78135	536	287 296	23.15167	73.21202
502	252 004	22.40536	70.85196	537	288 369	23.17326	73.28028
503	253 009	22.42766	70.92249	538	289 444	23.19483	73.34848
504	254 016	22.44994	70.99296	539	290 521	23.21637	73.41662
505	255 025	22.47221	71.06335	540	291 600	23.23790	73.48469
506	256 036	22.49444	71.13368	541	292 681	23.25941	73.55270
507	257 049	22.51666	71.20393	542	293 764	23.28089	73.62065
508	258 064	22.53886	71.27412	543	294 849	23.30236	73.68853
509	259 081	22.56103	71.34424	544	295 936	23.32381	73.75636
510	260 100	22.58318	71.41428	545	297 025	23.34524	73.82412
511	261 121	22.60531	71.48426	546	298 116	23.36664	73.89181
512	262 144	22.62742	71.55418	547	299 209	23.38803	73.95945
513	263 169	22.64950	71.62402	548	300 304	23.40940	74.02702
514	264 196	22.67157	71.69379	549	301 401	23.43075	74.09453
515	265 225	22.69361	71.76350	550	302 500	23.45208	74.16198
516	266 256	22.71563	71.83314	551	303 601	23.47339	74.22937
517	267 289	22.73763	71.90271	552	304 704	23.49468	74.29670
518	268 324	22.75961	71.97222	553	305 809	23.51595	74.36397
519	269 361	22.78157	72.04165	554	306 916	23.53720	74.43118
520	270 400	22.80351	72.11103	555	308 025	23.55844	74.49832
521	271 441	22.82542	72.18033	556	309 136	23.57965	74.56541
522	272 484	22.84732	72.24957	557	310 249	23.60085	74.63243
523	273 529	22.86919	72.31874	558	311 364	23.62202	74.69940
524	274 576	22.89105	72.38784	559	312 481	23.64318	74.76630

Table 6 (*continued*)

n	n²	√n	√10n	n	n²	√n	√10n
560	313 600	23.66432	74.83315	595	354 025	24.39262	77.13624
561	314 721	23.68544	74.89993	596	355 216	24.41311	77.20104
562	315 844	23.70654	74.96666	597	356 409	24.43358	77.26578
563	316 969	23.72762	75.03333	598	357 604	24.45404	77.33046
564	318 096	23.74868	75.09993	599	358 801	24.47448	77.39509
565	319 225	23.76973	75.16648	600	360 000	24.49490	77.45967
566	320 356	23.79075	75.23297	601	361 201	24.51530	77.52419
567	321 489	23.81176	75.29940	602	362 404	24.53569	77.58866
568	322 624	23.83275	75.36577	603	363 609	24.55606	77.65307
569	323 761	23.85372	75.43209	604	364 816	24.57641	77.71744
570	324 900	23.87467	75.49834	605	366 025	24.59675	77.78175
571	326 041	23.89561	75.56454	606	367 236	24.61707	77.84600
572	327 184	23.91652	75.63068	607	368 449	24.63737	77.91020
573	328 329	23.93742	75.69676	608	369 664	24.65766	77.97435
574	329 476	23.95830	75.76279	609	370 881	24.67793	78.03845
575	330 625	23.97916	75.82875	610	372 100	24.69818	78.10250
576	331 776	24.00000	75.89466	611	373 321	24.71841	78.16649
577	332 929	24.02082	75.96052	612	374 544	24.73863	78.23043
578	334 084	24.04163	76.02631	613	375 769	24.75884	78.29432
579	335 241	24.06242	76.09205	614	376 996	24.77902	78.35815
580	336 400	24.08319	76.15773	615	378 225	24.79919	78.42194
581	337 561	24.10394	76.22336	616	379 456	24.81935	78.48567
582	338 724	24.12468	76.28892	617	380 689	24.83948	78.54935
583	339 889	24.14539	76.35444	618	381 924	24.85961	78.61298
584	341 056	24.16609	76.41989	619	383 161	24.87971	78.67655
585	342 225	24.18677	76.48529	620	384 400	24.89980	78.74008
586	343 396	24.20744	76.55064	621	385 641	24.91987	78.80355
587	344 569	24.22808	76.61593	622	386 884	24.93993	78.86698
588	345 744	24.24871	76.68116	623	388 129	24.95997	78.93035
589	346 921	24.26932	76.74634	624	389 376	24.97999	78.99367
590	348 100	24.28992	76.81146	625	390 625	25.00000	79.05694
591	349 281	24.31049	76.87652	626	391 876	25.01999	79.12016
592	350 464	24.33105	76.94154	627	393 129	25.03997	79.18333
593	351 649	24.35159	77.00649	628	394 384	25.05993	79.24645
594	352 836	24.37212	77.07140	629	395 641	25.07987	79.30952

Table 6 (*continued*)

n	n^2	$\sqrt{n}$	$\sqrt{10n}$	n	n^2	$\sqrt{n}$	$\sqrt{10n}$
630	396 900	25.09980	79.37254	665	442 225	25.78759	81.54753
631	398 161	25.11971	79.43551	666	443 556	25.80698	81.60882
632	399 424	25.13961	79.49843	667	444 889	25.82634	81.67007
633	400 689	25.15949	79.56130	668	446 224	25.84570	81.73127
634	401 956	25.17936	79.62412	669	447 561	25.86503	81.79242
635	403 225	25.19921	79.68689	670	448 900	25.88436	81.85353
636	404 496	25.21904	79.74961	671	450 241	25.90367	81.91459
637	405 769	25.23886	79.81228	672	451 584	25.92296	81.97561
638	407 044	25.25866	79.87490	673	452 929	25.94224	82.03658
639	408 321	25.27845	79.93748	674	454 276	25.96151	82.09750
640	409 600	25.29822	80.00000	675	455 625	25.98076	82.15838
641	410 881	25.31798	80.06248	676	456 976	26.00000	82.21922
642	412 164	25.33772	80.12490	677	458 329	26.01922	82.28001
643	413 449	25.35744	80.18728	678	459 684	26.03843	82.34076
644	414 736	25.37716	80.24961	679	461 041	26.05763	82.40146
645	416 025	25.39685	80.31189	680	462 400	26.07681	82.46211
646	417 316	25.41653	80.37413	681	463 761	26.09598	82.52272
647	418 609	25.43619	80.43631	682	465 124	26.11513	82.58329
648	419 904	25.45584	80.49845	683	466 489	26.13427	82.64381
649	421 201	25.47548	80.56054	684	467 856	26.15339	82.70429
650	422 500	25.49510	80.62258	685	469 225	26.17250	82.76473
651	423 801	25.51470	80.68457	686	470 596	26.19160	82.82512
652	425 104	25.53429	80.74652	687	471 969	26.21068	82.88546
653	426 409	25.55386	80.80842	688	473 344	26.22975	82.94577
654	427 716	25.57342	80.87027	689	474 721	26.24881	83.00602
655	429 025	25.59297	80.93207	690	476 100	26.26785	83.06624
656	430 336	25.61250	80.99383	691	477 481	26.28688	83.12641
657	431 649	25.63201	81.05554	692	478 864	26.30589	83.18654
658	432 964	25.65151	81.11720	693	480 249	26.32489	83.24662
659	434 281	25.67100	81.17881	694	481 636	26.34388	83.30666
660	435 600	25.69047	81.24038	695	483 025	26.36285	83.36666
661	436 921	25.70992	81.30191	696	484 416	26.38181	83.42661
662	438 244	25.72936	81.36338	697	485 809	26.40076	83.48653
663	439 569	25.74879	81.42481	698	487 204	26.41969	83.54639
664	440 896	25.76820	81.48620	699	488 601	26.43861	83.60622

Table 6 (*continued*)

n	n^2	$\sqrt{n}$	$\sqrt{10n}$	n	n^2	$\sqrt{n}$	$\sqrt{10n}$
700	490 000	26.45751	83.66600	735	540 225	27.11088	85.73214
701	491 401	26.47640	83.72574	736	541 696	27.12932	85.79044
702	492 804	26.49528	83.78544	737	543 169	27.14774	85.84870
703	494 209	26.51415	83.84510	738	544 644	27.16616	85.90693
704	495 616	26.53300	83.90471	739	546 121	27.18455	85.96511
705	497 025	26.55184	83.96428	740	547 600	27.20294	86.02325
706	498 436	26.57066	84.02381	741	549 081	27.22132	86.08136
707	499 849	26.58947	84.08329	742	550 564	27.23968	86.13942
708	501 264	26.60827	84.14274	743	552 049	27.25803	86.19745
709	502 681	26.62705	84.20214	744	553 536	27.27636	86.25543
710	504 100	26.64583	84.26150	745	555 025	27.29469	86.31338
711	505 521	26.66458	84.32082	746	556 516	27.31300	86.37129
712	506 944	26.68333	84.38009	747	558 009	27.33130	86.42916
713	508 369	26.70206	84.43933	748	559 504	27.34959	86.48699
714	509 796	26.72078	84.49852	749	561 001	27.36786	86.54479
715	511 225	26.73948	84.55767	750	562 500	27.38613	86.60254
716	512 656	26.75818	84.61678	751	564 001	27.40438	86.66026
717	514 089	26.77686	84.67585	752	565 504	27.42262	86.71793
718	515 524	26.79552	84.73488	753	567 009	27.44085	86.77557
719	516 961	26.81418	84.79387	754	568 516	27.45906	86.83317
720	518 400	26.83282	84.85281	755	570 025	27.47726	86.89074
721	519 841	26.85144	84.91172	756	571 536	27.49545	86.94826
722	521 284	26.87006	84.97058	757	573 049	27.51363	87.00575
723	522 729	26.88866	85.02941	758	574 564	27.53180	87.06320
724	524 176	26.90725	85.08819	759	576 081	27.54995	87.12061
725	525 625	26.92582	85.14693	760	577 600	27.56810	87.17798
726	527 076	26.94439	85.20563	761	579 121	27.58623	87.23531
727	528 529	26.96294	85.26429	762	580 644	27.60435	87.29261
728	529 984	26.98148	85.32292	763	582 169	27.62245	87.34987
729	531 441	27.00000	85.38150	764	583 696	27.64055	87.40709
730	532 900	27.01851	85.44004	765	585 225	27.65863	87.46428
731	534 361	27.03701	85.49854	766	586 756	27.67671	87.52143
732	535 824	27.05550	85.55700	767	588 289	27.69476	87.57854
733	537 289	27.07397	85.61542	768	589 824	27.71281	87.63561
734	538 756	27.09243	85.67380	769	591 361	27.73085	87.69265

Table 6 (*continued*)

n	n^2	$\sqrt{n}$	$\sqrt{10n}$	n	n^2	$\sqrt{n}$	$\sqrt{10n}$
770	592 900	27.74887	87.74964	805	648 025	28.37252	89.72179
771	594 441	27.76689	87.80661	806	649 636	28.39014	89.77750
772	595 984	27.78489	87.86353	807	651 249	28.40775	89.83318
773	597 529	27.80288	87.92042	808	652 864	28.42534	89.88882
774	599 076	27.82086	87.97727	809	654 481	28.44293	89.94443
775	600 625	27.83882	88.03408	810	656 100	28.46050	90.00000
776	602 176	27.85678	88.09086	811	657 721	28.47806	90.05554
777	603 729	27.87472	88.14760	812	659 344	28.49561	90.11104
778	605 284	27.89265	88.20431	813	660 969	28.51315	90.16651
779	606 841	27.91057	88.26098	814	662 596	28.53069	90.22195
780	608 400	27.92848	88.31761	815	664 225	28.54820	90.27735
781	609 961	27.94638	88.37420	816	665 856	28.56571	90.33272
782	611 524	27.96426	88.43076	817	667 489	28.58321	90.38805
783	613 089	27.98214	88.48729	818	669 124	28.60070	90.44335
784	614 656	28.00000	88.54377	819	670 761	28.61818	90.49862
785	616 225	28.01785	88.60023	820	672 400	28.63564	90.55385
786	617 796	28.03569	88.65664	821	674 041	28.65310	90.60905
787	619 369	28.05352	88.71302	822	675 684	28.67054	90.66422
788	620 944	28.07134	88.76936	823	677 329	28.68798	90.71935
789	622 521	28.08914	88.82567	824	678 976	28.70540	90.77445
790	624 100	28.10694	88.88194	825	680 625	28.72281	90.82951
791	625 681	28.12472	88.93818	826	682 726	28.74022	90.88454
792	627 264	28.14249	88.99428	827	683 929	28.75761	90.93954
793	628 849	28.16026	89.05055	828	685 584	28.77499	90.99451
794	630 436	28.17801	89.10668	829	687 241	28.79236	91.04944
795	632 025	28.19574	89.16277	830	688 900	28.80972	91.10434
796	633 616	28.21347	89.21883	831	690 561	28.82707	91.15920
797	635 209	28.23119	89.27486	832	692 224	28.84441	91.21403
798	636 804	28.24889	89.33085	833	693 889	28.86174	91.26883
799	638 401	28.26659	89.38680	834	695 556	28.87906	91.32360
800	640 000	28.28472	89.44272	835	697 225	28.89637	91.37833
801	641 601	28.30194	89.49860	836	698 896	28.91366	91.43304
802	643 204	28.31960	89.55445	837	700 569	28.93095	91.48770
803	644 809	28.33725	89.61027	838	702 244	28.94823	91.54234
804	646 416	28.35489	89.66605	839	703 921	28.96550	91.59694

Table 6   (*continued*)

n	n^2	$\sqrt{n}$	$\sqrt{10n}$	n	n^2	$\sqrt{n}$	$\sqrt{10n}$
840	705 600	28.98275	91.65151	875	765 625	29.58040	93.54143
841	707 281	29.00000	91.70605	876	767 376	29.59730	93.59487
842	708 964	29.01724	91.76056	877	769 129	29.61419	93.64828
843	710 649	29.03446	91.81503	878	770 884	29.63106	93.70165
844	712 336	29.05168	91.86947	879	772 641	29.64793	93.75500
845	714 025	29.06888	91.92388	880	774 400	29.66479	93.80832
846	715 716	29.08608	91.97826	881	776 161	29.68164	93.86160
847	717 409	29.10326	92.03260	882	777 924	29.69848	93.91486
848	719 104	29.12044	92.08692	883	779 689	29.71532	93.96808
849	720 801	29.13760	92.14120	884	781 456	29.73214	94.02127
850	722 500	29.15476	92.19544	885	783 225	29.74895	94.07444
851	724 201	29.17190	92.24966	886	784 996	29.76575	94.12757
852	725 904	29.18904	92.30385	887	786 769	29.78255	94.18068
853	727 609	29.20616	92.35800	888	788 544	29.79933	94.23375
854	729 316	29.22328	92.41212	889	790 321	29.81610	94.28680
855	731 025	29.24038	92.46621	890	792 100	29.83287	94.33981
856	732 736	29.25748	92.52027	891	793 881	29.84962	94.39280
857	734 449	29.27456	92.57429	892	795 664	29.86637	94.44575
858	736 164	29.29164	92.62829	893	797 449	29.88311	94.49868
859	737 881	29.30870	92.68225	894	799 236	29.89983	94.55157
860	739 600	29.32576	92.73618	895	801 025	29.91655	94.60444
861	741 321	29.34280	92.79009	896	802 816	29.93326	94.65728
862	743 044	29.35984	92.84396	897	804 609	29.94996	94.71008
863	744 769	29.37686	92.89779	898	806 404	29.96665	94.76286
864	746 496	29.39388	92.95160	899	808 201	29.98333	94.81561
865	748 225	29.41088	93.00538	900	810 000	30.00000	94.86833
866	749 956	29.42788	93.05912	901	811 801	30.01666	94.92102
867	751 689	29.44486	93.11283	902	813 604	30.03331	94.97368
868	753 424	29.46184	93.16652	903	815 409	30.04996	95.02631
869	755 161	29.47881	93.22017	904	817 216	30.06659	95.07891
870	756 900	29.49576	93.27379	905	819 025	30.08322	95.13149
871	758 641	29.51271	93.32738	906	820 836	30.09983	95.18403
872	760 384	29.52965	93.38094	907	822 649	30.11644	95.23655
873	762 129	29.54657	93.43447	908	824 464	30.13304	95.28903
874	763 876	29.56349	93.48797	909	826 281	30.14963	95.34149

Table 6   (*continued*)

n	n^2	$\sqrt{n}$	$\sqrt{10n}$	n	n^2	$\sqrt{n}$	$\sqrt{10n}$
910	828 100	30.16621	95.39392	945	893 025	30.74085	97.21111
911	829 921	30.18278	95.44632	946	894 916	30.75711	97.26253
912	831 744	30.19934	95.49869	947	896 809	30.77337	97.31393
913	833 569	30.21589	95.55103	948	898 704	30.78961	97.36529
914	835 396	30.23243	95.60335	949	900 601	30.80584	97.41663
915	837 225	30.24897	95.65563	950	902 500	30.82207	97.46794
916	839 056	30.26549	95.70789	951	904 401	30.83829	97.51923
917	840 889	30.28201	95.76012	952	906 304	30.85450	97.57049
918	842 724	30.29851	95.81232	953	908 209	30.87070	97.62172
919	844 561	30.31501	95.86449	954	910 116	30.88689	97.67292
920	846 400	30.33150	95.91663	955	912 025	30.90307	97.72410
921	848 241	30.34798	95.96874	956	913 936	30.91925	97.77525
922	850 084	30.36445	96.02083	957	915 849	30.93542	97.82638
923	851 929	30.38092	96.07289	958	917 764	30.95158	97.87747
924	853 776	30.39737	96.12492	959	919 681	30.96773	97.92855
925	855 625	30.41381	96.17692	960	921 600	30.98387	97.97959
926	857 476	30.43025	96.22889	961	923 521	31.00000	98.03061
927	859 329	30.44667	96.28084	962	925 444	31.01612	98.08160
928	861 184	30.46309	96.33276	963	927 369	31.03224	98.13256
929	863 041	30.47950	96.38465	964	929 296	31.04835	98.18350
930	864 900	30.49590	96.43651	965	931 225	31.06445	98.23441
931	866 761	30.51229	96.48834	966	933 156	31.08054	98.28530
932	868 624	30.52868	96.54015	967	935 089	31.09662	98.33616
933	870 489	30.54505	96.59193	968	937 024	31.11270	98.38699
934	872 356	30.56141	96.64368	969	938 961	31.12876	98.43780
935	874 225	30.57777	96.69540	970	940 900	31.14482	98.48858
936	876 096	30.59412	96.74709	971	942 841	31.16087	98.53933
937	877 969	30.61046	96.79876	972	944 784	31.17691	98.59006
938	879 844	30.62679	96.85040	973	946 729	31.19295	98.64076
939	881 721	30.64311	96.90201	974	948 676	31.20897	98.69144
940	883 600	30.65942	96.95360	975	950 625	31.22499	98.74209
941	885 481	30.67572	97.00515	976	952 576	31.24100	98.79271
942	887 364	30.69202	97.05668	977	954 529	31.25700	98.84331
943	889 249	30.70831	97.10819	978	956 484	31.27299	98.89388
944	891 136	30.72458	97.15966	979	958 441	31.28898	98.94443

Table 6 (*continued*)

n	n^2	$\sqrt{n}$	$\sqrt{10n}$	n	n^2	$\sqrt{n}$	$\sqrt{10n}$
980	960 400	31.30495	98.99495	990	980 100	31.46427	99.49874
981	962 361	31.32092	99.04544	991	982 081	31.48015	99.54898
982	964 324	31.33688	99.09591	992	984 064	31.49603	99.59920
983	966 289	31.35283	99.14636	993	986 049	31.51190	99.64939
984	968 256	31.36877	99.19677	994	988 036	31.52777	99.69955
985	970 225	31.38471	99.24717	995	990 025	31.54362	99.74969
986	972 196	31.40064	99.29753	996	992 016	31.55947	99.79980
987	974 169	31.41656	99.34787	997	994 009	31.57531	99.84989
988	976 144	31.43247	99.39819	998	996 004	31.59114	99.89995
989	978 121	31.44837	99.44848	999	998 001	31.60696	99.94999
				1000	1000 000	31.62278	100.00000

Computed by J. Huang, Department of Statistics, University of Florida.

Table 7   *Critical Values for* $D = \max |F_0(x) - S_n(x)|$ *in the Kolmogorov–Smirnov Test for Goodness of Fit: One Sample*

Sample Size n	$a = .20$	$a = .15$	$a = .10$	$a = .05$	$a = .01$
1	.900	.925	.950	.975	.995
2	.684	.726	.776	.842	.929
3	.565	.597	.642	.708	.828
4	.494	.525	.564	.624	.733
5	.446	.474	.510	.565	.669
6	.410	.436	.470	.521	.618
7	.381	.405	.438	.486	.577
8	.358	.381	.411	.457	.543
9	.339	.360	.388	.432	.514
10	.322	.342	.368	.410	.490
11	.307	.326	.352	.391	.468
12	.295	.313	.338	.375	.450
13	.284	.302	.325	.361	.433
14	.274	.292	.314	.349	.418
15	.266	.283	.304	.338	.404
16	.258	.274	.295	.328	.392
17	.250	.266	.286	.318	.381
18	.244	.259	.278	.309	.371
19	.237	.252	.272	.301	.363
20	.231	.246	.264	.294	.356
25	.21	.22	.24	.27	.32
30	.19	.20	.22	.24	.29
35	.18	.19	.21	.23	.27
Over 35	$\dfrac{1.07}{\sqrt{n}}$	$\dfrac{1.14}{\sqrt{n}}$	$\dfrac{1.22}{\sqrt{n}}$	$\dfrac{1.36}{\sqrt{n}}$	$\dfrac{1.63}{\sqrt{n}}$

From "The Kolmogorov–Smirnov Test for Goodness of Fit." *Journal of the American Statistical Association*, Vol. 46 (1951), p. 70, by F. J. Massey, Jr. Reproduced by permission of the Editor.

Table 8 *Random Numbers*

Line/Col.	(1)	(2)	(3)	(4)	(5)	(6)	(7)	(8)	(9)	(10)	(11)	(12)	(13)	(14)
1	10480	15011	01536	02011	81647	91646	69179	14194	62590	36207	20969	99570	91291	90700
2	22368	46573	25595	85393	30995	89198	27982	53402	93965	34095	52666	19174	39615	99505
3	24130	48360	22527	97265	76393	64809	15179	24830	49340	32081	30680	19655	63348	58629
4	42167	93093	06243	61680	07856	16376	39440	53537	71341	57004	00849	74917	97758	16379
5	37570	39975	81837	16656	06121	91782	60468	81305	49684	60672	14110	06927	01263	54613
6	77921	06907	11008	42751	27756	53498	18602	70659	90655	15053	21916	81825	44394	42880
7	99562	72905	56420	69994	98872	31016	71194	18738	44013	48840	63213	21069	10634	12952
8	96301	91977	05463	07972	18876	20922	94595	56869	69014	60045	18425	84903	42508	32307
9	89579	14342	63661	10281	17453	18103	57740	84378	25331	12566	58678	44947	05585	56941
10	85475	36857	53342	53988	53060	59533	38867	62300	08158	17983	16439	11458	18593	64952
11	28918	69578	88231	33276	70997	79936	56865	05859	90106	31595	01547	85590	91610	78188
12	63553	40961	48235	03427	49626	69445	18663	72695	52180	20847	12234	90511	33703	90322
13	09429	93969	52636	92737	88974	33488	36320	17617	30015	08272	84115	27156	30613	74952
14	10365	61129	87529	85689	48237	52267	67689	93394	01511	26358	85104	20285	29975	89968
15	07119	97336	71048	08178	77233	13916	47564	81056	97735	85977	29372	74461	28551	90707
16	51085	12765	51821	51259	77452	16308	60756	92144	49442	53900	70960	63990	75601	40719
17	02368	21382	52404	60268	89368	19885	55322	44819	01188	65255	64835	44919	05944	55157
18	01011	54092	33362	94904	31273	04146	18594	29852	71585	85030	51132	01915	92747	64951
19	52162	53916	46369	58586	23216	14513	83149	98736	23495	64350	94738	17752	35156	35749
20	07056	97628	33787	09998	42698	06691	76988	13602	51851	46104	88916	19509	25625	58104
21	48663	91245	85828	14346	09172	30168	90229	04734	59193	22178	30421	61666	99904	32812
22	54164	58492	22421	74103	47070	25306	76468	26384	58151	06646	21524	15227	96909	44592
23	32639	32363	05597	24200	13363	38005	94342	28728	35806	06912	17012	64161	18296	22851
24	29334	27001	87637	87308	58731	00256	45834	15398	46557	41135	10367	07684	36188	18510
25	02488	33062	28834	07351	19731	92420	60952	61280	50001	67658	32586	86679	50720	94953

Abridged from *Handbook of Tables for Probability and Statistics*, Second Edition, edited by William H. Beyer, © The Chemical Rubber Co., 1968. Used by permission of the Chemical Rubber Co.

GLOSSARY OF COMMON STATISTICAL TERMS

Acceptance region   Set of values of a test statistic that imply acceptance of the null hypothesis.

Alternative hypothesis   Hypothesis to be accepted if the null hypothesis is rejected (also called the *research hypothesis*).

Analysis of variance   Method of data analysis useful in testing the equality of k population means. (See Chapter 12.)

Average deviation   Measure of data variation. (See Chapter 4.)

Biased estimator   Estimator whose probability distribution has a mean value that is not equal to the value of the estimated parameter.

Binomial experiment   Experiment involving n identical independent dichotomous trials. (See Chapter 5 for an exact description of a binomial experiment.)

Binomial random variable   Discrete random variable representing the number of successes in n identical independent trials. (For an exact definition of a binomial experiment, see Chapter 5.)

Chi-square   Test statistic used to test the null hypothesis of "independence" for the two classifications of a contingency table and used for a goodness-of-fit test. Also has many other statistical applications not discussed in this text. (See Chapters 7 and 9.)

Circle chart   Graphical method to describe data; also called a *pie chart*. (See Chapter 3.)

Class frequency   Number of observations falling in a class (referring to a frequency histogram).

Class intervals   Cells of a frequency histogram.

Class limit   Dividing point between two cells in a frequency histogram.

Coefficient of linear correlation   Measure of linear dependence between two random variables.

Confidence coefficient   Probability that an interval estimate (a confidence interval) will enclose the parameter of interest.

Confidence interval Concerned with interval estimation. Two numbers, computed from sample data, that form an interval estimate for some parameter. (Naturally, we would like the interval to enclose the parameter of interest.)

Contingency coefficient Nominal measure of association. (See Chapter 10.)

Contingency table Two-way table constructed for classifying count data. The entries in the table show the number of observations falling in the cells. The objective of an analysis is to determine whether the two directions of classification are dependent (contingent) upon one another.

Critical value of a test statistic Value (or values) that separate the rejection and the acceptance regions in a statistical test.

Degrees of freedom Parameter of Student's *t* and the chi-square probability distributions. Degrees of freedom measure the quantity of information available in normally distributed data for estimating the population variance, σ^2.

Deviation from the mean Distance between a sample observation and the sample mean, $\bar{X}$.

Discrete random variable Random variable that can assume only a finite number or a countable infinity of values.

Empirical rule Rule that describes the variability of data that possess a mound-shaped frequency distribution. (See Chapter 4.)

Error of estimation Distance between an estimate and the true value of the parameter estimated.

Estimate Number computed from sample data used to approximate a population parameter.

Estimator Rule that tells how to compute an estimate based on information contained in a sample. An estimator is usually given as a mathematical formula.

Event Outcome of an experiment.

Expected value of a random variable Mean value of the random variable based on its probability distribution (theoretical frequency distribution).

Experiment Process of making an observation.

Fisher's exact test Nonparametric test for count data. (See Chapter 9).

Frequency Number of observations falling in a cell or in classification category.

F statistic Test statistic used to compare variances from two normal populations. Used in the analysis of variance.

Gamma Proportional reduction in error measure of association for ordinal data. (See Chapter 10.)

Histogram Graphical method for describing a set of data. (See Chapter 3.)

Interval estimate Two numbers computed from the sample data. The interval formed by the numbers should enclose some parameter of interest. An interval estimate is usually called a *confidence interval*.

Interval estimator Rule that explains how to calculate from sample data the two numbers required for an interval estimate.

Kendall's *W* Measure of association for ordinal data. (See Chapter 10.)

Kolmogorov–Smirnov test Goodness-of-fit test. (See Chapter 7.)

Kruskal–Wallis test One-way analysis of variance by ranks. (See Chapter 12.)

Lambda Proportional reduction in error measure of association for nominal data. (See Chapter 10.)

Least squares Method of curve fitting that selects as the best-fitting curve the one that minimizes the sum of the squares of deviations of the data points from the fitted curve. (See Chapter 11.)

Lower confidence limit The smaller of the two numbers that form a confidence interval.

McNemar test Test of change in attitude or preference. (See Chapter 9.)

Mean Average of a set of measurements. The symbols $\bar{X}$ and μ denote the means of a sample and a population, respectively.

Median Middle measurement when a set is ordered according to numerical value. (See Chapter 4 for a more precise definition.)

Multinomial experiment Experiment that consists of n identical independent trials, where each trial can result in one of k (a fixed number) of outcomes.

Nonparametric methods Usually refers to statistical tests of hypotheses about population probability distributions—but not about specific parameters of the distributions.

Normal distribution Bell-shaped probability distribution. The curve possesses a specific mathematical formula. (See References of Chapter 5.)

Null hypothesis Hypothesis under test in a statistical test of an hypothesis.

Paired-difference test Statistical test for the comparison of two population means. The test is based on paired observations, one from each of the two populations. (See Chapter 8.)

Parameter of a population Numerical descriptive measure of a population.

Parametric methods Statistical methods for estimating parameters or testing hypotheses about population parameters.

Pearson's r   Measure of the linear dependence between two variables measured on an interval or ratio scale. (See Chapter 11.)

Percentiles   See Chapter 4 for definition.

Phi coefficient   Nominal measure of association useful in 2 × 2 tables. (See Chapter 10.)

Pie chart   Graphical method for describing data; also called a *circle chart*. (See Chapter 3.)

Point estimate   *See* Estimate.

Population   Set of measurements, existing or conceptual, that is of interest to the experimenter. Samples are selected from the population.

Probability   As a practical matter, we think of the probability of an event as a measure of one's belief that the event will occur when the experiment is conducted once. The exact definition, giving a quantitative measure of this belief, is subject to debate. The relative frequency concept is most widely accepted.

Probability distribution, continuous   Smooth curve that gives the theoretical frequency distribution for the continuous random variable. An area under the curve over an interval is proportional to the probability that the random variable will fall in the interval.

Probability distribution, discrete   Listing, a mathematical formula, or a histogram that gives the probability associated with each value of the random variable.

Proportional reduction in error   Method for interpreting the strength of the relationship between two variables.

Quartiles   See Chapter 4 for definition.

Random sample   Sample of n measurements selected in such a way that every different sample of n elements in the population has an equal probability of being selected.

Random variable   Random variable is associated with an experiment. Its values are numerical events that cannot be predicted with certainty.

Range of a set of measurements   Difference between the largest and smallest members of the set. (See Chapter 4 for a more precise definition.)

Rank correlation coefficient   Rank correlation coefficient is a coefficient of linear correlation between two random variables that is based on the ranks of the measurements, not their actual values. (See coefficient of linear correlation.)

Real class limits   Interval endpoints for grouped data that form nonoverlapping, contiguous classes of equal width.

Regression line   Line fit to data points using the method of least squares.

Rejection region   Set of values of a test statistic that indicates rejection of the null hypothesis.

Relative frequency   Class frequency divided by the total number of measurements.

Research hypothesis   See *Alternative hypothesis.*

Sample   Subset of measurements selected from a population.

Sign test   Nonparametric statistical test used to compare two populations. (See Chapter 8.)

Spearman's rho   Measure of the linear correlation between ranked values of two random variables. ρ^2, the coefficient of determination, has a proportional reduction in error interpretation. (See Chapter 10.)

Standard deviation   Measure of data variation. (See Chapter 4.)

Standardized normal distribution   Normal distribution with mean and standard deviation equal to 0 and 1, respectively. The standardized normal variable is denoted by the symbol z.

Statistical map   Graphical method for describing data. (See Chapter 3.)

Student's *t* distribution   Particular symmetric mound-shaped distribution that possesses more spread than the standard normal probability distribution.

Test statistic   Function of the sample measurements, used as a decision maker in a test of an hypothesis.

Type I error   Rejecting the null hypothesis when it is true.

Type II error   Accepting the null hypothesis when it is false and the alternative hypothesis is true.

Unbiased estimator   Estimator that has a probability distribution with mean equal to the estimated parameter.

Upper Confidence limit   The larger of the two numbers that form a confidence interval.

Variance   Measure of data variation. (See Chapter 4.)

z statistic   Standardized normal random variable that is frequently used as a test statistic.

ANSWERS TO EXERCISES

Chapter 2

1. Quantitative: can be measured according to degree of racism, thus can be ordered.

5. An infant mortality rate is calculated by first counting the number of infant deaths over a long period of time and then dividing this number by the total number of infants born during the same period. This division can result in infinitely many values, not just integer values.

7. ratio

9. No, because the distances between the ranks are unknown and are not necessarily the same for both categories.

10. Not only can we make absolute comparisons (as with an interval scale), but we can also make valid relative comparisons.

13. (a) $18/48 = .375$ (b) $48/59 = .814$ (c) $(11/107)100 = 10.3\%$
    (d) $2/48 = .042$ (e) $6/59 = .102$ (f) Yes, because we could compare categories more easily.

14. | | |
     |------|------|
     | 740 | 640 |
     | 680 | 440 |
     | 1280 | 790 |
     | 100 | 1280 |

15. Yes, for example the number of students enrolled in universities could be rounded to the nearest 100.

16. 11−12

17. | | | |
     |------|-----|-------|
     | 0.0 | 1.2 | 139.8 |
     | .7 | 1.6 | 17.4 |

18. | | | |
     |-----|------|-----|
     | 100 | 7900 | 200 |
     | 0 | 0 | 100 |

19. 20
     20
     10
     10
     10

20
20
10
10
20

22. *Discrete*: a quantitative variable that can take on a countable number of values. *Continuous*: a quantitative variable that can assume the infinitely large number of values in a line interval.

24. (a)   15,000/16,000 = .9375   (b)   16,000/15,000 = 1.0670
    (c)   15,000/31,000 = .4839   (d)   16,000/31,000 = .5161

25.   4/1116 = .00358

26.   to examine trends across time

27.

number	ones	tens	hundreds
6.5	6	10	0
14.3	14	10	0
78.7	79	80	100
100.5	100	100	100
149.5	150	150	100
698.1	698	700	700
1028.7	1029	1030	1000
1499.5	1500	1500	1500

28.   6.45–6.55
       14.25–14.35
       78.65–78.75
       100.45–100.55
       149.45–149.55
       698.05–698.15
       1028.65–1028.75
       1499.45–1499.55

30.   percent = proportion × 100

31.   A rate expresses the number of occurrences on a per-unit basis; a ratio expresses the number of occurrences in one group relative to the number in another group.

33.   1.611
       2.760
       5.952
       7.576
       14.269
       17.100
       2.205

34. | 160.5–161.5 | 1426.5–1427.5 |
    |---|---|
    | 275.5–276.5 | 1709.5–1710.5 |
    | 594.5–595.5 | 219.5–220.5 |
    | 757.5–758.5 | |

35. | 161 | 1427 |
    |---|---|
    | 276 | 1710 |
    | 595 | 220 |
    | 758 | |

Chapter 3

2. Both principles are violated.

3. Exclusiveness is violated. Inclusiveness is followed.

4. Exclusiveness is violated; for example, a physics major can be classified under physics and physical sciences. Inclusiveness is followed.

5.

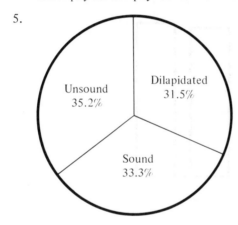

6.

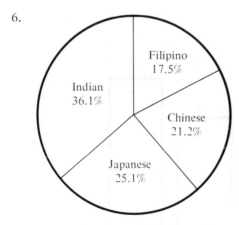

8.

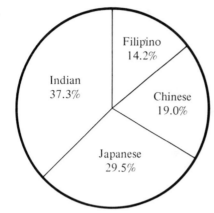

9.

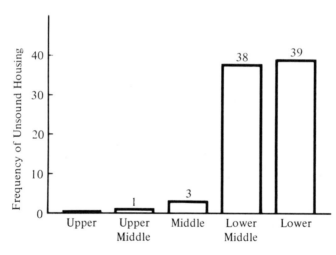

10.

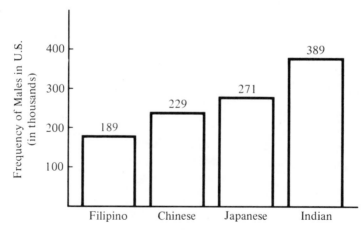

11.

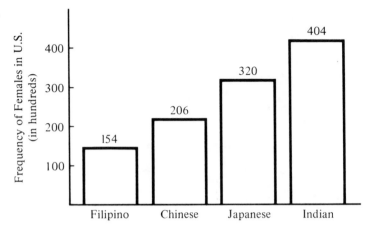

13. *relative frequency, males*

$2/311 = .0064$

0

$2/311 = .0064$

$13/311 = .0418$

$13/311 = .0418$

$29/311 = .0932$

$72/311 = .2315$

$64/311 = .2058$

$68/311 = .2186$

$42/311 = .1350$

$5/311 = .0161$

$1/311 = .0032$

14. *relative frequency, females*

$2/319 = .0063$

$5/319 = .0157$

$3/319 = .0094$

$12/319 = .0376$

$28/319 = .0878$

$32/319 = .1003$

$70/319 = .2194$

$74/319 = .2320$

$77/319 = .2414$

$16/319 = .0502$

0

0

15.   4

16.

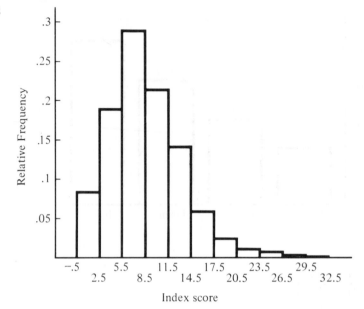

17. *relative frequency*

2/137 = .0146	13/137 = .0949
10/137 = .0730	12/137 = .0876
12/137 = .0876	2/137 = .0146
16/137 = .1168	4/137 = .0292
27/137 = .1971	1/137 = .0073
37/137 = .2701	1/137 = .0073

18.

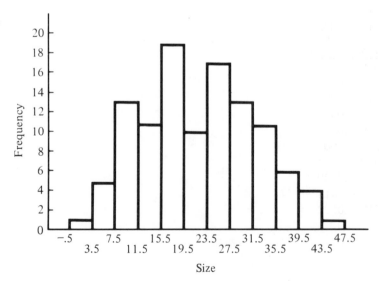

19.

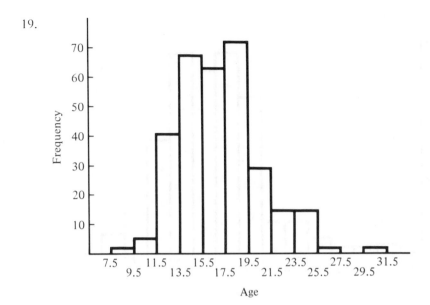

20.

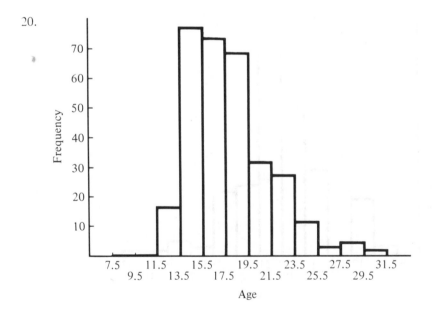

21.

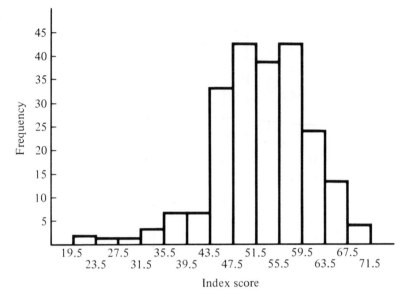

23.

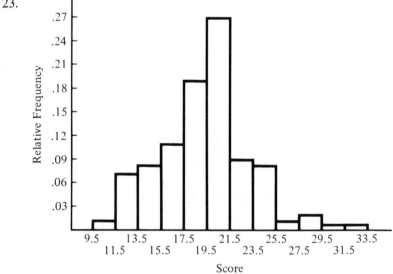

24. $-.5 - 3.5$

25.

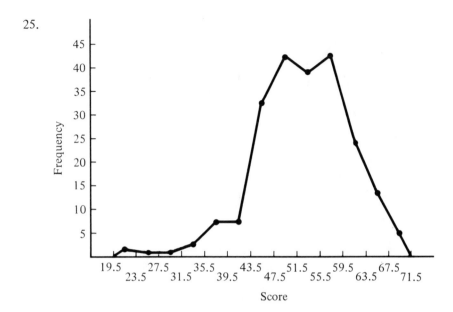

26.

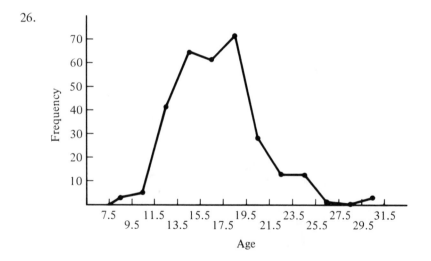

27.   no

28.

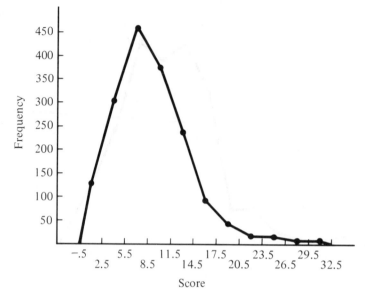

29.

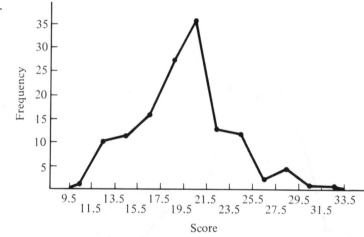

30.

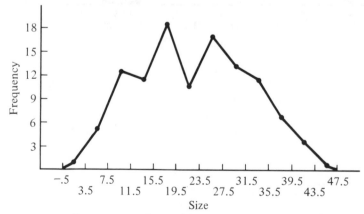

31. | *cumulative relative frequency, males* |
    | --- |
    | .9998 |
    | .9934 |
    | .9934 |
    | .9870 |
    | .9452 |
    | .9034 |
    | .8102 |
    | .5787 |
    | .3729 |
    | .1543 |
    | .0193 |
    | .0032 |

32. | *cumulative relative frequency, females* |
    | --- |
    | 1.0010 |
    | .9938 |
    | .9781 |
    | .9687 |
    | .9311 |
    | .8433 |
    | .7430 |
    | .5236 |
    | .2916 |
    | .0502 |
    | 0 |
    | 0 |

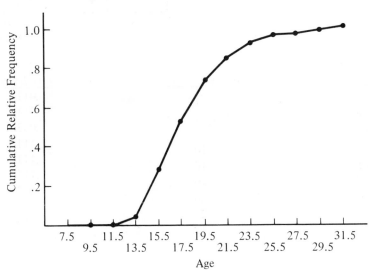

34.

cumulative relative frequency
1.000
.999
.997
.992
.985
.960
.907
.766
.544
.263
.079

35.

cumulative relative frequency
1.0001
.9855
.9125
.8249
.7081
.5110
.2409
.1460
.0584
.0438
.0146
.0073

36.

cumulative relative frequency
1.0000
.9910
.9550
.9009
.8018
.6847
.5315
.4414
.2702
.1711
.054
.009

43.

45.

murder rate	*number of cities*
23.5–26.5	1
20.5–23.5	1
17.5–20.5	3
14.5–17.5	10
11.5–14.5	10
8.5–11.5	10
5.5–8.5	20
2.5–5.5	23
–.5–2.5	12

46.

relative frequency
.011
.011
.033
.111
.111
.111
.222
.256
.133

47.

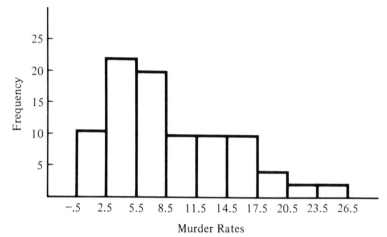

48.

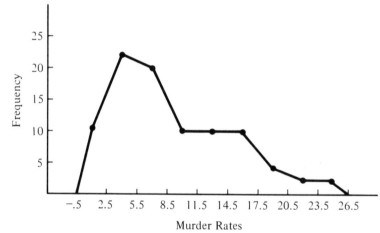

52.

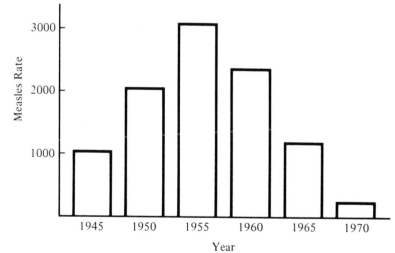

53. | cumulative relative frequency |
    | --- |
    | 1.0000 |
    | .9953 |
    | .9953 |
    | .9811 |
    | .9770 |
    | .9245 |
    | .8915 |
    | .8491 |
    | .8066 |
    | .7075 |
    | .5377 |
    | .3915 |
    | .2736 |
    | .1415 |
    | .0755 |
    | .0472 |

.0142
.0047

55. | cumulative relative frequency |
    | --- |
    | 1.000 |
    | .999 |
    | .998 |
    | .994 |
    | .989 |
    | .973 |
    | .946 |
    | .897 |
    | .808 |
    | .678 |
    | .477 |
    | .263 |
    | .050 |

Chapter 4

1. 45

2. 4

3. 12, 9.915

4. 85, 85.250

5. 45

6. 4.5

7. 15.068

8. 82.726

9. 46.2

10. 4.9

11. 19.405

12. 19.405

13. 19.405

14. 22.257

15. 29.731

16. 22.1, 50%; 17.5, crude mode; 22.257, mean.

17. 19.568

18. 21.419, 50%; 20.5, crude mode; 19.405, mean.

19. 87.89%

20. 97.43%

21. 95.69%

22. No, the categories must be nonoverlapping.

23. 4.604

24. 4.485

25. 2.3

26. 2.3

27. 2.3/4.9 = .469

28. 2.3/4.5 = .511

29. $\sqrt{74.4} = 8.626$

30. $\sqrt{8.767} = 2.961$

31. $\sqrt{12.45} = 3.528$

32. $\sqrt{69.26} = 8.322$

33. $\sqrt{12.45} = 3.528$

34. $\sqrt{21.88} = 4.678$

35. $\sqrt{16.82} = 4.101$

36. $\sqrt{99.058} = 9.953$

37. 6

38. 13

39. (a) 12.7   11.8
         12.2   17.7
         18.1   14.6
         15.7   16.9
         13.4   10.3
    (b) There is no mode.   (c) 14   (d) 14.347   (e) 2.287
    (f) 2.287

40. 7.87

41. 4.433

43. (a) 65   (b) 56.9, $\sqrt{305.4} = 17.476$ (c) 65/4 = 16.25

44. $62.27, $138.73

45. $2.143, \sqrt{3.143} = 1.773, 1.5$

46. $1.968, \sqrt{7.32} = 2.706$

47. Because the range/4 = .150, we have reason to doubt the answer. $s = .132$

48. No, since scores cannot be greater than 100, the distribution is skewed to the right.

49. $15.88, \sqrt{12.56} = 3.54$

50. 14.46

51. 15.318

52. 14.4, mean; 15, median; 16, mode.

53. $\sqrt{13.971} = 3.738$

54. 3.5

55. 16, mode; 15, median; 14.87, mean.

56. $\sqrt{15.838} = 3.98$

57. 3.75

Chapter 5

1. A discrete random variable is one that can assume only a countable number of values; a continuous random variable can assume the infinitely large number of values in a line interval.

3. The empirical approach to probability relies on the repetition of an experiment a large number of times. If the experiment is conducted N times of which n outcomes result in an event, then the probability of that event is approximately n/N.

6. 120, 1

8. $(.5)^{10} = .000977$ $\qquad$ $(.5)^{10} = .000977$ $\qquad$ yes

9. $5(.5)^5 + (.5)^5 = .1875$ $\qquad$ yes

11. $np, \sqrt{npq}$

12. $500, \sqrt{250} = 15.81$

13. yes

14. $30, \sqrt{29.7} = 5.45$

15. yes, yes

16. (a) $\quad$ .3413 $\qquad$ (b) $\quad$ .4772 $\qquad$ (c) $\quad$ .4236 $\qquad$ (d) $\quad$ .0692
    (e) $\quad$ .2886 $\qquad$ (f) $\quad$ .4236

17. (a) .1538   (b) .6916   (c) .1359   (d) .0215

18. (a) .8537   (b) .0267   (c) .7488   (d) .6181

19. .67

20. .167

21. −.433

22. .1191

23. 17.36%, 3.01%

24. Parameters are numerical descriptive measures of populations. Statistics are numerical descriptive measures of samples.

25. A random sample of n measurements from a population is one that gives every different sample of size n from the population an equal probability of selection.

29. 10

32. (a) 72.87%   (b) 31.02%   (c) 98.90%

33. $z = (X - \mu)/\sigma$

34. (a) .4332   (b) .4641

35. (a) .4938   (b) .4332

36. (a) .2881   (b) .5762

37. (a) .9500   (b) .9902

38. (a) .5119   (b) .0267
39. $z = .524$
40. $z = 1.645$
41. $z = 1.96$
42. $z = -.5$
43. $z = .75$
44. .4649
45. $\mu = 20, \sigma = 3.46$
46. yes
47. $z = -.583$
48. $z = .56$; yes
49. $\mu = 30, \sigma = 4.58$
50. $\mu = 3500, \sigma = 32.4$
51. no, 2600 lies approximately 28 standard deviations below the mean.
52. $z = 2.33$
53. $z = 1.28$
54. .1335, approximately 114
55. .0475

Chapter 6

1. A point estimate is a single number (or point) used to estimate a population parameter. The error of estimation is the positive distance between a single estimate and the true parameter.

2. 25.6, ±1.48

3. 6.3, ±.648

4. 4.2, ±.392

6. With 95% confidence the population mean lies between $\bar{X} - 1.96 \, \sigma/\sqrt{n}$ and $\bar{X} + 1.96 \, \sigma/\sqrt{n}$.

7. \$.75 ± \$.09

8. 35 ± 1.952

9. 50/180 = .278

10. $2\sqrt{.001115} = .0668$

11. .2, ±.08

12. .685 ± .0455

13. .258 ± .0357

14. .0767 ± .0253

15. .0767 ± .0396

16. 676

17. 117

18. 1067

19. The variance can be estimated by using a value of s^2 obtained from a prior study or by using a range estimate of s to compute s^2.

21. No, the value would not be an estimate but an exact value.

23. An unbiased estimator, $\hat{\theta}$, of a parameter θ has an expected value equal to θ; the distribution of estimates obtained by using the estimator $\hat{\theta}$ is centered on θ. A biased estimator does not have an expected value equal to the parameter it is estimating.

24. (1) The distribution of estimates should be centered about the parameter estimated; (2) the standard deviation of the distribution of estimates is as small as possible.

25. .10, ±2(.013)

26. 120 ± 1.078

27. 120 ± 1.419

28. $.45, \pm .0315$

29. $.45 \pm .0406$

31. 78

32. 400

33. 384, no

Chapter 7

1. null hypothesis, alternative hypothesis, test statistic, rejection region

2. Yes, rejection region is greater than 89.465 and less than 62.535.

3. Yes, rejection region is greater than 51.76 and less than 28.24.

4. Insufficient evidence to reject; rejection region is greater than 358.2 and less than 337.8.

5. Reject; rejection region is greater than 8.11 and less than 5.89.

6. Insufficient evidence to reject; rejection region is greater than 8.36 and less than 5.64.

7. Reject; rejection region is greater than $54.01 and less than $49.98.

8. Provided the sample size is 30 or more, we can use the sample standard deviation, S, in place of σ.

9. $|z| > 2.33$; i.e., reject if $z > 2.33$ or $z < -2.33$

10. .10

11. .05

12. $\chi^2 = 77.34$; reject

13. $\chi^2 = 95.89$; reject

14. yes

15. $D = .216$; insufficient evidence to reject

16. $D = .366$; insufficient evidence to reject

18. For a given sample size, α and β are inversely related.

19. Both α and β decrease.

20. Yes, one-tailed; rejection region is greater than 526.00.

21. Insufficient evidence to reject; rejection region is greater than 372.1 and less than 347.9.

22. $\chi^2 = 9.375$; yes

23. $\chi^2 = 10$; reject

24. $D = .333$; reject. *Note:* .333 is greater than the critical value for $n = 20$ and $n = 25$.

25. Yes, reject; rejection region is greater than 526.

26. Reject; $X = 133$ lies 4.67 standard deviations above $\mu = 100$.

27. Insufficient evidence to reject; $\bar{X} = 6.2$ lies 1.38 standard deviations above $\mu = 5.0$.

Chapter 8

1. $1.4, \pm .6924$

2. $z = 6.18$; reject

3. $z = -2.508$; reject

4. $z = 3.147$; reject

5. $z = 1.109$; insufficient evidence to reject

6. $z = 1.67$; insufficient evidence to reject

7. $.29, \pm .139$

8. $z = 3.966$; reject

9. $z = 3.33$; reject

10. We have an independent random sample from each of two populations. We then utilize the ranks to compute U. Both sample sizes should be 10 or more.

11. $z = 2.20$; reject

13. Yes, we could randomly pair the measurements after the data are obtained and perform a paired-difference test; however, the unpaired analysis would likely detect a difference when it exists. *Note:* If the data are paired initially, we must perform a paired analysis.

14. to filter out variability in sample data

17. $z = 11/\sqrt{7.5} = 4.017$; reject

18. $z = 2.33$; reject

19. $z = -1.5/\sqrt{2.75} = -.904$; insufficient evidence to reject

20. $z = 1.79$; insufficient evidence to reject

21. The rejection region using z is not appropriate for the test statistic, $(\bar{X} - \mu_0)/(s/\sqrt{n})$, when n is less than 30.

22. $\bar{X} = 638.6, s = \sqrt{162.3}, t = -11.4/\sqrt{32.46} = -2,00$; insufficient evidence to reject.

23. $t = -1.75/\sqrt{.6696} = -2.139$; insufficient evidence to reject

24. Assume that both populations are normal and possess the same variance.

25. $t = 2.2$; no, insufficient evidence to reject $H_0: \mu_d = 0$

26. $t = 1.712$; no, insufficient evidence to reject

27. $H_a: \mu_1 - \mu_2 \neq 0, H_0: \mu_1 - \mu_2 = 0, z = -2.295$; reject

28. $z = -7.2$; reject $H_0: \mu_1 - \mu_2 = 0$ for $\alpha = .01$

29. $z = -3.78$; reject

30. $z = 2.462$; yes, reject

31. $z = 2.623$; yes, reject H_0 for $\alpha = .01$

32. $z = 3.23$; yes, reject H_0 for $\alpha = .01$

33. when the assumptions of normality or equality of the two population variances are violated

34. $z = -.898$; insufficient evidence to reject

35. $z = .6324$; insufficient evidence to reject

36. (a) $z = 1.807$; insufficient evidence to reject
    (b) $t = 1.434$; insufficient evidence to reject

37. $z = 9.016$; yes, reject

38. To use the z test, the sample size must be 30 or more. Smaller samples require the t test.

40. We have drawn an independent random sample from a normal population with mean μ and variance σ^2.

Chapter 9

1.

	Murder Rate			
	0–6	7–12	13–18	19–24
North	2.7	38.9	69.6	100.0
South	54.1	27.8	17.4	0
West	43.2	33.3	13.0	0
	100.0	100.0	100.0	100.0
	$n = 37$	$n = 18$	$n = 23$	$n = 2$

4. need an expected cell count of 5 or more for all cells

5. $\chi^2 = 6.1$; reject H_0

6. $\chi^2 = 12.816$; reject H_0

7. to find out if two variables are related, so one variable can be predicted based on knowledge of the other variable

9. $\chi^2 = 4.0$; insufficient evidence to reject

10. $\chi^2 = 33.45$; reject

11. for 2×2 tables

12. (a) Chi-square test of independence can be used to detect dependence between any type of cross-classified data. (b) Fisher's exact test is used for analyzing the relation between two variables when the data are nominal or ordinal and the sample sizes are small. (c) The McNemar test is used to determine if there is a significant change over a time period. The data are sampled at two different points in time.

13. $\chi^2_M = 9.0$; reject

14. $P(1, 4) + P(0, 5) = .0863$; insufficient evidence to reject

15. $P(1, 4) + P(0, 5) = .1002$; insufficient evidence to reject

16. $\chi^2_M = 30$; reject

17. Yes, the expected cell count of lower-class males and females is less than 5.

18. $\chi^2 = 2.56$; insufficient evidence to reject

19. $\chi^2 = 17.675$; reject

20. $z = .215 \sqrt{.002495} = 4.30$; reject

21. no, $\chi^2 = 2.872$; insufficient evidence to reject, $\alpha = .05$

22. 4

23. yes, $\chi^2 = 11.626$; reject

Chapter 10

1. $\Phi = \sqrt{.044} = .210$

2. $\Phi = \sqrt{.111} = .333$

3. The contingency coefficient can be used for tables larger than a 2×2.

4. .707

5. $\sqrt{.042} = .206, .206/.707 = .291$

6. C_{adj} enables one to compare tables of different sizes.

7. the difference between the number of elements misclassified under the "no association" rule and the "perfect association" rule

8. The independent variable contributes no information for the prediction of the dependent variable.

9. The dependent variable can be predicted exactly, based upon knowledge of the independent variable.

10. $\text{PRE} = (E_1 - E_2)/E_1$

11. is a measure of association for nominal data

12. λ_r is computed using the row variable as the dependent variable. λ_c is computed when the column variable is the dependent variable.

13. .222

14. 0

15. .111

16. Both λ_r and λ_c have a proportional reduction in error interpretation.

19. $\gamma = .487$

20. Gamma is preferred when the data are measured on an ordinal scale. Lambda is for nominal data.

21. $\gamma = .6939$

22. ρ is a proportional reduction in error measure of association and W is not; however, W can be used to measure association among two or more sets of ranks.

23. $\rho = .87$. *Note:* the highest test score received the rank of 1.

24. $t = .87\sqrt{.2431} = .429$; yes, yes

25. $\rho = -.5874$. *Note:* the lowest rating and shortest distance received the rank of 1.

26. $t = -2.295$; reject, $\alpha = .05$, yes

27. no

28. $W = .9365$; yes

30. $\Phi = \sqrt{.145} = .381, C = \sqrt{.127} = .356, C_{max} = .707, C_{adj} = .356/.707 = .504$

31. because it has a proportional reduction in error interpretation

32. It can be used for tables larger than 2×2.

33. $.707, \sqrt{2/3} = .816, \sqrt{3/4} = .866, \sqrt{4/5} = .894, \sqrt{5/6} = .913$

36. λ_r assumes the row variable is the dependent variable. λ_c is computed assuming the column variable is the dependent variable.

37. 0

38. $\gamma = .378$. There is a positive relationship between degree of perceived interest and self-esteem.

39. $W = .7144$, yes

40. γ, because both variables can be measured on an ordinal scale

41. $\rho = .9534$, $t = .9534\sqrt{197.7403} = 13.407$; reject

Chapter 11

1. the line running through the points of a scatter diagram

2. a plot of data corresponding to measurements on two variables

3. A freehand regression line gives a prediction equation, however, there could be different prediction equations from different "eyeball" fits to the same data.

4.

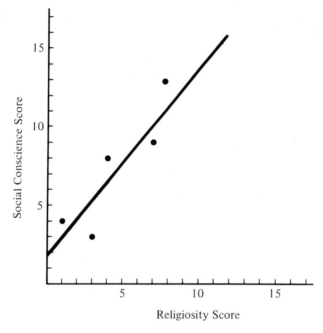

5. the value of Y at the point where the line crosses the Y axis ($X = 0$)

6. No, because ordinal data are only ranked, there is no meaning to the distance between the ranks.

7. $b = 1.12, a = 2.14$

8.

X	Predicted Y
7	9.98
4	6.62
3	5.50
12	15.58
8	11.10
1	3.26

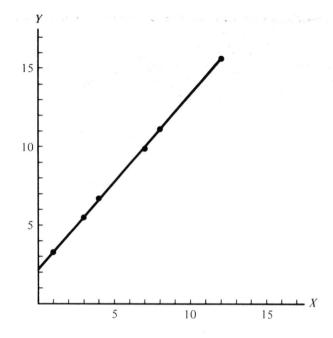

9. $a = 0, b = .938$

10.

z_x	Predicted z_y
.323	.303
−.506	−.475
−.782	−.734
1.701	1.596
.598	.561
−1.333	−1.250

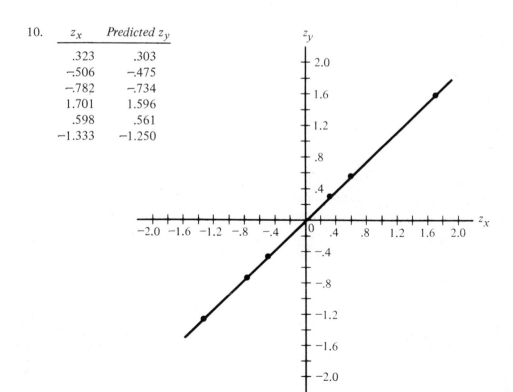

11. $r = .938$

12. $r^2 = .8798$

13. r^2 determines the proportion of the total variability in the Y values that can be accounted for by using the model. r is the slope of the regression line $z_y = rz_x$ and can be used for prediction purposes and for information concerning the strength of the relation between X and Y.

14. $\hat{Y} = a + bX$

15. No

16. If the scatter of Y values for corresponding X values is uniform across values of X, the Y values are homoscedastic.

17. $t = .8\sqrt{272.2} = 13.198$; reject H_0: $r = 0$; thus r is significantly different from zero

18. $r = 49.15/\sqrt{3448.228} = .837$

19. $t = 4.839$; reject

20. .70

21. Grade-point average does not necessarily reflect a student's IQ.

22. to know what to expect for the value of r—positive, negative, or no relation

23. the trend line running through the points of a scatter diagram

24. r can take on values between -1 and 1

25. The slope (b) is the increase in Y that corresponds to a 1-unit increase in X.

26. The method of least squares chooses those values for a and b that make SSE a minimum.

27. r is the slope of the standardized least-squares prediction equation; b is the slope of the least-squares prediction equation calculated from non-standardized data.

28. so that his results can be compared to other studies, since b depends on the scale of the original measurements

29. $-.80$ indicates z_y decreases .80 per 1-unit increase in z_x.
    $+.80$ indicates z_y increases .80 per 1-unit increase in z_x.

30. The proportional reduction in the error sum of squares is the coefficient of determination, r^2. r^2 is the fraction of the total variation in the Y values attributable to X

31. $Y = 3.62 + 1.45X$, 45.67

32. $r = \sqrt{.976} = .988$, $r^2 = .976$

33. To be homoscedastic, the scatter of Y values for corresponding X values is uniform across values of X. To be linear, the trend line must be straight. To be normal, the data must be randomly sampled from a normal population.

34.

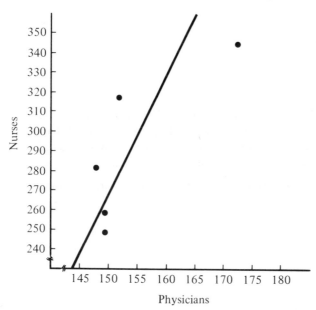

36. $Y = -248.9 + 3.5X$

37. $r = \sqrt{.682} = .826$

38. $r^2 = .682$

39. $t = 2.54$; insufficient evidence to reject

Chapter 12

1. Assume that the five populations are approximately normal with a common variance σ^2.

2. s_W^2 represents a combined estimate of the common variance σ^2 and measures the variability of the observations "within" the populations. s_B^2 estimates the variability among the sample means for the populations.

3. s_B^2/s_W^2 is the F ratio and it is the decision maker used to test equality of the population means.

4. $F = 5.909/.6779 = 8.7166$; reject

5. $F = 1.107/.38 = 2.913$; reject

6. .if the scientist suspects that the measurements were not drawn from normal populations with a common variance

7. $H = 10.7485$; reject

8. $H = 19.572$; reject

9. $F = .776$; insufficient evidence to reject

10. $F = 19.642$; reject

11. $H = 13.763$; reject

12. $F = 9.626$; reject

13. $H = 15.808$; reject

14.   The Kruskal–Wallis procedure is more appropriate than the analysis of variance of Section 12.3 for cases in which it cannot be assumed that the data were obtained from normal populations with a common variance. The Kruskal–Wallis method is also used with ordinal data.

16.   yes, when the normality assumption or the equality-of-variance assumption does not hold

17. $F = 16.86$; reject

18. $H = 20.186$; reject

INDEX

ESSENTIAL MATHEMATICAL SYMBOLS AND RELATIONSHIPS

refined mode (grouped data) $\quad M_0 = L + \left(\dfrac{D_1}{D_1 + D_2}\right) i$

median (grouped data) $\quad M_d = L + \dfrac{i}{f_m} \, (.5n - cf_b)$

population mean $\quad \mu$

sample mean (ungrouped data) $\quad \bar{X} = \dfrac{\Sigma X}{n}$

sample mean (grouped data) $\quad \bar{X} = \dfrac{\Sigma f X}{n}$

population variance $\quad \sigma^2$

sample variance (ungrouped data) $\quad s^2 = \dfrac{\Sigma(X - \bar{X})^2}{n - 1} = \dfrac{1}{n - 1}\left[\Sigma X^2 - \dfrac{(\Sigma X)^2}{n}\right]$

sample variance (grouped data) $\quad s^2 = \dfrac{\Sigma f(X - \bar{X})^2}{n - 1} = \dfrac{1}{n - 1}\left[\Sigma f X^2 - \dfrac{(\Sigma f X)^2}{n}\right]$

population standard deviation $\quad \sigma$

sample standard deviation $\quad s$

population proportion (binomial population) $\quad p$

sample proportion $\quad \hat{p} = \dfrac{X}{n}$

large-sample confidence interval for μ $\quad \bar{X} \pm z_{\alpha/2} \dfrac{\sigma}{\sqrt{n}}$

large-sample confidence interval for $(\mu_1 - \mu_2)$ $\quad (\bar{X}_1 - \bar{X}_2) \pm z_{\alpha/2} \sqrt{\dfrac{\sigma_1^2}{n_1} + \dfrac{\sigma_2^2}{n_2}}$

large-sample confidence interval for p $\quad \hat{p} \pm z_{\alpha/2} \sqrt{\dfrac{pq}{n}}$

large-sample confidence interval for $(p_1 - p_2)$ $\quad (\hat{p}_1 - \hat{p}_2) \pm z_{\alpha/2} \sqrt{\dfrac{p_1 q_1}{n_1} + \dfrac{p_2 q_2}{n_2}}$